Quantum Information

Important concepts from the diverse fields of physics, mathematics, engineering, and computer science coalesce in this foundational text on the cutting-edge field of quantum information. Designed for undergraduate and graduate students with any STEM background, and written by a highly experienced author team, this textbook draws on quantum mechanics, number theory, computer science technologies, and more, to delve deeply into learning about qubits, the building blocks of quantum information, and how they are used in quantum computing and quantum algorithms. The pedagogical structure of the chapters features exercises after each section as well as focus boxes, giving students the benefit of additional background and applications without losing sight of the big picture. Recommended further reading and answers to select exercises further support learning. Written in approachable and conversational prose, this text offers a comprehensive treatment of the exciting field of quantum information while remaining accessible to students and researchers within all STEM disciplines.

Asma Al-Qasimi is a researcher in quantum information and optics at the University of Rochester and she obtained her degree in physics and mathematics from the University of Toronto. She is a former APS Blewett Fellow and has received awards for her contributions to the teaching of undergraduate quantum mechanics.

Daniel F. V. James is Professor of Theoretical Physics at the University of Toronto and has been working in quantum information science and technologies for nearly 30 years. His achievements include the first tomographic measurement of an entangled state, the first experimental realization of deterministic quantum teleportation, and an early proof of principle demonstration of Shor's algorithm.

Quantum Information

A First Course

ASMA AL-QASIMI
University of Rochester, New York

DANIEL F. V. JAMES
University of Toronto

CAMBRIDGE
UNIVERSITY PRESS

Shaftesbury Road, Cambridge CB2 8EA, United Kingdom

One Liberty Plaza, 20th Floor, New York, NY 10006, USA

477 Williamstown Road, Port Melbourne, VIC 3207, Australia

314–321, 3rd Floor, Plot 3, Splendor Forum, Jasola District Centre,
New Delhi – 110025, India

103 Penang Road, #05–06/07, Visioncrest Commercial, Singapore 238467

Cambridge University Press is part of Cambridge University Press & Assessment,
a department of the University of Cambridge.

We share the University's mission to contribute to society through the pursuit of
education, learning and research at the highest international levels of excellence.

www.cambridge.org
Information on this title: www.cambridge.org/highereducation/isbn/9781009514729
DOI: 10.1017/9781009514750

First published 2025

Printed in the United Kingdom by CPI Group Ltd, Croydon CR0 4YY

Cover illustration: enot-poloskun / E+ / Getty Images.

A catalogue record for this publication is available from the British Library

A Cataloging-in-Publication data record for this book is available from the Library of Congress

ISBN 978-1-009-51472-9 Hardback

Additional resources for this publication at www.cambridge.org/al-qasimi-1e

For our children

Contents

Preface

This is a book that presents an introduction to quantum information for a one-semester course (if the instructor is selective) or a two-semester course (if the entire book is covered), aimed at second- and third-year undergraduate students in physics, mathematics, computer science, and engineering. It takes a pedagogical approach to the subject and tries to be as self-contained as possible, making minimal assumptions about the physics or mathematical background of the student. In fact, quantum physics is not even a prerequisite; whatever topics we need from quantum physics are introduced in the book. It does assume knowledge of high school algebra, trigonometry, geometry, matrices and calculus, probability theory, and some basic concepts from physics, such as forces, energy, and electricity. Thus, arguably the book is accessible to highly motivated high school students or to professionals in the financial services industry with good STEM backgrounds (who might like to know more about what they are investing in). Of course, having a background in first-year calculus and/or university-level linear algebra would be an asset, but it is not essential.

One point we would like to stress is that the subject of this book is mathematical, meaning that to learn, one has to solve problems. With this in mind, the end-of-section exercises are designed to reinforce concepts learned in the corresponding sections as well as to teach new concepts; they are there to complement the text and be part of the learning experience. The end-of-chapter exercises are usually designed to stretch the students more, as well as to get them to combine several learned skills. In order to help the students gain confidence that they are on the right track when solving problems, we include the solutions to most of the odd-numbered exercises, which are to be found at the end of the book. We also include a glossary of terms one might encounter in quantum information (the first time a glossary word is mentioned in the text it will appear in **bold face**).

Our Pedagogical Approach: Notes on Boxed Material

Considering the diverse background of our audience, there may be gaps in individuals' preparation which we aim to fill by placing explanations of some background concepts in boxes. The reason for not including them in the main body of the text is to make the presentation as friendly and easy to understand as possible and to avoid overloading the main text. In addition, by compartmentalizing certain concepts in boxes, we highlight them, saving students from rummaging through a lot of text to find a supplementary or background concept. In many textbooks, presenting material in boxes implies that it isn't necessary to understand the text; in other words, it is optional and can be skipped. However, this is *not* always the case in this book.

To elaborate, there are four types of boxes in this book, which we have indicated with icons:

- ⚓ Physics Tips
- ✚ Mathematics Tips
- 🚨 Looking a Little Bit Further
- 🐚 Interesting Facts

The first two (Physics and Mathematics Tips) are essential to the understanding of the material. If the student is sufficiently comfortable with the contents of a given box, he/she may skip reading it, but with the understanding that the material in the book depends on it, and the exercises in the book assume familiarity. With regards to the "Looking a Little Bit Further" boxes, they can be considered optional, and the exercises in the book do not depend on them. However, we leave it to the discretion of individual instructors if they would like to assign them to stretch the students a little bit more. The "Interesting Facts" box is also not required reading, but offers facts that might be of interest to a student of quantum information. For example, it answers questions like: How does the science fiction version of teleportation differ from teleportation in science fact? How did the human fascination and skill in encrypting secret messages develop, and where does quantum power come in the picture? Where do words like "algorithm" and "tomography" come from?

Some background and extra material that needs more space is included in the appendices of the book, such as the topics of solving differential equations, probabilities, and the singular value decomposition.

Organization of Material

In the introductory chapter of the book (Chapter 1), the field of quantum information and quantum computing is introduced, with some intuitive examples of the computational power that it promises. Chapter 2 focuses on the mathematics and physics of a qubit, the smallest system that can be used as a building block for quantum computers; the chapter also introduces the precepts from quantum mechanics and the various operations that can be applied on the qubit. In this chapter, both the *ket* and *density matrix* representation of the qubit are covered, as well as unitary evolutions. Chapter 3 is all about *two* qubits, with a focus on entanglement, the non-classical property and the star resource in quantum information and quantum computing. Chapter 4 teaches the basics of quantum computing: how to store and manipulate numbers in quantum computers, as well as various simple quantum algorithms. Shor's algorithm, arguably the "killer app" of all quantum algorithms, is the highlight of Chapter 5. Here, the necessary mathematical skills to appreciate it are built. Moreover, RSA encryption, which is made vulnerable by Shor's algorithm, is also covered. In Chapter 6 we begin our discussion of how to build a quantum computer by discussing the requirements that must be placed on a single two-level quantum system for it to be a viable qubit. The abstract concepts are illustrated by four example qubit technologies: photon polarization, spin-1/2 particles, atomic ions, and electric circuits with Josephson junctions in Chapter 7. This discussion is continued in Chapter 8, where the physical implementation of two-qubit gates is discussed. In Chapter 9 we introduce a very important tool needed in quantum device development, namely quantum state tomography. Chapter 10 describes some interesting paradigm shifts in quantum computing – psuedo-pure states, non-deterministic gates, and cluster-state quantum computing. In Chapter 11 we look at decoherence, the effect that destroys quantum behaviour on a large scale, including killing Schrödinger's long-suffering cat, and which also is the curse of quantum computing; but we conclude by describing the important cures for decoherence, namely decoherence-free spaces and error-correcting codes. In Chapter 12 we describe other interesting quantum technologies allied to quantum computing, such as quantum key distribution and quantum metrology.

We put every effort into writing the material in this book as best as we can to make it as helpful as possible to the starting student of quantum information. Being human, we are not infallible. We welcome questions and comments to improve on this pedagogical endeavour to teach the subject.

Acknowledgements

We acknowledge with gratitude Joseph H. Eberly, Greg Gbur, John F. James, Robert Joynt, José Morales Escalante, Tzu-Chieh Wei, and all the anonymous reviewers for their helpful comments on the manuscript. We are also grateful for the invaluable editorial help of Nicholas Gibbons, Tineke Bryson, and the other team members at Cambridge University Press.

1 Welcome to Quantum Computing

The main purpose of this book is to familiarize you with the foundations of quantum computing, which is a means for information storage and processing that exploits unique quantum mechanical phenomena, as opposed to the classical physics which drives your current personal electronics. In this chapter, we will discuss the fundamental differences between classical and quantum physics and give a few examples that demonstrate why **quantum computers** can give us an entirely different type of data processing power from what we are used to. The quantum mechanical notation for describing states will also be introduced, which we will build on in the following chapters.

Learning Objectives

- Understand the difference between state and configuration in classical systems versus quantum systems.
- Understand the difference between chosen and inherent randomness.
- Be introduced to the ket representation of quantum states and the mathematics behind it.
- Become familiar with quantum entanglement.
- Become aware of some of the advantages that quantum computers offer.

1.1 The Great Divorce: Observables and Configurations

1.1.1 The Classical State Versus the Quantum State

Classical physics is based on the implicit assumption that you can measure the complete state (arrangement and motion of the constituent elements) of the system you are interested in. For example, suppose you have a collection of particles (see Fig. 1.1) whizzing around, colliding, sticking together, and generally doing what particles do. Each of them has a *position* and a *momentum*, as well as intrinsic properties such as *mass* (and charge if you are dealing with electric and magnetic forces) or other traits (if more exotic forces are involved), all of which can be measured by some means.

Positions can be measured simply using a ruler and maybe a protractor to find the distance and direction from some reference point; velocities can be determined by measuring the positions at some short interval of time later, or perhaps by some more direct means such as Doppler velocimetry; and so on. The actual practical means of performing the measurements is not so important here, but the implicit assumption that such measurements *can* be made with arbitrary precision and accuracy, and they constitute a complete characterization of the system, is central to classical mechanics. With knowledge of the nature of forces acting on the various particles, one can then use Newton's second

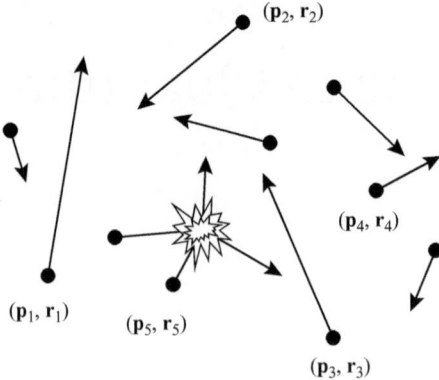

Figure 1.1   Classical particles. Each particle has a position, a momentum, and a mass associated with it.

law of motion, i.e., the rate of change of a given particle's momentum is equal to the sum of all the forces exerted on that particle by all the other particles, to write down a set of mathematical equations involving the dynamic parameters, i.e.,

$$\frac{d\mathbf{r}_i}{dt} = \mathbf{p}_i/m_i,$$

$$\frac{d\mathbf{p}_i}{dt} = \sum_{\substack{j=1 \\ j \neq i}}^{N} \mathbf{F}_{ij}, \tag{1.1}$$

where $\mathbf{r}_i$ is the position vector of the i-th particle, $\mathbf{p}_i/m_i$, the momentum divided by the mass, is its velocity, and $\mathbf{F}_{ij}$ is the force applied by the j-th particle on the i-th particle.

Solving these mathematical equations (assuming that such a solution exists) is of course an onerous, perhaps intractable task of mathematics; nevertheless, in principle, by measuring the positions, momenta, masses, etc. of all the particles in the system, one can predict their future positions and momenta. This is really what physics is about; you want to know what is going to happen next.

It seems intuitive that the complete collection of measured properties *is* the **configuration** of the system. What else could it be? But consider: an automobile has a mass, a position, a velocity, and power supplied by its engine, and one might accurately predict much about its future based on these quantities; but to fully understand the automobile, especially when it declines to start, one must look *under the hood* at all the complex parts that make it function. Could the same be true of particles at a more fundamental level?

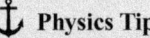 **Physics Tips 1.1**

Classical Objects Versus Quantum Objects. Classical physics laws, such as the ones developed by Issac Newton, describe *classical* objects. These objects tend to be big macroscopic objects such as a chair, an ice cream cone, a football, etc. On the other hand, quantum physics laws, developed by the famous physicists of a hundred years ago (Einstein, Bohr, Schrödinger, Born, etc.), describe the behaviour of tiny sub-microscopic entities, such as atoms, molecules, electrons, **photons**, etc.

The entities that can display quantum behaviour are called *quantum* objects. They are, therefore, the focus of **quantum information** and computation research as potential building blocks for building powerful quantum computers. Classical objects are, in fact, also governed by the laws of quantum mechanics at a fundamental level. It's just that a system's random interactions with its surroundings – generically called **decoherence** – disrupt the delicate quantum nature, leaving the one configuration with an immunity from decoherence to emerge, namely classical dynamics. The key to practical quantum technologies is building devices that are, in one way or another, immune to decoherence.

This issue was at the heart of the development of quantum mechanics a hundred years ago. The pioneers of this discipline found it necessary to discard the comfortable assumption that the measured properties of a system and the configuration of the system were the same. At the fundamental level, there *is* more going on "under the hood", and one must divorce the happy classical marriage between observational properties and dynamic configurations.

In quantum mechanics, the configuration, or **quantum state** of the system, whose mathematical formulation we will be discussing shortly, can be predicted by solving a different mathematical equation, called **Schrödinger's equation**, after its discoverer. Given a set of initial conditions, and external stimuli, the solution of this equation tells us the state of the system at a later time. In other words, we can predict the future, just as in classical mechanics. However, when you measure the system, the result is no longer directly determined by the state you have predicted. *All that the state does is give you a probability distribution for the random outcomes of your measurement.*

1.1.2 Randomness: Classical Versus Quantum

Random behaviour in science is usually an expression of our ignorance or inability to perform the sufficiently intricate calculations of complicated systems. For example, a dealer shuffling a pack of cards, a referee tossing a coin at the start of a sports match, or all of the molecules in the atmosphere being heated by sunlight or cooled by the oceans to produce tomorrow's weather are, in principle, predictable using Newtonian mechanics and the knowledge of the initial configuration and external forces. Since it is impractical to perform such difficult calculations, we accept our ignorance and resort to describing the outcomes as random events governed by a probability distribution to quantify the likelihood of various outcomes. However, unlike **classical randomness**, **quantum randomness** is an *inherent* randomness that is an inevitable characteristic of quantum mechanics.

Let us now focus on a simple case: flipping a coin, with the possible outcomes being Heads or Tails, as can be seen in Fig. 1.2. If you flip the coin in the air, you will have a **probability** associated with getting Heads and a probability associated with getting Tails. You cannot be in a situation where it is Heads and Tails simultaneously. As mentioned, in classical physics, conceivably you could work out the detailed dynamics of the coin based on how much force and how much torque was applied while flipping it, the air resistance, the coefficient of restitution of the surface on which the coin might bounce, and all other relevant parameters, and predict exactly what the result will be. However, all that trouble is not usually worth it; we choose to stay ignorant about the detailed dynamical description and go with a probabilistic analysis instead. During its flight, except for isolated instants when it happens to be exactly on its edge due to it spinning, the coin will always have either the Heads or the Tails side uppermost, but in any interval that can be perceived without sophisticated apparatus capable of tracking its dynamics on timescales shorter than one-tenth of a second or so, the coin is equally likely

Figure 1.2   Outcomes of a coin flip. H stands for Heads (left) and T for Tails (right).

to be Heads or Tails. When the coin's flight stops as it falls on the floor, either of these two outcomes is equally possible.

On the other hand, if this was a quantum mechanical problem, the coin is *both* Heads and Tails at the same time; each outcome has a probability associated with it which is constantly changing in time. Only when we measure the coin does it *project* into one or other of the two possible outcomes.

This may seem like hair-splitting of little importance to those of us not directly engaged in consideration of the philosophical foundations of the nature of reality; indeed, in the case of a single system with two possible outcomes, the distinction between a classical probabilistic description and a quantum mechanical description has no practical implications. But, as we will see, when we have two or more **quantum coins** involved, the quantum mechanical nature of the system has truly profound implications, both philosophically and technologically.

1.1.3   Mathematics of the Space of Possibilities

How can we formulate a mathematical description of this quantum mechanical state, in which the coin can be in two configurations at the same time? We take inspiration from the idea of vectors and, specifically, how any vector can be thought of as a linear combination of other vectors (see Fig. 1.3). Specifically, in a two-dimensional vector space (e.g., the page of this book), any vector $\mathbf{v}$ may be written as a combination of two basic vectors $\mathbf{e}_x$ and $\mathbf{e}_y$. Thus we have

$$\mathbf{v} = a\mathbf{e}_x + b\mathbf{e}_y, \tag{1.2}$$

where a and b are real numbers, called the components of $\mathbf{v}$ in the x- and y-directions, respectively. It is usually convenient to assume that the two basic vectors are perpendicular (or **orthogonal**) to each other, so that $\mathbf{e}_x \cdot \mathbf{e}_y = 0$, and are of unit length; i.e., $\mathbf{e}_x \cdot \mathbf{e}_x = \mathbf{e}_y \cdot \mathbf{e}_y = 1$ (here $\cdot$ is the dot (or scalar) product of two vectors). If this is not the case, one can always use any two non-parallel vectors to construct a pair of orthogonal unit vectors that satisfy these conditions (see Exercise 2).

We construct the configuration of our *quantum* coin in an analogous manner. Instead of the two directions, e_x and e_y, we have two possible configurations for the coin: Heads and Tails. Since our "space of configurations" has a completely different physical interpretation from a vector on the plane of our page, we use a different notation for these vectors: Heads is represented by the symbol (called a **ket**) $|H\rangle$ and Tails by the ket $|T\rangle$, and an arbitrary state, $|\phi\rangle$, by the combination (read: *ket phi equals alpha ket H plus beta ket T*):

$$|\varphi\rangle = \alpha|H\rangle + \beta|T\rangle. \tag{1.3}$$

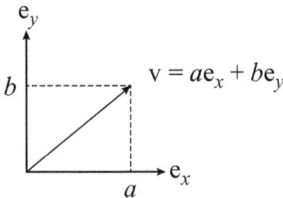

Figure 1.3 Two-dimensional space. A vector in a two-dimensional space can be broken down into two vectors, one pointing in the x-direction, and the other in the y-direction.

⊕ **Mathematics Tips 1.2**

Complex Numbers

- In physics and mathematics, it is sometimes necessary to work with complex numbers, which are numbers that are expressed in terms of the imaginary unit $i = \sqrt{-1}$ (sometimes the letter j is used instead, but we will use i in this book). A complex number, z, has the form $x + iy$, where x and y are real numbers; x is called the *real* part (Re{z}) and y is called the *imaginary* part (Im{z}).
- The *absolute value* or *modulus* of z, denoted $|z|$, is $\sqrt{x^2 + y^2}$.
- Sometimes it is easier to work with exponentials, so z is written as $z = |z| \exp(i\theta)$, where $x = |z| \cos\theta$ and $y = |z| \sin\theta$, where the identity $\exp(i\theta) = \cos\theta + i\sin\theta$ is used. The angle θ is called the *argument* of z (denoted Arg{z}), or, in physics contexts, the **phase** of z.
- The *complex conjugate* of a complex number $z = x + iy$ is $z^* = x - iy$; i.e., the imaginary component changes sign. In terms of exponentials $z^* = |z| \exp(-i\theta)$. Note that $|z|^2 = z^*z$, $x = (z + z^*)/2$ and $y = (z - z^*)/2i$. Mathematicians sometimes use an overbar $\bar{z}$ to denote the complex conjugate, so it's important to check what convention is being used.

Another subtle difference between geometric vectors such as **v** and quantum mechanical states is that the components of $|\varphi\rangle$, α and β, are in general *complex* numbers that give the weight or significance of each component. Their exact significance will be discussed shortly. In physics and engineering, complex numbers are usually employed as a mathematical short-cut to avoid a lot of tedious manipulations with trigonometric identities or as a slick way to perform definite integrals by Cauchy's theorem; no mensurable physical quantity has a complex value, and classical physical systems can be described without resorting to complex numbers if one really wants to. However, it turns out that the formulation of the quantum mechanical state (in itself, something that can only be estimated by indirect means and not measured directly) requires the use of complex amplitudes: attempts to formulate quantum theory with real-valued amplitudes have been in vain.

Often we will resort to writing the vectors/kets as column vectors:

$$|\varphi\rangle = \alpha|H\rangle + \beta|T\rangle = \begin{bmatrix} \alpha \\ \beta \end{bmatrix}. \tag{1.4}$$

In writing the state as a column vector, the **basis** kets $|H\rangle$ and $|T\rangle$ are implicit:

$$|H\rangle = \begin{bmatrix} 1 \\ 0 \end{bmatrix} \quad |T\rangle = \begin{bmatrix} 0 \\ 1 \end{bmatrix}. \tag{1.5}$$

We can always write $|\varphi\rangle$ using a different set of basis vectors, but then the components α and β will also change, just like the components of **v** would be different if we chose different axes. In general, we will not be dealing with the simple two-dimensional space corresponding to a system with only two possible measurement outcomes; rather it is an abstract space (in the linear algebra sense) that can be of any number of dimensions, even infinite. Such a space is called the **Hilbert space**, named after the famous mathematician David Hilbert.

Accompanying each state vector, or ket, there is a corresponding **bra**, which is most conveniently defined using the column-vector representation. The bra is formed by taking the complex conjugate and, simultaneously, the transpose (i.e., rows and columns are switched) of the ket. This operation, which is called "taking the **Hermitian adjoint**" (named after the mathematician Charles Hermite) is denoted by a "dagger" (†) superscript:

$$|\varphi\rangle^{\dagger} = \langle\varphi| = \alpha^*\langle H| + \beta^*\langle T| = \begin{bmatrix} \alpha^* & \beta^* \end{bmatrix}. \tag{1.6}$$

⊕ Mathematics Tips 1.3

Products of Two State Vectors. The analogue of the dot product of two geometric vectors is the scalar or inner product between two state vectors, which is defined by transforming one of the kets into a bra, and then forming a *bracket* (which is the origin of the words "bra" and "ket"). The result is a single complex number. For example, the product of $|\varphi\rangle = \alpha|H\rangle + \beta|T\rangle$ and $|\psi\rangle = \gamma|H\rangle + \delta|T\rangle$ is

$$\langle\varphi|\psi\rangle = \begin{bmatrix} \alpha^* & \beta^* \end{bmatrix} \cdot \begin{bmatrix} \gamma \\ \delta \end{bmatrix} = \alpha^*\gamma + \beta^*\delta. \tag{1.7}$$

We will also encounter the *outer product* of two states, which is a matrix, defined like this:

$$|\psi\rangle\langle\varphi| = \begin{bmatrix} \gamma \\ \delta \end{bmatrix} \cdot \begin{bmatrix} \alpha^* & \beta^* \end{bmatrix} = \begin{bmatrix} \gamma\alpha^* & \gamma\beta^* \\ \delta\alpha^* & \delta\beta^* \end{bmatrix} \tag{1.8}$$

Note that the trace of the outer product (i.e., the sum of all the elements of the leading diagonal of the matrix, denoted $\text{Tr}\{|\psi\rangle\langle\varphi|\}$) is equal to the inner product $\langle\varphi|\psi\rangle$.

State vectors are usually defined so that they are **normalized**, in the sense that $\langle\varphi|\varphi\rangle = 1$. In this case, the interpretation of the **probability amplitudes** is straightforward: the absolute value squared of the inner product between the state and the basis state $|H\rangle$ is the *probability* of measuring our quantum coin to be "Heads"; and similarly the absolute value squared of the inner product between the state and $|T\rangle$ is the probability for it to be "Tails":

$$P_H = |\langle H|\varphi\rangle|^2 = |\alpha|^2, \tag{1.9}$$

$$P_T = |\langle T|\varphi\rangle|^2 = |\beta|^2. \tag{1.10}$$

On their own, α and β are called *probability amplitudes*; it is when you take their absolute values and square them that you get the *probability* for obtaining the state they correspond to, Heads or Tails, upon measurement. These probabilities (and average values for measurement outcomes which can be inferred from the probabilities) are the quantities that have experimental implications; the probability amplitudes cannot be measured directly (though they can be inferred from experimental data by a process called **quantum state tomography**; see Chapter 9). Indeed, if we were to multiply each of the probability amplitudes by a common (or "global") phase factor $\exp(i\phi)$, all the probability amplitudes and averages would not be changed in the slightest.

Generally it is much easier to work with a basis that is orthogonal than one that is not orthogonal; in situations like that of coin flipping, where Heads and Tails are mutually exclusive, they are naturally represented with orthogonal kets for each outcome. Moreover, for probability amplitudes to be meaningful, we need this basis to be normalized. That is why the basis that describes the quantum coin is constructed so that the following relations hold:

$$\langle H|T\rangle = 0, \tag{1.11}$$

$$\langle H|H\rangle = \langle T|T\rangle = 1. \tag{1.12}$$

The interpretation of the ket representation of the quantum mechanical coin that we discussed above (about α and β being probability amplitudes corresponding to finding the state to be Heads or Tails) is due to a footnote placed by the physicist Max Born in one of his papers, which eventually won him the Nobel Prize! It is known as the **Born rule** and is the best explanation we have out there and is an accepted interpretation, so it has the status of an *axiom in quantum mechanics*.

1.1.4 A Pair of Quantum Coins

Now, if you think about the previous discussion, except for the use of some fancy notation, what is really special about the quantum coin? The situation seems identical to the classical coin case: you flip a coin, whether quantum or classical, and then you measure it: you have a probability associated with "Heads" and another probability associated with "Tails". Non-classical behaviour only really becomes interesting when we have two or more quantum coins.

⚓ Physics Tips 1.4

Information Cannot Travel Faster than the Speed of Light. Imagine you wake up and make the decision to wear black socks today rather than white socks; nothing other than your own free will determines your choice. Being a keen participant in social media, you post a photo of today's choice of socks for your friend Angharad who lives far away. The distance the information has travelled, L, will always be less than the time t since you made your decision multiplied by c, the speed of light, the maximum speed at which information can travel; thus, defining the quantity $\Delta S^2 = c^2 t^2 - L^2$, we must have $\Delta S^2 > 0$. The principle of relativity then tells us that ΔS^2 is independent of any observer's speed relative to you and Angharad, and the order of the two events (your choice, and Angharad receiving the news) will always occur in that order.

Suppose, however, the latest social media app is able to send the information at a speed v which is faster than that of light. In that case, the distance the message travels in time t will be $vt > ct$, and so the quantity ΔS^2 will become negative, which in turn implies that one can contrive a situation in which Angharad is moving sufficiently fast that she receives the news of your choice *before* you even woke up and decided what to wear. Thus, the principle of causality – the effect must always occur after its cause – implies that information must travel slower than light.

There are four possible mutually exclusive configurations for two quantum coins:

$$|HH\rangle, \quad |HT\rangle, \quad |TH\rangle, \text{ and } |TT\rangle, \tag{1.13}$$

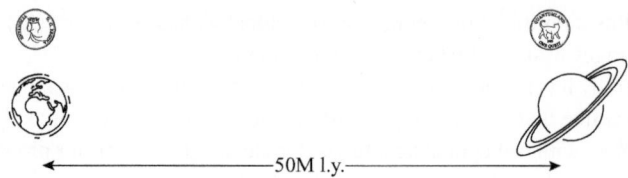

Figure 1.4   Two quantum coins. One coin is on planet Earth, while the other one is on another planet 50 million light years away. The coins are prepared in the state given in (1.15).

where $|HT\rangle$, for example, corresponds to the first coin being Heads and the second Tails, and so on. Here, the general possible configuration of two coins is represented as

$$|\psi\rangle = \alpha|HH\rangle + \beta|HT\rangle + \gamma|TH\rangle + \delta|TT\rangle, \qquad (1.14)$$

where, again, the coefficients are complex numbers, and the sum of the squares of their amplitudes is equal to unity.

We can engineer a special case of the state representation in Eq. (1.14), where we have equal probabilities of being in state $|HH\rangle$ and $|TT\rangle$ (and zero probabilities for the other two possible configurations):

$$|\psi\rangle = \frac{1}{\sqrt{2}}\{|HH\rangle + |TT\rangle\}. \qquad (1.15)$$

What this representation tells us is that both coins are the same, being either both Heads or both Tails. Look at the scenario in Fig. 1.4. Here we have a coin on Earth and another on a hypothetical planet 50 million light years away, the state of both coins together being represented by Eq. (1.15). If we measure the quantum coin on Earth and obtain $|H\rangle$, then immediately (and *not* 50 million years later) the configuration of the second coin will become $|H\rangle$! Likewise, if our measurement of the first coin yields $|T\rangle$, instantaneously (no delay!) the second coin is rendered into $|T\rangle$. "Hmm…" you might be thinking at this point: isn't this just the same as arranging two coins to be either both Heads or both Tails, and keeping the choice secret; then you send one coin to the distant planet and give the other to your aunty with strict instructions not to look at it until you say so. Then, when she looks at her coin, she immediately knows whether the coin sent to the distant planet was Heads or Tails. That sounds very similar, no?

But the quantum correlation is even stranger than you think: with *quantum* coins we have more options than just measuring in Heads or Tails. We can also measure to see if they are in one of the states that is *Heads and Tails*; for example:

$$|+\rangle = (|H\rangle + |T\rangle)/\sqrt{2},$$
$$|-\rangle = (|H\rangle - |T\rangle)/\sqrt{2}. \qquad (1.16)$$

It turns out (see Exercise 11) that

$$\frac{1}{\sqrt{2}}\{|HH\rangle + |TT\rangle\} = \frac{1}{\sqrt{2}}\{|++\rangle + |--\rangle\}. \qquad (1.17)$$

In other words, the complete correlation between measurements is the same, even in this basis: not only did the far away coin "know" the outcome of the measurement beforehand, it "knew" what measurement was going to be made as well (the same will not be true of your aunty's measurements).

In 1935, Albert Einstein, Boris Podolsky, and Nathan Rosen published their concern about this weird implication of quantum theory in a paper in *Physical Review*, which is famous among quantum physicists as the *EPR paper*, in reference to the initials of the authors. That is why this phenomenon, where a measurement in a far away place can instantaneously affect a system it is not in communication with, is sometimes referred to as the **EPR dilemma**. How does the second coin instantaneously "know" that anyone is doing anything to the first coin?!

In the same year as the EPR paper, Schrödinger wrote a paper that explains the consequences of states like (1.15), where he introduced the term **entanglement**. Entanglement, he pronounced, is *the* characteristic trait of quantum physics.

This implication of **quantum entanglement** strongly suggests we do not have the complete description of the system. The apparent randomness discovered by Born is, in fact, just like the coin-tossing randomness: a deterministic system, which is too complicated to fully analyze, and there are other **hidden variables** that will predetermine the outcome of any measurement of our quantum systems and complete the picture, and immediately all the quantum weirdness will disappear (we recover all certainty, just like we do if we work out all the detailed dynamical description of a classical coin). In 1964, the physicist John Stewart Bell proved that, in general, the theory of quantum mechanics cannot be described by these alleged hidden variables (we will elaborate more on this in Chapter 3). He came up with a testable way to demonstrate this result: a set of inequalities, dependent on specific experimental scenarios, that have to be satisfied for the existence of hidden variables to be possible to explain these weird outcomes. Today, we have come a long way in terms of experiments, and Bell's result is still standing strong: quantum systems violate the **Bell inequalities**, implying a hidden variable description is not applicable. In other words, quantum mechanical weirdness is inherent and unavoidable. Fascinating as this weirdness may be, the big question in this book is: How does this actually help us build a computer?

⚓ Looking a Little Bit Further 1.5

When Do We Have *Quantum* Entanglement? In this section, to define quantum entanglement, we first presented the idea that only quantum systems can be in superposition states, and second, we followed up on that by saying that, as a consequence of being in a superposition state, when we have at least two systems/"coins", entanglement materializes itself in quantum systems and not in classical systems. As they stand, these statements are not entirely true. We have to be careful: *Yes, some sort of superposition states are needed to have quantum entanglement*, but we have to be careful when we say which.

To elaborate, when we say that classical systems cannot be present in superposition states, we mean things like coins, whose states cannot be Heads and Tails at the same time, or a football that cannot be in the living room and the garden at the same time, or the poor kitty that cannot be both dead and alive at the same time. However, there are other classical systems that can be in superposition states. Think about a wave of polarized light, for example. Say that it only has **polarization** components perpendicular to the direction it is propagating in, so if it is propagating in the z-direction, then it has polarization components in the x–y plane; it can be x-polarized, y-polarized, or be in a superposition of both! Moreover, one can connive to arrange for two light waves to have their polarization degrees of freedom (represented mathematically by a vector space) entangled with some of their other (such as position-related) degrees of freedom, and that will result in **classical entanglement**! Yes, it will! In fact, if you search scientific journals, you will find articles on this type of entanglement. The thing is: entanglement is a consequence of a system having the ability to be in a superposition state, and there are classical systems that can be in such states. That's why, on its own, entanglement is not a uniquely quantum property.

In this book, we are not concerned about classical entanglement; we are only interested in quantum entanglement, and it is the property that we want to exploit to build powerful quantum computers. It is what Einstein referred to as *spooky action at a distance*. We will introduce appropriate mathematical representations (see *density matrices* in Chapters 2 and 3) and appropriate mathematical tools (*concurrence* in Chapter 3) to determine if our quantum states are entangled or not.

One of the most distinguishing properties of useful quantum entanglement is that it has to be between spatially separated systems; for example, entanglement between the degrees of freedom of two separated atoms or ions. We are not interested in entanglement between degrees of freedom of *one* atom, such as when its motional and energy degrees of freedom intermingle. In fact, many physicists interested in studying quantum power disparage classical entanglement being given the name because it is equivalent to the entanglement of degrees of freedom of *one* quantum system. Here, we will deal with entanglement between separate quantum systems, and when we say the system is classical, we will mean that this kind of entanglement is absent. Everything we do will be based on the assumption that we are dealing with genuine scenarios involving *quantum entanglement*.

Exercises 1.1.5

Unless otherwise stated, we assume that $\langle H|H \rangle = \langle T|T \rangle = 1$ and $\langle T|H \rangle = 0$ in the exercises in this book.

1. Given two non-parallel vectors $\mathbf{u}$ and $\mathbf{v}$, where θ is the angle between them:
   (a) Find the projection of $\mathbf{v}$ on $\mathbf{u}$.
   (b) Subtract this projection from $\mathbf{v}$.
   (c) Calculate the inner product between $\mathbf{u}$ and the result in (b). What can you conclude about these two vectors?
2. The vectors $\mathbf{v}_1$ and $\mathbf{v}_2$ are real-valued vectors in a two-dimensional plane. Assuming $\mathbf{v}_1$ and $\mathbf{v}_2$ are not parallel, find a pair of basis vectors $\mathbf{e}_x$ and $\mathbf{e}_y$, expressed in terms of these non-parallel vectors, which satisfy $\mathbf{e}_x \cdot \mathbf{e}_y = 0$.
3. Orthogonalize the following pairs of vectors:
   (a) $\mathbf{v}_1 = \mathbf{e}_x$, $\mathbf{v}_2 = \frac{1}{\sqrt{2}}\mathbf{e}_x + \frac{1}{\sqrt{2}}\mathbf{e}_y$.
   (b) $\mathbf{v}_1 = \frac{1}{\sqrt{2}}\mathbf{e}_x - \frac{1}{\sqrt{2}}\mathbf{e}_y$, $\mathbf{v}_2 = \frac{1}{\sqrt{3}}\mathbf{e}_x + \frac{2}{\sqrt{3}}\mathbf{e}_y$.
   (c) $\mathbf{v}_1 = \frac{1}{\sqrt{2}}\mathbf{e}_x - \frac{1}{\sqrt{2}}\mathbf{e}_y$, $\mathbf{v}_2 = \frac{1}{\sqrt{2}}\mathbf{e}_x + \frac{1}{\sqrt{2}}\mathbf{e}_y$.
4. Normalize the results in Exercise 3.
5. Write the following *kets* as column vectors:
   (a) $|H\rangle$.
   (b) $|T\rangle$.
   (c) $(4|H\rangle + 3|T\rangle)/5$.
   (d) $\frac{1}{\sqrt{3}}|H\rangle - i\sqrt{\frac{2}{3}}|T\rangle$, where i is the imaginary number $\sqrt{-1}$.
6. Find the *bra* representation (row vectors) by taking the *conjugate transpose* of each of the column vectors obtained in the previous problem. Then, calculate the scalar product between each bra and its corresponding ket (using matrix multiplication rules). (The answer should be a single number.)

7. Find which of the following pairs of states are orthogonal and normalized:
   (a) $(|H\rangle + |T\rangle)/\sqrt{2}$ and $(|H\rangle - |T\rangle)/\sqrt{2}$.
   (b) $(|H\rangle + |T\rangle)/\sqrt{2}$ and $(|H\rangle + i|T\rangle)/\sqrt{2}$.
   (c) $(\sqrt{3}|H\rangle + \sqrt{4}|T\rangle)/\sqrt{5}$ and $(\sqrt{4}|H\rangle - \sqrt{3}|T\rangle)/\sqrt{5}$.
   (d) $(5|H\rangle + i12|T\rangle)/13$ and $(i12|H\rangle + 5|T\rangle)/13$.

8. Consider two non-orthogonal normalized states of a single quantum coin, $|\chi\rangle$ and $|\psi\rangle$. Find a general expression for two normalized orthogonal states.

9. Orthonormalize the following pairs of kets:
   (a) $|\psi\rangle = \frac{i}{\sqrt{2}}|H\rangle + \frac{(1+i)}{2}|T\rangle$, $|\phi\rangle = \frac{1}{\sqrt{2}}|H\rangle + \frac{1}{\sqrt{2}}|T\rangle$.
   (b) $|\psi\rangle = \frac{\exp(-i\theta)}{\sqrt{2}}|H\rangle + \frac{1}{\sqrt{2}}|T\rangle$, $|\phi\rangle = |T\rangle$.

10. *Change of Basis.* In this problem, we will switch from using the $\{|H\rangle, |T\rangle\}$ orthonormal basis to the new one $\{|+\rangle, |-\rangle\}$. In terms of the vectors of the new basis, the vectors of the old basis are given by

$$|H\rangle = \frac{1}{\sqrt{2}}|+\rangle + \frac{1}{\sqrt{2}}|-\rangle,$$

$$|T\rangle = \frac{1}{\sqrt{2}}|+\rangle - \frac{1}{\sqrt{2}}|-\rangle.$$

Write the states in Exercise 5 in terms of this new basis, then find the scalar product of the states in the new basis. What changes and what does not change about the states in this new representation?

11. Prove Eq. (1.17).

12. Find the probability amplitudes as well as the probabilities of obtaining Heads and Tails in each of the states given in Exercise 5.

13. Normalize the state $|\psi\rangle = 2|H\rangle + 5|T\rangle$. Upon measuring a quantum coin in state $|\psi\rangle$, what is the probability of obtaining Heads? What about the probability of obtaining Tails?

14. Find the matrix representation of the *outer product* $|\psi\rangle\langle\phi|$ where
   (a) $|\psi\rangle = |H\rangle$ and $|\phi\rangle = |H\rangle$,
   (b) $|\psi\rangle = |T\rangle$ and $|\phi\rangle = |H\rangle$,
   (c) $|\psi\rangle = (|H\rangle + i|T\rangle)/\sqrt{2}$ and $|\phi\rangle = |H\rangle$,
   (d) $|\psi\rangle = (|H\rangle + i|T\rangle)/\sqrt{2}$ and $|\phi\rangle = (|H\rangle + i|T\rangle)/\sqrt{2}$.

15. Find the inner product of the following single quantum coin states:

$$|\psi\rangle = |H\rangle \text{ and } |\phi\rangle = \frac{1}{\sqrt{2}}|H\rangle + \frac{1}{\sqrt{2}}|T\rangle.$$

16. Find the inner product of the following two quantum coin states:

$$|\psi\rangle = |HT\rangle \quad \text{and} \quad |\phi\rangle = (|HH\rangle - |HT\rangle)/\sqrt{2}.$$

17. The following state is written in the $\{|HH\rangle, |HT\rangle, |TH\rangle, |TT\rangle\}$ basis. Find the ket representation of the state

$$|\varphi\rangle = \frac{1}{\sqrt{3}} \begin{bmatrix} \sqrt{2} \\ 0 \\ 0 \\ 1 \end{bmatrix}.$$

18. Given the state $|\psi\rangle = \frac{1}{\sqrt{2}}\{|HT\rangle + |TH\rangle\}$, what can be said about the second coin if the measurement result of the first coin yields Heads?

19. Given the state $|\psi\rangle = \frac{1}{\sqrt{2}}\{|HT\rangle + |HH\rangle\}$, what can be said about the second coin if the measurement result of the first coin yields Heads?

1.2   How Do Quantum Coins Help You Build a Better Computer?

1.2.1   Quantum Memories are BIG!

A computer is a device that processes information. First, we must be able to store the information in a computer memory so that it can be conveniently accessed and processed electronically. For this purpose, information must be disassembled into manageable chunks. For example, a database of student grades, a shopping list, or the collected works of Shakespeare can all be broken down into individual letters, each of which is assigned a number, which, in turn, can be represented as a series of base 2, or binary digits, or **bits** (from **BI**nary digi**T**). By saying that a bit is of base 2, we mean it has only 2 possible values, say 0 and 1. That is why binary numbers are written as series of 0s and 1s. Compare this to the decimal, or base 10, number system, where each digit has 10 possible values: 0, 1, 2, 3, 4, 5, 6, 7, 8, and 9.

If we only have one bit, the only possible numbers that may be stored will be 0 and 1; if we are allowed two bits, then the possible numbers will be 00, 01, 10, and 11, i.e., $2^2 = 4$ possibilities (corresponding to the numbers 0, 1, 2, 3 in the decimal number system). Similarly, if we are allowed three digits, then the possible numbers will be 000, 001, 010, 011, 100, 101, 110, and 111; i.e., $2^3 = 8$ possibilities. Do you notice a pattern? In the binary system, when using n digits to represent a character, one is able to write a representation for 2^n characters.

In the English language, there are 26 letters in the alphabet; we need to distinguish the upper-case letters from the lower case letters; that makes 52 characters. How about the numbers from 0 to 9, punctuation marks, and a character representing a *space*? Or non-printing characters corresponding to commands needed to control a computer or printer, like start a new page or skip a line? In all, 256 individual characters are assigned code numbers in the widely used American Standard Code for Information Interchange (**ASCII**), equivalent to eight bits (or one **byte**) of information per character. Thus, for example, the letter "A" is assigned the ASCII code number 65, which corresponds to the eight-digit number 01000001 in binary. When we count all the characters, including spaces, in the phrase "To be or not to be", we see it requires 18 bytes of memory; since each byte is 8 bits, this comes out to be equal to 144 bits. The binary representation of this phrase is: 01010100 01101111 00100000 01100010 01100101 00100000 01101111 01110010 00100000 01101110 01101111 01110100 00100000 01110100 01101111 00100000 01100010 01100101.

 Physics Tips 1.6

Physical Implementation of Classical Bits. When building a *classical* computer, each bit requires a *physical implementation*, more precisely a physical system that has *two* possible configurations. For example, in a semiconductor integrated circuit, a *dynamic random access memory* (DRAM) cell usually consists of a capacitor, which can be either discharged or charged, representing the two possible values of a bit; more permanent memory elements are *static random access memory*

(SRAM) cells, consisting of four transistors, again using voltages to encode bits of data. Other more permanent storage devices are magnetic drives, where small magnetic domains are aligned with specific magnetization to indicate a 0 or 1 (although these are nowadays falling out of fashion). Another interesting technology (at least from the physics perspective) is optical data storage, in which the reflection of a **laser** spot indicates a 0 or 1; but, like magnetic storage, CDs and DVDs look like their days are numbered. The earliest external memory technology, which some of us are old enough to remember, was paper tapes with holes in, where the absence or presence of a hole would record one bit of information.

Let's consider a simple (but not very practical) implementation of a digital memory in which each bit is represented by a coin, with "Heads" standing for "0" and "Tails" standing for "1". Thus the letter "A" would be represented by eight coins in a line, displaying Heads-Tails-Heads-Heads-Heads-Heads-Heads-Tails.

What about quantum coins: can they do any better? As we have seen, the state of a single quantum coin is

$$|\psi_1\rangle = \alpha_0|H\rangle + \alpha_1|T\rangle, \tag{1.18}$$

where α_0 and α_1 are probability amplitudes (we have resorted to using subscripts, since otherwise we will rapidly run out of Greek letters).

Stepping up, the state of two quantum coins will be

$$|\psi_2\rangle = \alpha_0|HH\rangle + \alpha_1|HT\rangle + \alpha_2|TH\rangle + \alpha_3|TT\rangle; \tag{1.19}$$

and for three coins:

$$\begin{aligned}|\psi_3\rangle &= \alpha_0|HHH\rangle + \alpha_1|HHT\rangle + \alpha_2|HTH\rangle \\ &+ \alpha_3|HTT\rangle + \alpha_4|THH\rangle + \alpha_5|THT\rangle \\ &+ \alpha_6|TTH\rangle + \alpha_7|TTT\rangle. \end{aligned} \tag{1.20}$$

You can start to see a pattern: each time you add another quantum coin, the number of complex numbers (i.e., the probability amplitudes α_n) required to specify the state is doubled.

Eight quantum coins have a total of $2^8 = 256$ possible configurations: $|HHHHHHHH\rangle$, $|HHHHHHHT\rangle$, $|HHHHHHTH\rangle$, ..., $|TTTTTTTT\rangle$, and thus the quantum state of a set of eight quantum coins is, in general, a combination of these 256 configurations, each of which represents a letter: thus, while eight classical coins can store a single letter, eight quantum coins can store all 256 letters simultaneously, each with its own weighting factor.

Extending this argument, it has been estimated that the total data storage capacity world-wide – that is, all personal computers, iPhones, government databases, all the published books in all the libraries, and so on – is about 10^{24} bits of information. Using the simple argument above, this information could all be stored in the quantum state of 80 quantum coins. We hasten to emphasise this example is just to illustrate the potential of quantum superposition for data storage, and, as we will see, this is *not* how the memory registers of quantum computers are configured during algorithms.

⊕ **Mathematics Tips 1.7**

Representation of the Hermitian Adjoint of Kets and Bras. What happens to the symbols inside a ket or a bra of multiple quantum coins upon taking their Hermitian adjoint can be ambiguous, so we will clarify this point here.

If we have the term $|AB\rangle$, for example, the corresponding bra will be $|AB\rangle^\dagger = \langle AB|$. If we have $\langle XYZ|$, the corresponding ket will be $\langle XYZ|^\dagger = |XYZ\rangle$. In other words, the symbols inside do not change order, just because the brackets do.

1.2.2 Quantum Computers are VERY Efficient

Another advantage of quantum data processing over its classical counterpart is its inherent high degree of **parallelism** – that is, to perform multiple operations on different parts of the stored data simultaneously. In classical computing, this is achieved by having multiple processor units; in the quantum realm it is native to the architecture. To elaborate, let us look at the three-coin state in (1.20) and say we flip the state of one of the coins; i.e., $|H\rangle$ becomes $|T\rangle$ and vice versa. By performing this operation, *all* the terms in (1.20) will be affected at the same time. Due to the nature of this configuration in being in a superposition state, you do not need to perform an operation on every single one of the eight terms; a single operation on the coin will affect them all. For example, if we flip the third coin, the state will change to be this:

$$|\psi_3\rangle_{\text{flipped}} = \alpha_0|HHT\rangle + \alpha_1|HHH\rangle + \alpha_2|HTT\rangle + \alpha_3|HTH\rangle$$
$$+ \alpha_4|THT\rangle + \alpha_5|THH\rangle + \alpha_6|TTT\rangle + \alpha_7|TTH\rangle. \tag{1.21}$$

If you compare (1.20) and (1.21), you notice that the probability amplitudes α_0 and α_1, α_2 and α_3, α_4 and α_5, and α_6 and α_7 have all been flipped, all by a single operation! The operation, known as the $\hat{X}$ operation (we will discuss it in more detail later), affects all of these coefficients; i.e., all of the probability amplitudes. In contrast, in the classical bit case, an operation can only be performed on bits one at a time. Considering a classical bit only provides one place to store information, this means that one has to perform an operation on every piece of data; but in the quantum case, in a multi-(quantum coin) system of size N, *an operation on one quantum coin can be equivalent to an operation of order 2^N classical single-bit operations.* Although this particular example is not of any immediately obvious utility, nevertheless it is a simple illustration of **quantum parallelism**, the key feature that enables the great efficiency of quantum computers.

1.2.3 A Very, Very Big "BUT": Measurement

So, at first glance this idea of using quantum systems, which can be in two configurations simultaneously, enables quantum computers to have very big memories, and be very efficient computationally. But, besides the enormous problems associated with physical implementation of these ideas, there is an unavoidable fundamental problem that arises when we try to access all the information that the quantum computer has stored and processed. When the time comes to measure the final configuration of our N quantum coins, we can only get N bits of classical information. Recall from the previous discussion that when you measure a superposition state of a quantum coin, the coin ends up in either "Heads" or "Tails", never anything in between. The superposition state, which stores all that information, has been irrevocably projected into one of the kets in the initial superposition description. This means that we can only extract a small amount of all that information.

Nevertheless, the promise of all that data storage and parallelism can, for a certain class of mathematical problems, overcome the shortcomings of the output of a quantum register. For example, suppose we are interested only in discovering some particular global property of a function $f(x)$ – say its period, or the value of the function argument x for which $f(x)$ takes some particular value. These are answers that only need a relatively modest number of bits to write down, but nevertheless need an awful lot of data storage and processing to discover: one needs to evaluate the function for a large number of arguments in order to discover the period; or we must search the database over and over again before finding the specific value of the argument we seek. *Tasks that can be performed efficiently in parallel, and require a relatively small amount of information which we can extract despite the limitations of quantum measurement, are the tasks at which a quantum computer can excel.*

Exercises 1.2.4

1. In the binary number system, work out how many whole numbers, starting from 0, can be expressed with three bits. *Hint: Each bit represents a digit.* Work out the binary representations and label each with the number it represents.
2. *Binary Code Versus Morse Code*
   (a) In the binary number system, each digit is represented by either 0 or 1. What is the maximum number of symbols (letters, numbers, etc.) that can be represented by five digits?
   (b) Morse code uses a series of dots and dashes to represent symbols. With a maximum of five dots and/or dashes, how many symbols can be represented?
   (c) With a maximum of five *digits*, which of these two coding systems would be more appropriate to use to represent symbols? Why?
3. How many bytes of memory does the expression "Quantum Information" require to be stored? How many bits does this correspond to?
4. How many quantum coins would we need to store the expression "Quantum Information"?
5. Using the normalization relation in two quantum coins, show that a two-(quantum coin) system depends on six real coefficients.
6. How big is the memory of a 10-bit classical computer? How about a 10-(quantum coin) quantum computer?
7. If we want to perform an operation on 100 bits, how many times do we have to apply the corresponding operator? How many times do we have to perform this operation if we have the most general 100-(quantum coin) superposition state? What if we have 100 quantum coins, but they are not in a superposition state?
8. A three-(quantum coin) state potentially has eight pieces of information. How much of this information can one retrieve upon measurement of the state?

1.3 CONCEPT CHECK AND EXERCISES

Concept Check

- In the classical description of systems, the configuration of a system and the measured quantities are the same. However, the two are not the same in the quantum case.
- The state of a quantum system gives us probabilities of the possible measurement outcomes. Quantum mechanical systems can be in two or more configurations at once, until the system is measured, when they collapse into being in a single configuration.
- Classical randomness is a measure of our ignorance, while quantum randomness is an inherent property of the system.
- The mathematical description of the quantum coin is analogous to that of vectors.

- A quantum coin has two possible configurations; mathematically its state is

$$|\varphi\rangle = \alpha|H\rangle + \beta|T\rangle = \begin{bmatrix} \alpha \\ \beta \end{bmatrix},$$

alternatively,

$$\langle\varphi| = \alpha^*\langle H| + \beta^*\langle T| = \begin{bmatrix} \alpha^* & \beta^* \end{bmatrix}.$$

- Implications for data processing (when the number of quantum coins is greater than 1): exponentially large data storage; massive parallelism; but limited output.

Chapter Exercises

1. Given three linearly independent vectors **a**, **b**, and **c**, how can they be used to construct three orthonormal vectors?

2. A **qutrit** is a quantum "coin" that has three possible measurement results. Given three different non-orthogonal states $|a\rangle$, $|b\rangle$, and $|c\rangle$ of a qutrit, find a general expression for three orthogonal and normalized states.

3. Given the general two-(quantum coin) states $|\psi\rangle = a|H\rangle + b|T\rangle$ and $|\phi\rangle = c|H\rangle + d|T\rangle$, find:
   (a) the matrix representation of each state,
   (b) the possible inner products, and
   (c) the possible outer products.

4. Assuming that the states $|\psi\rangle$ and $|\phi\rangle$ given in Exercise 3 are not normalized, normalize them. What are the probabilities of obtaining heads and tails for each state?

5. Given the following one-(quantum coin) operator: $\hat{X} = \begin{bmatrix} 0 & 1 \\ 1 & 0 \end{bmatrix}$.

   For a state that describes two quantum coins, if we want to apply $\hat{X}$ on both coins, we need to take the tensor product of two copies of $\hat{X}$ and then apply it on the state. In other words, we need to apply $\hat{X} \otimes \hat{X}$. Find the matrix representation of the two-(quantum coin) operator that will result in the states of the individual qubits to be flipped.

6. Given the quantum coin state $|\psi\rangle = \frac{1}{\sqrt{3}}|H\rangle + \sqrt{\frac{2}{3}}|T\rangle$, upon measurement, what is the most likely state that the quantum coin will collapse to? Explain.

7. What decimal number (base 10) has the binary representation 1000000000?

8. What is the maximum number of symbols that can be represented by n bits? How about a maximum of n dots and/or dashes in Morse code; what is the maximum number of symbols that can be represented in this case?

9. How many bits do we need to store a 100-character expression? How many quantum coins will suffice for the same purpose?

10. In mathematics, the symbol $\binom{n}{r}$ is read "n choose r", and it corresponds to the number of combinations of n objects taken r at a time, with a disregard for the order in which the objects are chosen. It appears in Pascal's triangle as well as in the binomial theorem, for example. In this exercise, we will define an n-(quantum coin) state in terms of this symbol as follows: $|\psi\rangle = \sum_{r=0}^{2^n-1} \binom{2^n-1}{r}|r\rangle$, where n and r are whole numbers, where $\langle r|s\rangle = 0$ unless $r = s$.

(a) Find the normalization constant for $|\psi\rangle$.

(b) Expand the state $|\psi\rangle$ for $n = 2$. Then, re-write the state so that the numbers in the kets, which are written in base 10, are re-written in base 2. What are the probabilities associated with each of the kets in the state description?

(c) Repeat (b) for $n = 3$.

11. Let us assume we have a fair classical coin, so that the probabilities of obtaining Heads and Tails are the same.

(a) What is the probability of obtaining Heads with one toss?

(b) If we toss the coin twice, what are the probabilities of not obtaining Heads at all, of obtaining one Head only, and of obtaining two Heads? *Hint: Binomial distribution.*

(c) What are the various probabilities if we now toss the coin three times?

(d) Now consider the quantum coin of equal superposition of Heads and Tails: $(|H\rangle + |T\rangle)/\sqrt{2}$. What happens if you have two quantum coins? What are the probability amplitudes associated with getting various numbers of Heads? How are they related to the probabilities associated with the classical coin?

(e) Repeat (d) for three quantum coins.

2 Quantum Mechanics for Quantum Computers

In this chapter, we will further explore the ideas that we learned about informally in Chapter 1. From now on, our quantum coins will be referred to as **qubits**, and their states, in keeping with our computer-science goals, will be labeled $|0\rangle$ and $|1\rangle$; i.e., they will be written as linear combinations of the elements of the **computational basis**: $\{|0\rangle, |1\rangle\}$. We will investigate the mathematics of the manipulation and control of single qubits and how to describe them when they are measured. We will also learn about quantum circuits and how they can be used to represent qubit operations, a skill we will need in later chapters.

Learning Objectives

- Be able to solve linear algebra problems that pertain to quantum information.
- Learn about the general properties of the one-qubit unitary operator and become familiar with common quantum gates.
- Learn about the Schrödinger equation and its role in studying the evolution of the quantum state.
- Become familiar with quantum circuit diagrams.
- Learn about the mathematics of quantum measurements.
- Be introduced to the density matrix representation of quantum systems and understand its role in representing randomness.

 Interesting Facts 2.1

The Invention of the Word "Qubit". According to the *Oxford English Dictionary*, the word "qubit" first appears in 1994 in a News and Views article in *Nature* **367**, pp. 513–514 written by Arthur Ekert; however, this article is describing a recent work of Benjamin Schumacher, subsequently published in *Physical Review A*, and it is he who is generally accepted as the originator of the term. Despite spirited philological rear-guard actions for other spellings, such as "qbit", the tide of "qubit" has proven irresistible, and we will use it in this book. Note that "qubit" is short for *quantum bit*.

2.1 Vectors and Matrices

This section mostly covers background information that is already assumed plus a few more necessary mathematical concepts. Depending on the level and background of the reader, it can be studied, reviewed, or skipped, keeping in mind that it contains material that is essential for continuing the learning journey in this book. Regardless of the situation, we recommend that students at least solve the exercises at the end of the section.

In this section, we will have a brief tour of one of the very important mathematical tools essential for understanding the control of single qubits in quantum information, namely *matrices*. We have already seen them in Chapter 1, but we will go over them in a bit more detail here. A matrix is a rectangular array of numbers, written in rows and columns. When dealing with single qubits we will be dealing mostly with three types of matrices: 1×2 row vectors, or bras:

$$\langle u | = \begin{bmatrix} r & s \end{bmatrix}, \tag{2.1}$$

2×1 column vectors, or kets:

$$|v\rangle = \begin{bmatrix} x \\ y \end{bmatrix}, \tag{2.2}$$

and 2×2 *square matrices*, also called operators:

$$\hat{M} = \begin{bmatrix} a & b \\ c & d \end{bmatrix}. \tag{2.3}$$

The numbers in these matrices – they are called *elements* – are all assumed to be complex in this book. Often the elements are referred to by subscripts, thus M_{12} is the element of the matrix $\hat{M}$ in row 1 and column 2; i.e., $M_{12} = b$. The row index is always followed by the column index: this is a good thing to remember.

2.1.1 How to Add and Multiply Matrices

There are rules about how to do arithmetic with these arrays. Adding two matrices together is just a question of adding the corresponding numbers together. For example, suppose we have two matrices

$$\hat{M} = \begin{bmatrix} a & b \\ c & d \end{bmatrix}$$

and

$$\hat{L} = \begin{bmatrix} e & f \\ g & h \end{bmatrix};$$

then

$$\hat{M} + \hat{L} = \begin{bmatrix} a+e & b+f \\ c+g & d+h \end{bmatrix}.$$

You can only add two matrices that have the same number of rows and the same number of columns. Multiplying a matrix by a number is also simple: you multiply each element of the array by the number.

Multiplying two matrices together is a bit more involved. It involves row by column multiplication. It is done by multiplying the first element of the row with the first element of the column, the second element of the row with the second element of the column, and so on. Then, all the products are added together. That resulting sum becomes the element of the new matrix such that its row number is the same as the one from the first matrix involved in the matrix multiplication, and its column number is the same as the number of the column in the second matrix involved in the multiplication, as can be seen in the subscripts in Eq. (2.4). If the matrix product of $\hat{L}$ and $\hat{M}$ is $\hat{P}$, then

$$P_{ij} = \sum_k L_{ik} M_{kj}, \tag{2.4}$$

where, as before, we denote the element in the i-th row and the j-th column of the matrix $\hat{P}$ by P_{ij}. The sum over k runs from 1 to the number of columns in $\hat{L}$, which must be equal to the number of rows in matrix $\hat{M}$; if these two numbers are not the same, the product of the two matrices is not defined. The product matrix $\hat{P}$ has the same number of rows as $\hat{L}$ and the same number of columns as $\hat{M}$.

In general $\hat{L}\hat{M}$ does not equal $\hat{M}\hat{L}$, even if they are both square matrices with the same number of rows and columns. If it happens that $\hat{L}\hat{M} = \hat{M}\hat{L}$ then $\hat{L}$ and $\hat{M}$ are called *commuting* matrices (the word is used in the sense of "to swap", rather than "to be stuck in traffic twice a day").

At first this appears a rather complicated way of doing things, but there is a method in it. For example, in the 2×2 matrices used above:

$$\hat{M}\hat{L} = \begin{bmatrix} ae + bg & af + bh \\ ce + dg & cf + dg \end{bmatrix}. \tag{2.5}$$

We can also multiply a 2×2 matrix $\hat{M}$ by a 2×1 column vector $|v\rangle$:

$$\hat{M}|v\rangle = \begin{bmatrix} a & b \\ c & d \end{bmatrix} \begin{bmatrix} x \\ y \end{bmatrix} = \begin{bmatrix} ax + by \\ cx + dy \end{bmatrix}, \tag{2.6}$$

and similarly a 1×2 row vector $\langle u|$ can multiply a 2×2 matrix $\hat{M}$:

$$\langle u|\hat{M} = \begin{bmatrix} r & s \end{bmatrix} \begin{bmatrix} a & b \\ c & d \end{bmatrix} = \begin{bmatrix} ar + cs & br + ds \end{bmatrix}. \tag{2.7}$$

We cannot multiply $\hat{M}\langle u|$ or $|v\rangle\hat{M}$: in both cases the number of columns of the first does not match the number of rows of the second.

Two other important examples we have already seen in Chapter 1: if we multiply a row vector (i.e., a 1×2 matrix) and a column vector (which is a 2×1 matrix) we obtain a 1×1 matrix, i.e., just a single number (or a *scalar*):

$$\langle u|v\rangle = \begin{bmatrix} r & s \end{bmatrix} \begin{bmatrix} x \\ y \end{bmatrix} = rx + sy. \tag{2.8}$$

This is the scalar product we encountered in the Products of Two State Vectors (Mathematics Tips 1.3). Similarly, multiplying a column vector (2×1) and a row vector (1×2) gives us a 2×2 matrix:

$$|v\rangle\langle u| = \begin{bmatrix} x \\ y \end{bmatrix} \begin{bmatrix} r & s \end{bmatrix} = \begin{bmatrix} xr & xs \\ yr & ys \end{bmatrix}. \tag{2.9}$$

This is the outer product which we also saw before.

One important matrix is the identity, $\hat{I}$, which is a square matrix with 1 along all of the top-left to bottom-right diagonal, and 0 everywhere else. For the 2×2 matrices we are particularly interested in,

$$\hat{I} = \begin{bmatrix} 1 & 0 \\ 0 & 1 \end{bmatrix}. \tag{2.10}$$

As the name suggests, $\hat{I}\hat{M} = \hat{M}\hat{I} = \hat{M}$, $\hat{I}|v\rangle = |v\rangle$ and $\langle u|\hat{I} = \langle u|$; but $\hat{I}\langle u|$ and $|v\rangle\hat{I}$ just do not exist, as we noted before, because the number of columns of the matrix on the left must equal the number of rows on the matrix to the right.

Multiplying a square matrix $\hat{M}$ can be reversed by the *inverse* matrix $\hat{M}^{-1}$, given by

$$\hat{M}^{-1} = \frac{1}{ad - bc} \begin{bmatrix} d & -b \\ -c & a \end{bmatrix}. \tag{2.11}$$

Multiplying $\hat{M}^{-1}$ by $\hat{M}$ we find

$$\hat{M}^{-1}\hat{M} = \frac{1}{ad - bc} \begin{bmatrix} d & -b \\ -c & a \end{bmatrix} \begin{bmatrix} a & b \\ c & d \end{bmatrix} = \frac{1}{ad - bc} \begin{bmatrix} da - bc & db - bd \\ -ca + ac & -cd + ad \end{bmatrix} = \begin{bmatrix} 1 & 0 \\ 0 & 1 \end{bmatrix}. \tag{2.12}$$

The quantity

$$D = ad - bc \tag{2.13}$$

is called the *determinant* of the matrix $\hat{M}$, or Det$\{\hat{M}\}$. It turns out to be quite an important quantity which keeps cropping up. In the case of the inverse matrix, there is an implicit assumption that $D \neq 0$ (otherwise, how can we divide by it?). If $D = 0$ the matrix is called *singular*, and an inverse just does not exist.

⊕ Mathematics Tips 2.2

Cayley–Hamilton Theorem for 2×2 **Matrices.** We can calculate functions of matrices just like we can calculate a function of a regular number. For example, $\hat{M}^2 = \hat{M}\hat{M}$, as one would expect. Using Eq. (2.5),

$$\hat{M}^2 = \begin{bmatrix} a^2 + bc & ab + bd \\ ca + dc & cb + d^2 \end{bmatrix} = (a + d) \begin{bmatrix} a & b \\ c & d \end{bmatrix} - (ad - bc) \begin{bmatrix} 1 & 0 \\ 0 & 1 \end{bmatrix} = T\hat{M} - D\hat{I}, \tag{2.14}$$

where the *Trace* T is defined by

$$T = \text{Tr}\{\hat{M}\} = a + d. \tag{2.15}$$

In general, the Trace of a square matrix is the sum of all the elements along the top-left to bottom-right diagonal; i.e., it equals the sum of the diagonal elements of the matrix, and is another quantity we will encounter a lot. Note that the formula Eq. (2.14) applies only to 2×2 matrices; it is a special case of the *Cayley–Hamilton* theorem, which generalizes things for square matrices of any size. We will be seeing examples with other functions a little later.

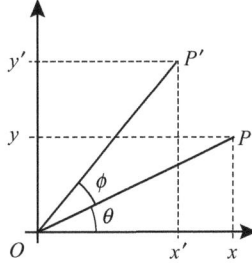

Figure 2.1   Rotation of vector $\overrightarrow{OP}$ by an angle ϕ in the x–y plane.

Rotations and Orthogonal Matrices

Let us illustrate all this with a simple example of a two-dimensional vector with real (rather than complex) elements. The position vector $\overrightarrow{OP}$ of a point in the x–y plane fits this description. A rotation changes the direction of a vector but keeps its length intact. In the figure the original vector $\overrightarrow{OP}$ has length ℓ and makes an angle θ with the x-axis; thus, its two components are $x = \ell \cos \theta$ and $y = \ell \sin \theta$. The rotated vector $\overrightarrow{OP'}$ has the same length but makes an angle $\theta' = \theta + \phi$ with the x-axis, ϕ being the angle of rotation. Thus, the components of $\overrightarrow{OP'}$ are $x' = \ell \cos(\theta + \phi)$ and $y' = \ell \sin(\theta + \phi)$. Using the trigonometry formulas:

$$\cos(\theta + \phi) = \cos\theta \cos\phi - \sin\theta \sin\phi,$$

$$\sin(\theta + \phi) = \sin\theta \cos\phi + \cos\theta \sin\phi,$$

we see that $x' = x \cos\phi - y \sin\phi$, and $y' = x \sin\phi + y \cos\phi$, or, in matrix form:

$$\begin{bmatrix} x' \\ y' \end{bmatrix} = \begin{bmatrix} \cos\phi & -\sin\phi \\ \sin\phi & \cos\phi \end{bmatrix} \begin{bmatrix} x \\ y \end{bmatrix}.$$

The 2×2 matrix in this formula is the rotation matrix for a rotation in the x–y plane about the z-axis (which is coming out of the origin perpendicular to this page) by an angle ϕ; we will denote it by $\hat{R}_z(\phi)$.

The rotation matrix has some interesting special properties. In particular, if we calculate its determinant, we find $D = \cos^2\phi + \sin^2\phi = 1$, and thus the inverse matrix is, using Eq. (2.11),

$$\hat{R}_z(\phi)^{-1} = \begin{bmatrix} \cos\phi & \sin\phi \\ -\sin\phi & \cos\phi \end{bmatrix} = \hat{R}_z(\phi)^T, \tag{2.16}$$

where the superscript T denotes the *transpose* of the matrix, i.e., the matrix formed from swapping the rows and columns:

$$\begin{bmatrix} a & b \\ c & d \end{bmatrix}^T = \begin{bmatrix} a & c \\ b & d \end{bmatrix}. \tag{2.17}$$

Matrices whose transpose is equal to their inverse are called *orthogonal* matrices.

A similar type of transformation of a matrix which we will encounter a great deal is the *Hermitian conjugate* (as a mark of its popularity, it has several other names: *Hermitian adjoint*, *Hermitian transpose* or *conjugate transpose*; all mean the same thing), in which the matrix it transposed and, at the same time, each element is replaced by its complex conjugate. This operation is denoted by a *dagger* symbol, $\hat{M}^{\dagger}$:

$$\begin{bmatrix} a & b \\ c & d \end{bmatrix}^{\dagger} = \begin{bmatrix} a^* & c^* \\ b^* & d^* \end{bmatrix}. \tag{2.18}$$

✦ Mathematics Tips 2.3

Special Types of Matrices. The elements of our generic 2×2 matrix $\hat{M}$ are four complex numbers, and performing even basic matrix arithmetic with them can get quite complicated. Fortunately, we often end up dealing with simpler matrices; here is a brief outline of some of the more common.

1. **Symmetric matrices** have the property that $\hat{\mathscr{S}}^T = \hat{\mathscr{S}}$, or

$$\begin{bmatrix} a & b \\ c & d \end{bmatrix} = \begin{bmatrix} a & c \\ b & d \end{bmatrix}, \tag{2.19}$$

i.e., $b = c$.

2. **Hermitian matrices** have the property that $\hat{\mathscr{H}}^{\dagger} = \hat{\mathscr{H}}$, or

$$\begin{bmatrix} a & b \\ c & d \end{bmatrix} = \begin{bmatrix} a^* & c^* \\ b^* & d^* \end{bmatrix}. \tag{2.20}$$

In other words, a and d are both real, and $c = b^*$. As we will see, a lot of important matrices in quantum mechanics are Hermitian.

3. **Orthogonal matrices**, as we saw above, have the property that $\hat{\mathscr{O}}^{-1} = \hat{\mathscr{O}}^T$, or

$$\frac{1}{ad - bc} \begin{bmatrix} d & -b \\ -c & a \end{bmatrix} = \begin{bmatrix} a & c \\ b & d \end{bmatrix}. \tag{2.21}$$

If we calculate the determinant of the matrices on the two sides of the equation, we find that $D = (ad - bc) = \pm 1$; and comparing corresponding elements, $d = Da$ and $b = -Dc$. It turns out that any symmetric matrix can be written in terms of an orthogonal matrix $\hat{\mathscr{O}}$ and a diagonal matrix $\hat{\Lambda}$ by the relation $\hat{\mathscr{S}} = \hat{\mathscr{O}}^T \hat{\Lambda} \hat{\mathscr{O}}$.

4. **Unitary matrices** have the property that $\hat{U}^{-1} = \hat{U}^{\dagger}$, or

$$\frac{1}{ad - bc} \begin{bmatrix} d & -b \\ -c & a \end{bmatrix} = \begin{bmatrix} a^* & c^* \\ b^* & d^* \end{bmatrix}. \tag{2.22}$$

This implies that $|ad - bc| = 1$, and so the determinant D is just a phase factor $\exp(i\phi)$ and $d = Da$ and $b = -Dc^*$. Any Hermitian matrix can be written in terms of a unitary matrix $\hat{U}$ and a real-valued diagonal matrix $\hat{\Lambda}$ by the relation $\hat{\mathscr{H}} = \hat{U}^{\dagger} \hat{\Lambda} \hat{U}$.

Unitary matrices crop up a lot, and all the basic operations needed to make a quantum device are of this class.

2.1.2 Eigenvalues and Eigenvectors

When a vector is multiplied by a matrix $\hat{M}$ and the resulting vector is simply a scalar multiple of the original vector, then that specific vector is called an *eigenvector* of the matrix. The scalar that multiplies it after this transformation is the corresponding *eigenvalue*. In other words, the effect of $\hat{M}$ on its eigenvectors is to stretch or shrink them (or maybe just leave them unchanged); the vectors do not get rotated. Using our compact notation, the relation is as follows:

$$\hat{M}|\mu\rangle = \mu|\mu\rangle, \tag{2.23}$$

where we have used the eigenvalue as a label for its corresponding eigenvector (we do that a lot in quantum mechanics). Only certain vectors have this property, and we have to do a bit of algebra to find these vectors and their eigenvalues.

 Interesting Facts 2.4

What Does "Eigen" Mean? The German word *eigen* is an adjective that has several meanings, such as: "own", "particular", or "specific". Its use in the present context originated with the mathematician David Hilbert, who used it in a paper he wrote in German on the theory of integral equations in 1904 (related ideas had been explored by mathematicians for nearly two centuries before this, but they used different words). Contemporary English-speaking scientists translated

Hilbert's term as "proper", and the phrases *proper value* or *proper function* can be found in some early English-language books on quantum mechanics in the 1930s. The use of the German word in English-language publications was popularized by Paul Dirac in his famous book *Principles of Quantum Mechanics* (1930), where he pointed out that the English word "proper" already had a lot of other meanings in physics, so using the German term would be more distinctive. For a more thorough discussion see Jeff Miller, "Earliest Uses of Some Words of Mathematics" https://mathshistory.st-andrews.ac.uk/Miller/mathword/.

Note that *any* square matrix $\hat{M}$ has eigenvalues and eigenvectors; but the right eigenvectors $\hat{M}|v\rangle = \mu|v\rangle$ and the left eigenvectors $\langle u|\hat{M} = \mu\langle u|$ turn out to be different from one another (i.e., in general $\langle u|^\dagger \neq |v\rangle$). To keep things simple, we only deal with Hermitian matrices, represented by $\hat{\mathcal{H}}$, in this section (where both the right and left eigenvectors are the same); these turn out to be the most important class of matrices for quantum mechanics anyway.

To find the eigenvalues and eigenvectors of the matrix $\hat{\mathcal{H}}$ we must solve the following equation:

$$\begin{bmatrix} a & b \\ b^* & d \end{bmatrix} \begin{bmatrix} x \\ y \end{bmatrix} = \mu \begin{bmatrix} x \\ y \end{bmatrix}, \tag{2.24}$$

where x and y are the components of the eigenvector, and μ is the eigenvalue. Multiplying the matrices and vectors out, this matrix equation is equivalent to a pair of algebraic equations we must solve together:

$$ax + by = \mu x, \qquad b^*x + dy = \mu y. \tag{2.25}$$

Solving the first of these equations we get $y = (\mu - a)x/b$; substituting this into the second, cancelling the common factor of x and rearranging we find

$$\mu^2 - T\mu + D = 0. \tag{2.26}$$

Here we have used the determinant $D = ad - |b|^2$ and the trace $T = a+d$ once again. This polynomial equation is generically known as the *characteristic equation*; the degree of the polynomial is equal to the dimension of the matrix: thus 2×2 matrices have quadratic characteristic equations.

Using the quadratic formula, the two roots of the characteristic equation are the two eigenvalues of $\hat{\mathcal{H}}$:

$$\mu_+ = (T + R)/2 \quad \text{and} \quad \mu_- = (T - R)/2, \tag{2.27}$$

where $R = \sqrt{T^2 - 4D}$. Note that

$$T = \mu_+ + \mu_- \tag{2.28}$$

and

$$D = \mu_+\mu_-. \tag{2.29}$$

For larger matrices, the determinant is equal to the product of all the eigenvalues (there is an involved formula for the determinant in terms of the elements of the matrix as well), and the trace is equal to the sum of all the eigenvalues (which is equal to the sum of all the elements of the matrix along the top-left to bottom-right diagonal of the array).

The two normalized eigenvectors are

$$|\mu_+\rangle = \frac{1}{\sqrt{2R(R+S)}} \begin{bmatrix} R+S \\ 2b^* \end{bmatrix},$$

$$|\mu_-\rangle = \frac{1}{\sqrt{2R(R-S)}} \begin{bmatrix} R-S \\ -2b^* \end{bmatrix}, \tag{2.30}$$

where $S = a - d = T - 2d = \sqrt{R^2 - 4|b|^2}$. To show this is so, let us multiply everything out (we will just do the $|\mu_+\rangle$ case, and we will leave out the $1/\sqrt{2R(R+S)}$ normalizing factor, which will be a common factor multiplying both sides):

$$\begin{bmatrix} a & b \\ b^* & d \end{bmatrix} \begin{bmatrix} S+R \\ 2b^* \end{bmatrix} = \begin{bmatrix} aR + aS + 2|b|^2 \\ b^*(R+S+2d) \end{bmatrix}. \tag{2.31}$$

The first term is $aR + aS + 2|b|^2 = aR + a(T - 2d) + 2|b|^2 = a(T + R) - 2ad + 2|b|^2 = 2(a\mu_+ - D) = 2(\mu_+^2 - d\mu_+)$ (we used the quadratic characteristic equation to do this bit) $= 2\mu_+(\mu_+ - d) = \mu_+(T + R - 2d) = \mu_+(S + R)$; the second term is $b^*(R + S + 2d) = b^*(R + T) = \mu_+ 2b^*$. Very similar steps can be used to verify that $|\mu_-\rangle$ is also an eigenvector with eigenvalue $\mu_- = (T - R)/2$. The normalization factors $1/\sqrt{2R(R+S)}$ and $1/\sqrt{2R(R-S)}$ ensure that $\langle\mu+|\mu+\rangle = 1$ and $\langle\mu-|\mu-\rangle = 1$.

If we take the scalar product of $\langle\mu_+|\mu_-\rangle$, something rather interesting happens:

$$\begin{aligned} \langle\mu_+|\mu_-\rangle &= \frac{1}{\sqrt{2R(R+S)}\sqrt{2R(R-S)}} \begin{bmatrix} R+S & 2b \end{bmatrix} \begin{bmatrix} R-S \\ -2b^* \end{bmatrix} \\ &= \frac{(R+S)(R-S) - 4|b|^2}{2R\sqrt{(R+S)(R-S)}} \\ &= \frac{R^2 - S^2 - 4|b|^2}{2R\sqrt{R^2 - S^2}} = 0, \end{aligned} \tag{2.32}$$

where we have used $R = \sqrt{S^2 + 4|b|^2}$. In words: the vectors $|\mu_+\rangle$ and $|\mu_-\rangle$ are orthogonal, and we can use them as a basis for the two-dimensional space.

One immediate consequence of this orthogonality is that we can construct a 2×2 unitary matrix by using the each of the two eigenvectors as columns, as follows:

$$\hat{U} = \frac{1}{\sqrt{2R}} \begin{bmatrix} \sqrt{R-S} & \sqrt{R+S} \\ -2b^*/\sqrt{R-S} & 2b^*/\sqrt{R+S} \end{bmatrix}, \tag{2.33}$$

which has the useful property

$$\hat{U}^\dagger \hat{\mathcal{H}} \hat{U} = \begin{bmatrix} \mu_- & 0 \\ 0 & \mu_+ \end{bmatrix}. \tag{2.34}$$

In other words, the unitary matrix diagonalizes the matrix $\hat{\mathcal{H}}$.

2.1.3 Projectors and Functions of Matrices

Using the results we obtained for $\hat{\mathcal{H}}$, we define two **projectors** using the outer products of the eigenvectors:

$$\hat{\Pi}_+ = |\mu_+\rangle\langle\mu_+| = \frac{1}{2R(R+S)}\begin{bmatrix} (R+S)^2 & 2b(R+S) \\ 2b^*(R+S) & 4|b|^2 \end{bmatrix},$$

$$\hat{\Pi}_- = |\mu_-\rangle\langle\mu_-| = \frac{1}{2R(R-S)}\begin{bmatrix} (R-S)^2 & -2b(R-S) \\ -2b^*(R-S) & 4|b|^2 \end{bmatrix}. \tag{2.35}$$

A projector $\hat{\Pi}_v = |v\rangle\langle v|$ is an operator that transforms an arbitrary vector $|\varphi\rangle$ onto the fixed vector $|v\rangle$, multiplied by the inner product $\langle v|\varphi\rangle$. It is the analogue for quantum states of the projection of vectors we saw in the exercises for Section 1.1 Any projector has the property $\hat{\Pi}\hat{\Pi} = \hat{\Pi}$. For two-level systems, the projectors $\hat{\Pi}_\pm$ introduced above have the following properties: $\hat{\Pi}_+^2 = \hat{\Pi}_+$, $\hat{\Pi}_-^2 = \hat{\Pi}_-$, and $\hat{\Pi}_-\hat{\Pi}_+ = \hat{\Pi}_+\hat{\Pi}_- = 0$. Furthermore, after a bit of algebra, you can show:

$$\hat{\Pi}_+ + \hat{\Pi}_- = \hat{I},$$

$$\mu_+\hat{\Pi}_+ + \mu_-\hat{\Pi}_- = \hat{\mathcal{H}}. \tag{2.36}$$

These two equations can be re-written in the following form:

$$\hat{\Pi}_+ = \frac{-\mu_-\hat{I} + \hat{\mathcal{H}}}{\mu_+ - \mu_-},$$

$$\hat{\Pi}_- = \frac{\mu_+\hat{I} - \hat{\mathcal{H}}}{\mu_+ - \mu_-}. \tag{2.37}$$

If $f(x)$ is some function of a number x, then the function of a matrix is defined by

$$f(\hat{\mathcal{H}}) = f(\mu_+)\hat{\Pi}_+ + f(\mu_-)\hat{\Pi}_-. \tag{2.38}$$

Using Eq. (2.37), this may be written as

$$f(\hat{\mathcal{H}}) = \frac{f(\mu_+) + f(\mu_-)}{2}\hat{I} + \frac{f(\mu_+) - f(\mu_1)}{\mu_+ - \mu_-}\left(\hat{\mathcal{H}} - \frac{T}{2}\hat{I}\right)$$

$$= \frac{f(\mu_+)\mu_- - f(\mu_1)\mu_+}{\mu_+ - \mu_-}\hat{I} + \frac{f(\mu_+) - f(\mu_1)}{\mu_+ - \mu_-}\hat{\mathcal{H}}. \tag{2.39}$$

Exercises 2.1.4

1. Given the two matrices

$$\hat{M} = \begin{bmatrix} 1 & 2 \\ 3 & 4 \end{bmatrix}, \quad \hat{L} = \begin{bmatrix} 5 & 6 \\ 7 & 8 \end{bmatrix},$$

the vector: $|v\rangle = \begin{bmatrix} i \\ 3 \end{bmatrix}$, and the 2×2 identity matrix $\hat{I}$, find:

(a) L_{21}, (b) M_{11}, (c) $3\hat{M}$, (d) $\hat{M} - \hat{L}$, (e) $\hat{M}\hat{L}$, (f) $\hat{L}\hat{M}$, (g) $\langle v|\hat{M}$,
(h) $\hat{L}|v\rangle$, (i) $\langle v|v\rangle$, (j) $|v\rangle\langle v|$, (k) $\hat{I}|v\rangle$, (l) $\hat{I}\hat{M}$, (m) $\hat{M}^{-1}$.

2. Given the following three matrices:

$$\hat{X} = \begin{bmatrix} 0 & 1 \\ 1 & 0 \end{bmatrix}, \quad \hat{Y} = \begin{bmatrix} 0 & -i \\ i & 0 \end{bmatrix}, \quad \hat{W} = \begin{bmatrix} 2 & 1 \\ 1 & 2 \end{bmatrix},$$

which one of the following pairs commutes? (If matrices $\hat{A}$ and $\hat{B}$ commute, it means that $\hat{A}\hat{B} = \hat{B}\hat{A}$.)

(a) $\hat{X}$, $\hat{Y}$.

(b) $\hat{X}$, $\hat{W}$.

(c) $\hat{Y}$, $\hat{W}$.

3. Given the following vector in the x–y plane: $|v\rangle = \begin{bmatrix} 1 \\ 1 \end{bmatrix}$.

(a) If it is rotated $45°$, what are its new coordinates?

(b) By how much does the length of the new vector change?

(c) What is the matrix that will reverse this rotation; i.e., what is the inverse of the rotation matrix described in (a)?

4. What is the absolute value or the modulus of the exponential $\exp(i\theta)$? What about $\exp(if(n)\theta)$, where $f(n)$ is a general function of a variable n?

5. For a general 2×2 matrix

$$\begin{bmatrix} a & b \\ c & d \end{bmatrix},$$

work out the properties of the matrix elements and re-write the matrix to reflect these properties for the following cases in which the matrix is

(a) symmetric,

(b) Hermitian,

(c) orthogonal,

(d) unitary.

6. For each of the following Hermitian matrices, find their eigenvalues and their corresponding normalized eigenvectors:

$$(a)\ \hat{\mathscr{H}}_1 = \begin{bmatrix} 1 & 1 \\ 1 & 1 \end{bmatrix},\ (b)\ \hat{\mathscr{H}}_2 = \begin{bmatrix} 1 & i \\ -i & 1 \end{bmatrix},\ (c)\ \hat{\mathscr{H}}_3 = \begin{bmatrix} 41 & 4+3i \\ 4-3i & 17 \end{bmatrix}.$$

7. Show that $|\mu_-\rangle$ defined by Eq. (2.30) is an eigenvector of the Hermitian matrix

$$\hat{\mathscr{H}} = \begin{bmatrix} a & b \\ b^* & d \end{bmatrix}.$$

8. Show that $\langle \mu_+|\mu_+\rangle = 1$ and $\langle \mu_-|\mu_-\rangle = 1$, where $|\mu_+\rangle$ and $|\mu_-\rangle$ are defined by Eq. (2.30).

9. Show that if we used Eq. (2.37) to define the projectors for an arbitrary matrix, i.e., we replace the Hermitian matrix $\hat{\mathscr{H}}$ with the generic 2×2 matrix $\hat{M}$, then these projectors also have the properties $\hat{\Pi}_+^2 = \hat{\Pi}_+$, $\hat{\Pi}_-^2 = \hat{\Pi}_-$ and $\hat{\Pi}_-\hat{\Pi}_+ = \hat{\Pi}_+\hat{\Pi}_- = 0$. *Hint: You might find it useful to use the Cayley–Hamilton formula, Eq. (2.14).*

10. Consider the matrix

$$\hat{M} = \frac{1}{25} \begin{bmatrix} 11 & -48i \\ 48i & 39 \end{bmatrix}.$$

(a) Which of the following terms apply to $\hat{M}$?

i. symmetric,

ii. Hermitian,

iii. orthogonal,

iv. unitary.

    (b) What are the eigenvalues of $\hat{M}$?

    (c) Find the matrices $\cos(\pi \hat{M})$ and $\sin(\pi \hat{M})$.

11. For the matrix

$$\hat{\mathcal{H}} = \begin{bmatrix} 56 & 15 + 36i \\ 15 - 36i & -48 \end{bmatrix},$$

find:

(a) $f(\hat{\mathcal{H}}) = \hat{\mathcal{H}}^2$,

(b) $f(\hat{\mathcal{H}}) = \sqrt{\hat{\mathcal{H}}}$,

(c) $f(\hat{\mathcal{H}}) = 1/\hat{\mathcal{H}}$.

12. If $\hat{M}$ is a Hermitian operator, use Eq. (2.39) to show that

$$\exp(i\hat{M}) = e^{i\tau/2} \left\{ \cos(\Delta\mu/2)\hat{I} + i\frac{\sin(\Delta\mu/2)}{\Delta\mu/2} \left[\hat{M} - \frac{\tau}{2}\hat{I} \right] \right\},$$

where $\tau = \text{Tr}\{\hat{M}\}$ is the trace of $\hat{M}$ and $\Delta\mu = \sqrt{\tau^2 - 4\det\{\hat{M}\}}$ is the absolute value of the difference between the two eigenvalues of $\hat{M}$. Is this operator unitary?

2.2 Single Qubits

Now let us focus on the single qubit, the fundamental building block of a quantum computer, and discuss its quantum mechanics. Recall the most general (superposition) state of one qubit is given by

$$|\psi\rangle = \alpha|0\rangle + \beta|1\rangle, \tag{2.40}$$

where α and β are complex probability amplitudes, the components or weights of each vector in the basis. Initially, we assume that α and β are deterministic (i.e., not random) quantities; we will be studying the case were they can be random in Section 2.6. The probability of obtaining the result $|0\rangle$ upon measuring the system prepared in state $|\psi\rangle$ is $|\alpha|^2$, while $|\beta|^2$ gives the probability of obtaining $|1\rangle$. That is why it is not surprising that $|\alpha|^2 + |\beta|^2 = 1$, as they represent the only two possible measurement outcomes for the qubit. Consequently this also means that quantum mechanical states are normalized:

$$\langle\psi|\psi\rangle = \begin{bmatrix} \alpha^* & \beta^* \end{bmatrix} \cdot \begin{bmatrix} \alpha \\ \beta \end{bmatrix} = 1. \tag{2.41}$$

2.2.1 Operators and Gates

Qubits sitting around minding their own business is all very well, but what can we *do* to one of them to change things? In other words, how can we manipulate qubits to create quantum states we might want to exploit to build a quantum computer? Mathematically, the transformation of a qubit implies changing its state, so it changes from $|\psi\rangle$ to $|\psi'\rangle$, for example. Consequently, this means that the probability amplitudes will, in general, change: α to α' and β to β'.

To begin, we will assume that the transformation, represented by the 2×2 matrix $\hat{U}$, is *linear*. Its elements will transform the coefficients α and β in a linear fashion; i.e., without higher powers of the coefficients, such as α^2 or $\alpha\beta$, etc., involved:

$$\alpha' = U_{00}\alpha + U_{01}\beta,$$
$$\beta' = U_{10}\alpha + U_{11}\beta, \tag{2.42}$$

where each of the U_{ij} in (2.42) corresponds to the four matrix elements of $\hat{U}$. The linearity of quantum mechanics is a fundamental assumption, and it has, so far, proved consistent with all experiments that have tried to test it.

The matrix transformation that gives (2.42) looks like this:

$$\underbrace{\begin{bmatrix} \alpha' \\ \beta' \end{bmatrix}}_{|\psi'\rangle} = \underbrace{\begin{bmatrix} U_{00} & U_{01} \\ U_{10} & U_{11} \end{bmatrix}}_{\hat{U}} \times \underbrace{\begin{bmatrix} \alpha \\ \beta \end{bmatrix}}_{|\psi\rangle}.$$

In order to discover the necessary properties of $\hat{U}$ for it to be a valid quantum operation, let us start by taking the complex conjugate of (2.42):

$$\alpha'^* = U_{00}^*\alpha^* + U_{01}^*\beta^*,$$
$$\beta'^* = U_{10}^*\alpha^* + U_{11}^*\beta^*. \tag{2.43}$$

Now, let us write (2.43) in matrix form:

$$\underbrace{\begin{bmatrix} \alpha'^* & \beta'^* \end{bmatrix}}_{\langle\psi'|} = \underbrace{\begin{bmatrix} \alpha^* & \beta^* \end{bmatrix}}_{\langle\psi|} \times \underbrace{\begin{bmatrix} U_{00}^* & U_{10}^* \\ U_{01}^* & U_{11}^* \end{bmatrix}}_{\hat{U}^\dagger},$$

where $\hat{U}^\dagger$ is the Hermitian conjugate of $\hat{U}$.

As we already discussed, a quantum mechanical state is normalized. As a consequence of that we know that the coefficients of $|\psi'\rangle$ should give the following relation:

$$|\alpha'|^2 + |\beta'|^2 = 1. \tag{2.44}$$

This means that the inner product of $|\psi'\rangle$ has to be equal to 1: $\langle\psi'|\psi'\rangle = 1$. We know that $|\psi'\rangle = \hat{U}|\psi\rangle$, therefore

$$\langle\psi'|\psi'\rangle = \langle\psi|\hat{U}^\dagger\hat{U}|\psi\rangle = 1. \tag{2.45}$$

What can we conclude about the properties of $\hat{U}$ from (2.45)? Looking at this relation, we have $\hat{U}^\dagger\hat{U}$, which is a 2×2 matrix, from which we want to get insight into these properties. First note that $\hat{U}$ cannot depend on the properties of $|\psi\rangle$, $|\psi'\rangle$, or whatever the state of the quantum system is. Otherwise, quantum mechanics would not be linear. Therefore, in order to find out what $\hat{U}^\dagger\hat{U}$ is, we choose whatever convenient values for α and β we want that will reveal the matrix elements of $\hat{U}^\dagger\hat{U}$. The matrix equation that we get from (2.45) is

$$\underbrace{\begin{bmatrix} \alpha^* & \beta^* \end{bmatrix}}_{\langle\psi|} \times \underbrace{\begin{bmatrix} A & B \\ C & D \end{bmatrix}}_{\hat{U}^\dagger\hat{U}} \times \underbrace{\begin{bmatrix} \alpha \\ \beta \end{bmatrix}}_{|\psi\rangle} = 1.$$

Next what we need to do is to substitute in sets of values for (α, β), one at a time, to reveal the values of the matrix elements, A, B, C, and D. If we try the following four different sets of values: $(1, 0)$, $(0, 1)$, $(\frac{1}{\sqrt{2}}, \frac{1}{\sqrt{2}})$, $(\frac{1}{\sqrt{2}}, \frac{i}{\sqrt{2}})$, where i is the imaginary unit, we find that

$$\begin{bmatrix} A & B \\ C & D \end{bmatrix} = \begin{bmatrix} 1 & 0 \\ 0 & 1 \end{bmatrix}.$$

This means that

$$\hat{U}^\dagger \hat{U} = \hat{I}, \tag{2.46}$$

where $\hat{I}$ is the identity operator, and for the product of two matrices to be equal to that, one of them has to be the inverse of the other. Therefore,

$$\hat{U}^\dagger = \hat{U}^{-1}, \tag{2.47}$$

i.e., $\hat{U}^\dagger$ has to be the inverse of $\hat{U}$. *All valid quantum operations on a **closed system** must be unitary*; i.e., must satisfy (2.47). By *closed* we mean that there are no outside influences or randomness that are contributing to changing the system, only the randomness that comes from quantum mechanics; there is no external extraction of information from the system (through interacting with an outside system). Ideally, we want to keep quantum systems as closed as possible if we want a good functional quantum computer. That is why these are the systems that we will focus on in this book, the dynamics of which is described by unitary operators. However, we will also cover some material on non-unitary actions that can occur to the system, whether external association or measurement.

Next, we will look at some examples of unitary operators that can be applied on the one-qubit systems, known in quantum computing as **gates**. First, let us look at a common basis that is used to describe any 2×2 matrix representing a gate. This basis is made of the identity matrix and the three **Pauli matrices**:

$$\hat{I} = \begin{bmatrix} 1 & 0 \\ 0 & 1 \end{bmatrix}, \; \hat{X} = \begin{bmatrix} 0 & 1 \\ 1 & 0 \end{bmatrix}, \; \hat{Y} = \begin{bmatrix} 0 & -i \\ i & 0 \end{bmatrix}, \text{ and } \hat{Z} = \begin{bmatrix} 1 & 0 \\ 0 & -1 \end{bmatrix}. \tag{2.48}$$

Note that the identity operator $\hat{I}$ and the three operators $\hat{X}$, $\hat{Y}$, and $\hat{Z}$ are also often denoted as $\hat{\sigma}_0$, $\hat{\sigma}_x$, $\hat{\sigma}_y$, and $\hat{\sigma}_z$ (or $\hat{\sigma}_0$, $\hat{\sigma}_1$, $\hat{\sigma}_2$, and $\hat{\sigma}_3$), respectively. We will use the notation in Eq. (2.48) except when it is more convenient to do summations over operators.

2.2.2 Bit-Flip Gate

The **bit-flip gate**, as the names suggests, flips a state $|0\rangle$ into the state $|1\rangle$, and vice versa. This can be achieved by applying the $\hat{X}$ operator on the state. For example,

$$\underbrace{\begin{bmatrix} 0 & 1 \\ 1 & 0 \end{bmatrix}}_{\hat{X}} \times \underbrace{\begin{bmatrix} 1 \\ 0 \end{bmatrix}}_{|0\rangle} = \underbrace{\begin{bmatrix} 0 \\ 1 \end{bmatrix}}_{|1\rangle}.$$

If we apply this operator on the general one-qubit state (2.40), the coefficients α and β will switch places, and $|\psi\rangle$ will become

$$|\psi'\rangle = \beta|0\rangle + \alpha|1\rangle. \tag{2.49}$$

For this reason, the $\hat{X}$ operator is called the *bit-flip gate*.

2.2.3 Phase-Flip Gate

 Physics Tips 2.5

Quantum Mechanics' Phases. A *phase* is a term that a quantum state acquires as time passes by. It is a complex number represented by an exponential of the form $e^{i\phi}$, where ϕ is the phase angle. If a phase is *relative*, then it will be multiplying one of the terms describing the quantum state. For example,

$$|\psi\rangle = \alpha|0\rangle + e^{i\phi}\beta|1\rangle. \tag{2.50}$$

If the phase multiplies all the terms describing the state, then it is a **global phase**, as shown here:

$$|\psi\rangle = e^{i\phi}\{\alpha|0\rangle + \beta|1\rangle\}. \tag{2.51}$$

Although they are important in some areas, in quantum computation we do not care about global phases. Global phases do not change the value of any matrix element of the form $\langle\psi|\hat{M}|\psi\rangle$, so they have no influence on any experimentally verifiable predictions the theory might make, so we can ignore them if we want. Look at this state where we factor out a phase:

$$\begin{aligned} |\psi\rangle &= e^{i\theta}\alpha|0\rangle + e^{i\phi}\beta|1\rangle \\ &= e^{i\theta}\{\alpha|0\rangle + e^{i(\phi-\theta)}\beta|1\rangle\}. \end{aligned} \tag{2.52}$$

We can ignore the global phase $e^{i\theta}$, but we cannot ignore the relative phase $e^{i(\phi-\theta)}$, obtaining

$$|\psi\rangle = \alpha|0\rangle + e^{i(\phi-\theta)}\beta|1\rangle. \tag{2.53}$$

The states described in (2.52) and (2.53) are considered the same in quantum mechanics.

Let us see what happens when we apply the $\hat{Z}$ operator on $|1\rangle$:

$$\underbrace{\begin{bmatrix} 1 & 0 \\ 0 & -1 \end{bmatrix}}_{\hat{Z}} \times \underbrace{\begin{bmatrix} 0 \\ 1 \end{bmatrix}}_{|1\rangle} = \underbrace{\begin{bmatrix} 0 \\ -1 \end{bmatrix}}_{-|1\rangle}.$$

Note that $-1 = e^{i\pi}$. Therefore $\hat{Z}|1\rangle = e^{i\pi}|1\rangle$. The **phase-flip gate** $\hat{Z}$ adds a π-phase to $|1\rangle$, but not to $|0\rangle$:

$$\begin{aligned} \hat{Z}|\psi\rangle &= \alpha\hat{Z}|0\rangle + \beta\hat{Z}|1\rangle \\ &= \alpha|0\rangle - \beta|1\rangle \\ &= \alpha|0\rangle + e^{i\pi}\beta|1\rangle. \end{aligned} \tag{2.54}$$

You can see in (2.54) that the phase-flip gate adds a relative phase of angle π to the state $|\psi\rangle$. Remember that relative phases matter.

2.2.4 Bit-Flip and Phase-Flip Gates Combined

If you combine both the bit-flip and the phase-flip gates, you will get both the states flipping and a relative phase added. The combined gates are equal to $i\hat{Y}$, where i is an overall phase, and it equals $e^{i\frac{\pi}{2}}$.

2.2.5 Arbitrary 2 × 2 Unitaries

Now we turn our attention to the matrix description of an arbitrary unitary gate. Let us start with a general 2 × 2 matrix:

$$\begin{bmatrix} a & b \\ c & d \end{bmatrix}. \tag{2.55}$$

If we take its Hermitian adjoint, we get

$$\begin{bmatrix} a & b \\ c & d \end{bmatrix}^{\dagger} = \begin{bmatrix} a^* & c^* \\ b^* & d^* \end{bmatrix}. \tag{2.56}$$

We can also write the inverse of this matrix as follows:

$$\begin{bmatrix} a & b \\ c & d \end{bmatrix}^{-1} = \frac{1}{ad - bc} \begin{bmatrix} d & -b \\ -c & a \end{bmatrix}, \tag{2.57}$$

where $ad - bc$ is the determinant of the matrix, and as long as it is not equal to 0, i.e., the matrix is not *singular*, (2.57) can be used to find the inverse of a 2 × 2 matrix. We know that if the matrix is unitary, then both (2.56) and (2.57) are equal. Using this fact and a little bit of algebra, we get

$$|ad - bc|^2 = 1. \tag{2.58}$$

Equation (2.58) is one of the properties of a unitary matrix, and it implies that the determinant of a unitary is of the form $e^{i\phi}$, where ϕ is a phase. However, since we do not care about the overall phase in quantum mechanics, it is safe to make the assumption that the determinant in this case is given by

$$ad - bc = 1. \tag{2.59}$$

Equation (2.59) makes it easier to compare the matrices in (2.56) and (2.57) to find that the following relations hold between the matrix elements: $a^* = d$ and $c^* = -b$.

This allows us to write our general unitary matrix as

$$\boxed{\hat{U} = \begin{bmatrix} a & b \\ -b^* & a^* \end{bmatrix}.} \tag{2.60}$$

Equation (2.60) can be a useful way to express a unitary. Using (2.59), we get the relation

$$|a|^2 + |b|^2 = 1. \tag{2.61}$$

2.2.6 Changing the Basis with a Unitary

Recall the computational basis $\{|0\rangle, |1\rangle\}$, which is analogous in two-dimensional space to the x- and y-axes, such that the orthonormality conditions $\langle 0|0\rangle = \langle 1|1\rangle = 1$, and $\langle 0|1\rangle = 0$ hold; i.e., if two vectors are the same, their inner product is 1. Otherwise, it is 0. We can use a unitary matrix to transform this basis into another basis $\{|0'\rangle, |1'\rangle\}$:

$$\begin{aligned} \hat{U}|0\rangle &= |0'\rangle, \\ \hat{U}|1\rangle &= |1'\rangle. \end{aligned} \tag{2.62}$$

Let us use the result in (2.60) to find $|0'\rangle$ and $|1'\rangle$:

$$\underbrace{\begin{bmatrix} a & b \\ -b^* & a^* \end{bmatrix}}_{\hat{U}} \times \underbrace{\begin{bmatrix} 1 \\ 0 \end{bmatrix}}_{|0\rangle} = \underbrace{\begin{bmatrix} a \\ -b^* \end{bmatrix}}_{|0'\rangle}, \tag{2.63}$$

and

$$\underbrace{\begin{bmatrix} a & b \\ -b^* & a^* \end{bmatrix}}_{\hat{U}} \times \underbrace{\begin{bmatrix} 0 \\ 1 \end{bmatrix}}_{|1\rangle} = \underbrace{\begin{bmatrix} b \\ a^* \end{bmatrix}}_{|1'\rangle}. \tag{2.64}$$

This means that, in terms of the old basis, the new one is given by

$$|0'\rangle = a|0\rangle - b^*|1\rangle,$$
$$|1'\rangle = b|0\rangle + a^*|1\rangle. \tag{2.65}$$

Moreover, the orthonormality relations are maintained such that $\langle 0'|0'\rangle = \langle 1'|1'\rangle = 1$ and $\langle 0'|1'\rangle = 0$, since this is the nature of unitary operators. This alternative way of thinking about unitary operations turns out to be quite useful when analyzing quantum circuits, as we will see.

✦ Mathematics Tips 2.6

An Algebraic Expression for $\hat{U}$. To find another useful expression for a general $\hat{U}$, we expand the matrix elements into their real and imaginary parts:

$$a = a_1 + ia_2,$$
$$b = b_1 + ib_2. \tag{2.66}$$

We use (2.66) to express $\hat{U}$ in this form:

$$\hat{U} = \begin{bmatrix} a_1 + ia_2 & b_1 + ib_2 \\ -b_1 + ib_2 & a_1 - ia_2 \end{bmatrix} = a_1 \underbrace{\begin{bmatrix} 1 & 0 \\ 0 & 1 \end{bmatrix}}_{\hat{I}} + ib_2 \underbrace{\begin{bmatrix} 0 & 1 \\ 1 & 0 \end{bmatrix}}_{\hat{X}} + ib_1 \underbrace{\begin{bmatrix} 0 & -i \\ i & 0 \end{bmatrix}}_{\hat{Y}} + ia_2 \underbrace{\begin{bmatrix} 1 & 0 \\ 0 & -1 \end{bmatrix}}_{\hat{Z}}. \tag{2.67}$$

Moreover, using (2.61), we get the following relation between the real and imaginary parts of a and b:

$$a_1^2 + a_2^2 + b_1^2 + b_2^2 = 1. \tag{2.68}$$

Let us define a real three-dimensional (3D) unit vector $n = (n_x, n_y, n_z)$, where $n_x^2 + n_y^2 + n_z^2 = 1$. We also know from trigonometry that $\cos^2(\theta/2) + \sin^2(\theta/2) = 1$ (the reason we use $(\theta/2)$ rather than just θ will be clear a little later). Now, we will re-write the latter identity in another equivalent way:

$$\cos^2(\theta/2) + \sin^2(\theta/2) = 1,$$
$$\cos^2(\theta/2) + (n_x^2 + n_y^2 + n_z^2)\sin^2(\theta/2) = 1,$$
$$\underbrace{\cos^2(\theta/2)}_{a_1^2} + \underbrace{n_z^2 \sin^2(\theta/2)}_{a_2^2} + \underbrace{n_y^2 \sin^2(\theta/2)}_{b_1^2} + \underbrace{n_x^2 \sin^2(\theta/2)}_{b_2^2} = 1. \tag{2.69}$$

Comparing the last line in (2.69) with (2.68), we can redefine the constants as follows:

$$a_1 = \cos(\theta/2),$$
$$(b_2, b_1, a_2) = \sin(\theta/2)(n_x, n_y, n_z). \tag{2.70}$$

Substituting (2.70) into (2.67), we get this useful compact equation for $\hat{U}$:

$$\boxed{\hat{U} = \cos(\theta/2)\,\hat{I} + i\sin(\theta/2)\,\mathbf{n}\cdot\hat{\boldsymbol{\sigma}},} \tag{2.71}$$

where the vector operator $\hat{\boldsymbol{\sigma}} = (\hat{\sigma}_x, \hat{\sigma}_y, \hat{\sigma}_z) = (\hat{X}, \hat{Y}, \hat{Z})$.

Now we can see the geometric significance of the angle θ and the unit vector $\mathbf{n}$. Consider the following transformation of the operator $\mathbf{A}\cdot\hat{\boldsymbol{\sigma}} = A_x\hat{X} + A_y\hat{Y} + A_z\hat{Z}$ (*any* operator whose trace is zero can be written in this form):

$$\left(\cos(\theta/2)\hat{I} + i\sin(\theta/2)\,\mathbf{n}\cdot\hat{\boldsymbol{\sigma}}\right)\mathbf{A}\cdot\hat{\boldsymbol{\sigma}}\left(\cos(\theta/2)\hat{I} - i\sin(\theta/2)\,\mathbf{n}\cdot\hat{\boldsymbol{\sigma}}\right) = \mathbf{A}'\cdot\hat{\boldsymbol{\sigma}}, \tag{2.72}$$

where

$$\mathbf{A}' = (\mathbf{n}\cdot\mathbf{A})\mathbf{n} + \cos\theta\,\{\mathbf{A} - (\mathbf{n}\cdot\mathbf{A})\mathbf{n}\} - \sin\theta\,(\mathbf{n}\times\mathbf{A}) \tag{2.73}$$

is the vector $\mathbf{A}$ rotated by an angle θ (*not* $\theta/2$ – this is why the factor of $1/2$ was incorporated when we introduced the angle $\theta/2$) about an axis of rotation parallel to the vector $\mathbf{n}$. While this might appear a bit obscure, the result comes up often.

Exercises 2.2.7

1. Show that the three Pauli operators ($\hat{X}$, $\hat{Y}$, and $\hat{Z}$) are unitary.
2. The elements of the computational basis are eigenvectors of which of the Pauli matrices?
3. Work out what happens to $|\psi\rangle = a|0\rangle + b|1\rangle$ when:
   (a) The phase-flip gate is applied on it followed by the bit-flip gate.
   (b) The bit-flip gate is applied on it followed by the phase-flip gate.
   (c) What is the difference between the answers in the two cases? Does it matter? Why or why not?
4. Given the following unitary operation (also known as the *Hadamard gate*):

$$\hat{H} = \frac{1}{\sqrt{2}}\begin{bmatrix} 1 & 1 \\ 1 & -1 \end{bmatrix},$$

   (a) What is its effect on $a|0\rangle + b|1\rangle$?
   (b) What about $|0\rangle$?
   (c) What about $|1\rangle$?
   (d) From your answers, what can you say $\hat{H}$ does?
5. An arbitrary phase shift gate $\hat{U}_z(\theta)$ is given by

$$\begin{bmatrix} \exp(-i\theta/2) & 0 \\ 0 & \exp(i\theta/2) \end{bmatrix}.$$

   (a) Show that $\hat{U}_z(\theta) = \cos(\theta/2)\hat{I} - i\sin(\theta/2)\hat{Z}$.
   (b) Show that, neglecting the global phase, an arbitrary unitary $\hat{U} = \hat{U}_z(\phi)\hat{H}\hat{U}_z(\theta)\hat{H}\hat{U}_z(\chi)$.

6. In deriving the properties of the matrix elements of a general 2×2 unitary matrix, we started with the matrix

$$\begin{bmatrix} a & b \\ c & d \end{bmatrix},$$

and we alluded to the fact that both the Hermitian adjoint and the inverse of this matrix are equal implies that $|ad - bc|^2 = 1$. Prove that this is the case.

7. Which of the following matrices are unitary (we assume θ is real)?

(a) $i \begin{bmatrix} \cos[\theta] & \sin[\theta] \\ \sin[\theta] & -\cos[\theta] \end{bmatrix}$.

(b) $\begin{bmatrix} \cos[\theta] & \sin[\theta] \\ -\sin[\theta] & \cos[\theta] \end{bmatrix}$.

(c) $\begin{bmatrix} \cos[\theta] & i\sin[\theta] \\ -i\sin[\theta] & \cos[\theta] \end{bmatrix}$.

(d) $\cos\theta \hat{I} + \sin\theta \hat{X}$.

8. Show that the new basis described in (2.65) is orthonormal.

9. (a) Given the following basis: $\{\frac{1}{\sqrt{2}}|0\rangle + \frac{1}{\sqrt{2}}|1\rangle, \frac{1}{\sqrt{2}}|0\rangle - \frac{1}{\sqrt{2}}|1\rangle\}$. Transform it into a new basis using the general unitary operator:

$$\hat{U} = \begin{bmatrix} a & b \\ -b* & a* \end{bmatrix}. \tag{2.74}$$

(b) Now, let us look at the following basis: $\left\{\frac{1}{\sqrt{2}}|0\rangle + \frac{1}{\sqrt{2}}|1\rangle, \frac{1}{\sqrt{2}}|0\rangle + \frac{i}{\sqrt{2}}|1\rangle\right\}$. Again, transform it using the general unitary operator.

(c) What is the difference between the two bases? What does $\hat{U}$ preserve or not preserve in both cases?

2.3 Time Evolution of Quantum States

Before we start talking about the evolution of a quantum state, let us equip ourselves with some mathematical tools. First, if a matrix $\hat{M}$ satisfies the following relation:

$$\hat{M} = \hat{M}^\dagger, \tag{2.75}$$

then the matrix is said to be *Hermitian*. An interesting property of such a matrix is that its diagonal elements and its eigenvalues are always real. On the other hand, if the matrix displays this property:

$$\hat{M} = -\hat{M}^\dagger, \tag{2.76}$$

then we can say that the matrix is **anti-Hermitian**. The next question is: if we have a matrix that is *not* Hermitian, is there something that we can do to it to make it so? The answer is, yes. Let us say that we have a matrix $\hat{N}$ such that $\hat{N} \neq \hat{N}^\dagger$. Now let us define another matrix in terms of $\hat{N}$ as follows:

$$\hat{L} = \hat{N}\hat{N}^\dagger. \tag{2.77}$$

⊕ Mathematics Tips 2.7

Transpose of a Product of Matrices. Given a product of N operators, represented by matrices $\hat{M}_1\hat{M}_2\hat{M}_3\cdots\hat{M}_{N-1}\hat{M}_N$, the transpose of this product can be shown to be $\left(\hat{M}_1\hat{M}_2\hat{M}_3\cdots\hat{M}_{N-1}\hat{M}_N\right)^T = \hat{M}_N^T\hat{M}_{N-1}^T\cdots\hat{M}_3^T\hat{M}_2^T\hat{M}_1^T$. Similarly, the Hermitian conjugate of a product of operators is the product of the Hermitian conjugate of the individual operators in reverse order. For example, given operators $\hat{A}$ and $\hat{B}$, the following relation holds:

$$\left(\hat{A}\hat{B}\right)^\dagger = \hat{B}^\dagger\hat{A}^\dagger. \tag{2.78}$$

Using the property (2.78), find $\hat{L}^\dagger$. You will find that $\hat{L} = \hat{L}^\dagger$. Now, how about if we are given an anti-Hermitian operator, how can we relate it to a Hermitian matrix? The simple answer is to multiply the Hermitian matrix by the imaginary number i or a multiple of it. Take, for example, the Hermitian matrix $\hat{L}$ in (2.77) and multiply it by i, then take its Hermitian conjugate. This is what you get:

$$\left(i\hat{L}\right)^\dagger = -i\hat{L}^\dagger$$
$$= -\left(i\hat{L}\right). \tag{2.79}$$

Also, recall the property of a unitary matrix $\hat{U}$:

$$\hat{U}\hat{U}^\dagger = \hat{I}. \tag{2.80}$$

If we take the derivative of (2.80), using the product rule of differentiation, we get the relation

$$\hat{U}\frac{d\hat{U}^\dagger}{dt} = -\frac{d\hat{U}}{dt}\hat{U}^\dagger. \tag{2.81}$$

This (2.81) relation shows that $\frac{d\hat{U}}{dt}\hat{U}^\dagger$ is an anti-Hermitian operator.

Now, let us get back to the subject of this section. A final state of a closed quantum system at anytime is obtained by a unitary transformation of the initial state, which we set to be at time $t = 0$. A final state is expressed as a function of time, and consequently the unitary evolution is also given by a matrix that is a function of time:

$$|\psi(t)\rangle = \hat{U}(t)|\psi(0)\rangle. \tag{2.82}$$

Taking the time derivative of both sides of the equation in (2.82), we get

$$\frac{d}{dt}|\psi(t)\rangle = \frac{d\hat{U}(t)}{dt}|\psi(0)\rangle$$
$$= \left(\frac{d\hat{U}(t)}{dt}\hat{U}^\dagger(t)\right)|\psi(t)\rangle, \tag{2.83}$$

where we used $|\psi(0)\rangle = \hat{U}^\dagger(t)|\psi(t)\rangle$. In (2.83), we have a differential equation for the state $|\psi(t)\rangle$. Look at the term between brackets in (2.83) and notice that it is an anti-Hermitian operator. Using our result in (2.79), we know that it should be proportional to $i\hat{\mathcal{H}}$, where $\hat{\mathcal{H}}$ is a Hermitian matrix. If we choose $-1/\hbar$ as the proportionality constant in this relationship, then we get

$$\frac{d\hat{U}(t)}{dt}\hat{U}^\dagger(t) = -\frac{i\hat{\mathcal{H}}}{\hbar}. \tag{2.84}$$

If we apply both sides on $|\psi(t)\rangle$ and use the result in (2.83), we get

$$\boxed{i\hbar\frac{d}{dt}|\psi(t)\rangle = \hat{\mathcal{H}}|\psi(t)\rangle.}$$

(2.85)

Equation (2.85) is the *Schrödinger equation* from quantum mechanics. The **Hamiltonian** $\hat{\mathcal{H}}$ is the operator that describes the dynamics of the system, which makes sense when you think of this equation as describing how $|\psi(t)\rangle$ evolves in time. This equation is confirming that time evolution has to be unitary, as long as there is no outside interference or measurement. Usually the eigenstates of $\hat{\mathcal{H}}$; i.e., the eigenvectors of its matrix representation, are chosen to be the states in the computational basis $\{|0\rangle, |1\rangle\}$.

⚓ **Physics Tips 2.8**

Planck's Constant. We have justified Schrödinger's equation, Eq. (2.85) purely on mathematical grounds, with the *Hamiltonian* $\hat{\mathcal{H}}$ being some Hermitian operator which happens to determine the dynamics of the system. To make a connection to real, experimentally accessible devices (and incidentally justify the sudden appearance of the constant $\hbar$) we are obliged to resort to physics.

The first hint of quantum mechanics was postulated by Max Planck in 1900 as a means to explain the spectrum of thermal radiation. His argument was based on a simple hypothesis – that the energy of a vibrating system (in his case, a mode of the electromagnetic field) must be in discrete chunks (or "quanta"), whose energy is proportional to the vibration frequency: $E = hf$, where f is the frequency (measured in cycles per second, also called hertz), and the constant of proportionality, Planck's constant, is $h \approx 6.62607015 \times 10^{-34}$ joules seconds (physical constants are not rational numbers, and their values can only be stated to the limit imposed by measurements). When we express things using the angular frequency $\omega = 2\pi f$ (this is done to cut down on the number of times you have to write out "2π") then the expression becomes $E = \hbar\omega$, where $\hbar = 1.05457182 \times 10^{-34}$ joules seconds. While at the time it was a bold step into the unknown, in hindsight we know Planck was on to the right idea because it correctly explained the experimentally measured spectrum of light from a thermal source (i.e., anything that is radiating because it is hot, like a light-bulb, a candle, or the Sun).

Back to Eq. (2.85): suppose we had a very simple situation in which the system is not really changing perceptibly, as would be the case if $|\psi(t)\rangle$ were an eigenstate of $\hat{\mathcal{H}}$, with an eigenvalue η. Then the Schrödinger equation has the simple form

$$i\hbar\frac{d}{dt}|\psi(t)\rangle = \eta|\psi(t)\rangle.$$

The solution is $|\psi(t)\rangle = \exp(-i\eta t/\hbar)|\psi(0)\rangle$: in other words, the quantum state is vibrating at frequency $\eta/\hbar$. Thus, according to Planck's famous hypothesis, η corresponds to the energy of the system. In classical mechanics, a more precise term for "total energy", when expressed as a function of position and momentum, is the Hamiltonian of the system, and the name is retained for the operator $\hat{\mathcal{H}}$ in the quantum formulation of mechanics.

Exercises 2.3.1

1. **Properties of Hermitian matrices**: Given a Hermitian matrix, $\hat{M}$; i.e., $\hat{M} = \hat{M}^{\dagger}$,
   (a) Prove that its diagonal elements and eigenvalues are real.
   (b) Prove that its eigenvectors are orthogonal.

2. Given a non-Hermitian matrix $\hat{N}$; i.e., $\hat{N} \neq \hat{N}^{\dagger}$, show that $\hat{N}\hat{N}^{\dagger}$ is Hermitian.

3. Given a differential equation of this form: $\frac{d}{dt}f(t) = cf(t)$, where c is a constant, the solution is $f(t) = \exp(ct)f(0)$. Using this, and provided that the Hamiltonian is independent of time, we can write down the solution for the Schrödinger equation in (2.85): $|\psi(t)\rangle = \exp(-i\hat{\mathcal{H}}t/\hbar)|\psi(0)\rangle$. Solve the Schrödinger equation for the following Hamiltonians and starting conditions and sketch a plot of the absolute value of both probability amplitudes for subsequent times:

(a) $\hat{\mathcal{H}} = \begin{bmatrix} \hbar\omega_0 & 0 \\ 0 & \hbar\omega_1 \end{bmatrix}$ $|\varphi(0)\rangle = |0\rangle$,

(b) $\hat{\mathcal{H}} = \begin{bmatrix} \hbar\omega_0 & 0 \\ 0 & \hbar\omega_1 \end{bmatrix}$ $|\varphi(0)\rangle = \frac{1}{\sqrt{2}}(|0\rangle + |1\rangle)$,

(c) $\hat{\mathcal{H}} = \begin{bmatrix} -\hbar\omega/2 & \hbar\Omega/2 \\ \hbar\Omega/2 & \hbar\omega/2 \end{bmatrix}$ $|\varphi(0)\rangle = |0\rangle$.

2.4 Quantum Circuit Diagrams

A very useful method to represent unitary operations in quantum information is to use **quantum circuit diagrams**. We will illustrate how to use them with the simple unitaries we have worked with so far. As we advance through the book, these circuits will become more and more involved. To start, a qubit is represented by a horizontal line.

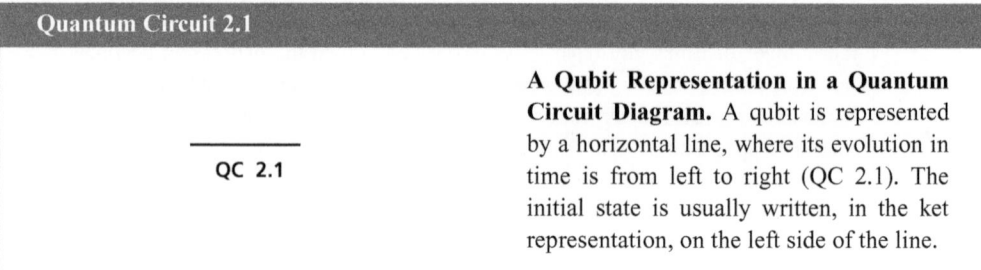

Quantum Circuit 2.1

QC 2.1

A Qubit Representation in a Quantum Circuit Diagram. A qubit is represented by a horizontal line, where its evolution in time is from left to right (QC 2.1). The initial state is usually written, in the ket representation, on the left side of the line.

Any operation that is performed on the qubit, say $\hat{U}$, is represented by a box.

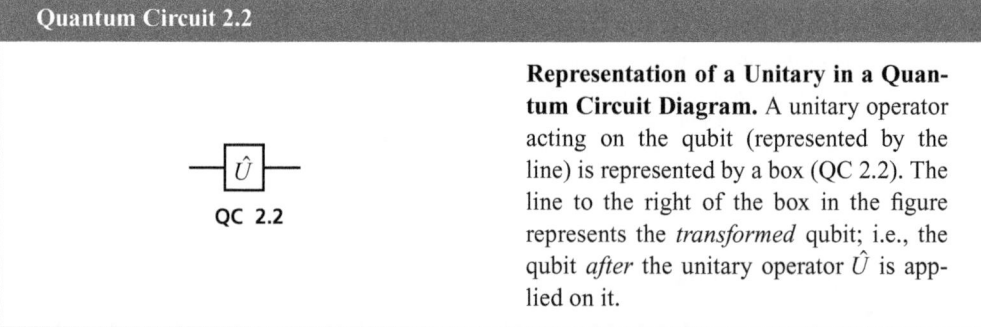

Quantum Circuit 2.2

$\hat{U}$

QC 2.2

Representation of a Unitary in a Quantum Circuit Diagram. A unitary operator acting on the qubit (represented by the line) is represented by a box (QC 2.2). The line to the right of the box in the figure represents the *transformed* qubit; i.e., the qubit *after* the unitary operator $\hat{U}$ is applied on it.

This sort of a diagram allows us to visualize very involved and elaborate sets of operators in a convenient way, keeping track of how the qubit transforms and when. Even though it might not look like it is essential when we are dealing with one or two simple operations on a qubit, when we look at more involved cases, where we deal with algorithms built from a series of operators applied at various times, especially in Chapter 4, you will appreciate the quantum circuit diagram representation even more.

Recall we discussed that a bit-flip operator, $\hat{X}$, followed by a phase-flip operator, $\hat{Z}$, is equivalent to applying $i\hat{Y}$. That is because $\hat{Z}\hat{X} = i\hat{Y}$. Notice that what is applied first, $\hat{X}$ here, is written on the right, and then what comes next, $\hat{Z}$, is written to its left, and so on. This is the rule of how you multiply several operators applied in series on the qubit of interest. Also, note that since the $i = e^{\frac{i\pi}{2}}$ is a global phase, it can be dropped, and we can say that $\hat{Z}\hat{X}$ is equivalent to $\hat{Y}$.

Quantum Circuit 2.3

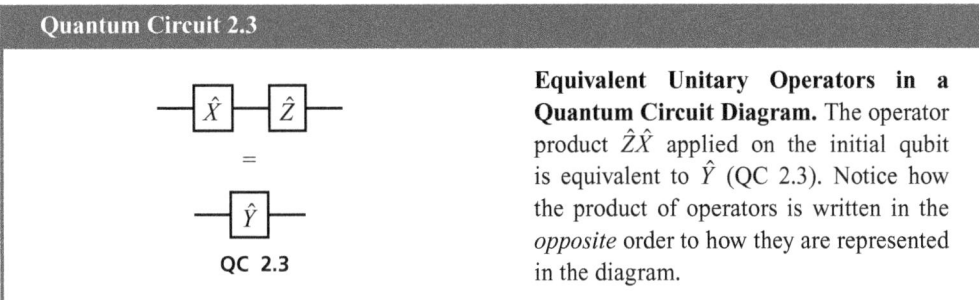

QC 2.3

Equivalent Unitary Operators in a Quantum Circuit Diagram. The operator product $\hat{Z}\hat{X}$ applied on the initial qubit is equivalent to $\hat{Y}$ (QC 2.3). Notice how the product of operators is written in the *opposite* order to how they are represented in the diagram.

Exercises 2.4.1

1. *True or False?* In a quantum circuit diagram, a qubit is represented by a vertical line.
2. *True or False?* In a quantum circuit diagram, the evolution of the qubit in time is from left to right.
3. Is the initial qubit represented by the line to the right or the one to the left of the unitary gate?
4. What Pauli matrix is the product $\hat{Y}\hat{Z}$ equivalent to? Express the relation using a quantum circuit diagram.
5. Given the qubit $|0\rangle$ that has the following operators applied on it (in the order in which they are applied): $\hat{A}, \hat{B}, \hat{C}, \hat{D}$:
   (a) Write down the final state in terms of the initial qubit and the operators applied.
   (b) Draw a quantum circuit diagram to illustrate the operations on $|0\rangle$.
6. Simplify QC 2.4:

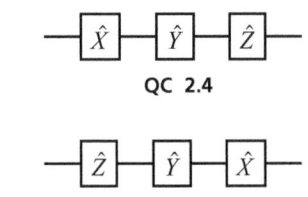

QC 2.4

7. Simplify QC 2.5:

$$\hat{Z} \quad \hat{Y} \quad \hat{X}$$

QC 2.5

2.5 Measurements and Non-Unitary Operations

When we measure a system, we remove the uncertainly as to its configuration: instead of being in an undetermined superposition of two or more possible configurations, it is rendered into a definite configuration, corresponding to the value of the quantity we just measured. This process is irreversible: if we start with a system in some superposition of "0" and "1", $|\varphi\rangle = \alpha|0\rangle + \beta|1\rangle$, then measure it and find that the result is "1", then all we can know about the values of α and β is that β is not zero; nothing else can be determined about the state, so reversing the **measurement** to re-create the initial state (assuming we did not know the values of α and β already) is impossible.

How can we describe this mathematically? The change a state undergoes due to measurement cannot be represented by a unitary matrix, because unitary matrices have an inverse, and thus describe processes that are reversible. Hence, measurements are non-unitary. The non-unitary operators $\hat{\Pi}_0 = |0\rangle\langle 0|$ and $\hat{\Pi}_1 = |1\rangle\langle 1|$ are called *projectors*, and they describe the outcome of measurements rather nicely:

$$\hat{\Pi}_0|\varphi\rangle = \alpha|0\rangle,$$
$$\hat{\Pi}_1|\varphi\rangle = \beta|1\rangle. \tag{2.86}$$

Note that the resultant states after measurement are not normalized (by virtue of the inconvenient factors α or β). As we have discussed above, these two outcomes are random, with probabilities given by $p_0 = \langle\varphi|\hat{\Pi}_0|\varphi\rangle$ and $p_1 = \langle\varphi|\hat{\Pi}_1|\varphi\rangle$, respectively. The correctly normalized post-measurement states of the qubits are

"0" measured: $|\varphi_{\text{after}}^{(0)}\rangle = \dfrac{1}{\sqrt{p_0}}\hat{\Pi}_0|\varphi\rangle$ (with probability $p_0 = \langle\varphi|\hat{\Pi}_0|\varphi\rangle$),

"1" measured: $|\varphi_{\text{after}}^{(1)}\rangle = \dfrac{1}{\sqrt{p_1}}\hat{\Pi}_1|\varphi\rangle$ (with probability $p_1 = \langle\varphi|\hat{\Pi}_1|\varphi\rangle$).

We are not limited to measurement in the computational basis, i.e., $\{|0\rangle, |1\rangle\}$. Classically, coins are either "Heads" or "Tails", never "25% Heads and 75% Tails". But in quantum mechanics you are free to measure whether the qubit is in one or the other of any pair of orthogonal states; the physical implementation may not be so straightforward, but in certain applications (such as Bell inequality measurements) such measurements are indispensable. (This possibility of this sort of measurement is not a distinctively quantum attribute: one can do such measurements with, for example, polarized light beams.) Further, ascribing the numerical value "0" to one outcome and "1" to the other is also arbitrary: one could pretty much give any two numerical values we like (or even non-numerical values, like "H" and "T", but the mathematical description of such measurements is a bit more tricky). How are such measurements described? Let the two possible outcomes be denoted ω_0 and ω_1, and their associated states $|\omega_0\rangle$ and $|\omega_1\rangle$; we assume the outcomes are distinct, so that $\langle\omega_0|\omega_1\rangle = 0$. Then a measurement yields the value ω_0 with probability $\langle\varphi|\hat{\Pi}_{\omega_0}|\varphi\rangle = |\langle\varphi|\omega_0\rangle|^2$ and the value ω_1 with probability $\langle\varphi|\hat{\Pi}_{\omega_1}|\varphi\rangle = |\langle\varphi|\omega_1\rangle|^2$. This measured quantity, which in the broader sense of quantum mechanics corresponds to a physically **observable** quantity, is represented by an operator $\hat{\Omega} = \sum_n \omega_n \hat{\Pi}_{\omega_n}$. Mathematically, we say we are measuring the variable represented by $\hat{\Omega}$, and that its eigenvalues and eigenvectors are represented by ω (which can take several different values) and $|\omega\rangle$, respectively.

Projectors such as $\hat{\Pi}_n = |n\rangle\langle n|$ have two distinctive properties, born of common sense: first, when you make a measurement, you must end up with *some* result. Mathematically, this implies $\sum_n \hat{\Pi}_n = \hat{I}$, where the sum is over all possible outcomes ($n = 0$ and $n = 1$ in the case of a single qubit, but the rule applies to systems with more configurations as well). Second, if your coin is Heads-up, and you do not toss it again, it will stay Heads-up, no matter how many times you look at it; it will be futile to try to see it in Tails-up, because it never will be. Quantumly, this can be expressed mathematically as follows:

$$\hat{\Pi}_n \cdot \hat{\Pi}_n = \hat{\Pi}_n,$$
$$\hat{\Pi}_m \cdot \hat{\Pi}_n = 0 \qquad (m \neq n).$$

We have seen that when we make a measurement on our system, assumed to be in some state $|\varphi\rangle$, of a quantity $\hat{\Omega}$, with eigenvalues ω_j $\{j = 1, 2, \dots\}$, and eigenvectors are $|\omega_j\rangle$ $\{j = 1, 2, \dots\}$, the possible measurement outcomes are ω_j, with probabilities $|\langle\varphi|\omega_j\rangle|^2$. Thus the average value of the measured quantity is

$$\begin{aligned}
\overline{\Omega} &= \sum_j \omega_j |\langle\varphi|\omega_j\rangle|^2 \\
&= \sum_j \langle\varphi|\omega_j\rangle \omega_j \langle\omega_j|\varphi\rangle \\
&= \langle\varphi| \left(\sum_j \omega_j |\omega_j\rangle\langle\omega_j| \right) |\varphi\rangle \\
&= \langle\varphi|\hat{\Omega}|\varphi\rangle \\
&= \text{Tr}\{|\varphi\rangle\langle\varphi|\hat{\Omega}\} \\
&= \text{Tr}\{\hat{\Pi}_\varphi \hat{\Omega}\},
\end{aligned} \tag{2.87}$$

where $\hat{\Pi}_\varphi = |\varphi\rangle\langle\varphi|$ is the projector for the system state $|\varphi\rangle$.

In quantum computing, qubits are often discarded after a measurement. They are discarded in the cases when measurements destroy the state and one no longer has access to them in the lab setting, such as what often happens upon measuring photons with the current lab technology by a photon getting absorbed by a detector. Another example is that of cooled ions that, upon measurement, heat up so they are not immediately available for re-use, having to be cooled again. That is why, when we represent measurements in a quantum circuit diagram, we have to remind ourselves that after a measurement we will no longer have a qubit.

Quantum Circuit 2.6

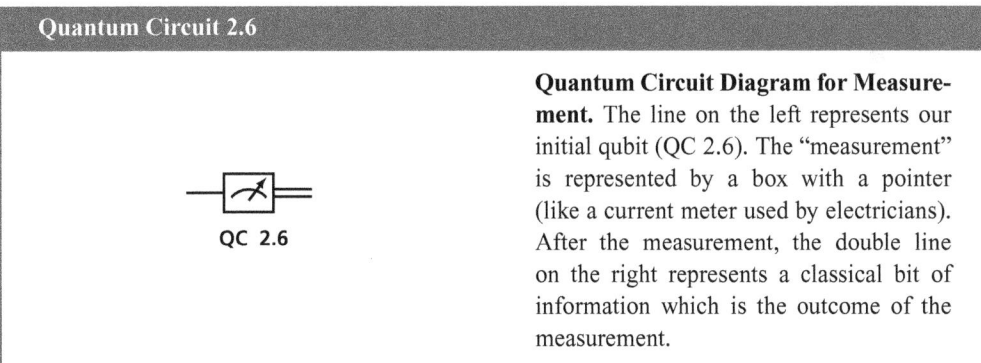

QC 2.6

Quantum Circuit Diagram for Measurement. The line on the left represents our initial qubit (QC 2.6). The "measurement" is represented by a box with a pointer (like a current meter used by electricians). After the measurement, the double line on the right represents a classical bit of information which is the outcome of the measurement.

Looking a Little Bit Further 2.9

Other Types of Quantum Measurement? The description of measurement in quantum mechanics is something of a minefield that the unwary quantum mechanic might regret straying into. In this book we will keep things as simple as possible and only deal with standard quantum measurements involving the projectors of eigenstates of the operator $\hat{\Omega}$ corresponding to an observable described above (and in most quantum mechanics textbooks). We will ignore entirely the very fascinating questions: "Why is quantum mechanics fundamentally random?"; "Why do the squared-absolute-values of the probability amplitudes give you probability?" and "How can quantum states collapse instantaneously?" Rather than open a can of worms that can't be closed, we adopt the pragmatic **operationalist interpretation** (also unknown as the "shut up and calculate" school): standard quantum measurement gives predictions entirely consistent with experiment, so there! Quantum information is already a challenging enough topic to master without having to worry about deeper interpretations.

That said, there are a number of different types of measurements that are occasionally discussed in quantum information, and which we would be remiss not to mention. In the language of functional analysis (the branch of mathematics in which functions are analyzed in a manner akin to vectors) the standard description of quantum measurement is called a *projection-valued measure* (PVM); in this context "measure" is an abstract mathematical notion that generalizes length or area. A generalization of such measurement, which offers some enticing theoretical possibilities, but is considerably more difficult to implement technologically, is the *positive operator-valued measure* (POVM). Here the collection of projectors $\hat{\Pi}_n$ used in PVMs is replaced by a set of operators $\hat{F}_i$ which are positive (i.e., $\langle \varphi | \hat{F}_i | \varphi \rangle > 0$ for all states $|\varphi\rangle$), and complete $\left(\sum_i \hat{F}_i = \hat{I} \right)$ but in general $\hat{F}_i \hat{F}_j \neq 0$ (for $i \neq j$); thus, the outcomes of this sort of measurement are no longer mutually exclusive.

Weak measurement is another type of generalized measurement for which the outcome, and consequently the output state, are not known precisely, but rather fall within some range of possible values.

Exercises 2.5.1

1. If the state $|\varphi\rangle = \alpha|0\rangle + \beta|1\rangle$ is measured, and the result "1" is obtained, what result is obtained if the same measurement is performed a second time?
2. A qubit is prepared in the state $|\varphi\rangle = 0.8|0\rangle + 0.6|1\rangle$ and then measured in the computational basis; i.e., the measurement operator is

$$\hat{\Omega} = \begin{bmatrix} 0 & 0 \\ 0 & 1 \end{bmatrix},$$

where

$$|0\rangle = \begin{bmatrix} 1 \\ 0 \end{bmatrix}, \quad \text{and} \quad |1\rangle = \begin{bmatrix} 0 \\ 1 \end{bmatrix}.$$

   (a) What are the possible outcomes of the measurement?
   (b) What are the corresponding probabilities?
   (c) What are the corresponding normalized states of the qubit after measurement?
   (d) Calculate the average value of the measurement operator $\overline{\Omega}$.
3. Repeat the analysis of Exercise 2, but with the new measurement operator $\hat{\Omega}' = \hat{H}\hat{\Omega}\hat{H}$, where $\hat{H}$ is the Hadamard operator.

4. If one of the two projectors that describe a measurement is given by

$$\frac{1}{2}\begin{bmatrix} 1 & -i \\ i & 1 \end{bmatrix},$$

what is the matrix representation of the other projector?

2.6 Random States and Density Operators

So far we have assumed that the probability amplitudes α and β are *deterministic* quantities. Determinism in this context means that if one does exactly the same thing to multiple identical systems (or to a single system multiple times), the resultant configurations of the system(s) will be identical each time; i.e., we get the same α and the same β. The opposite notion is *randomness*, i.e., the resultant configurations of our state $|\varphi\rangle$ will be different each time (different α and different β), due to various influences which are beyond our direct control.

To analyze random quantum systems we give each possible state of the system, now labelled $|\varphi_i\rangle$, an associated probability, p_i, whose value lies between 0 and 1 (i.e., $0 \leq p_i \leq 1$). The number of possible systems can be very large or even infinite, but we are *not* assuming that these different random states are linearly independent of each other, and since the system *must* be in *some* configuration, we require that $\sum_i p_i = 1$.

As we just discussed, there are two types of randomness that we are dealing with: we have a fundamentally *quantum* randomness required by the precepts of quantum mechanics describing the quantum state $|\varphi_i\rangle$, where measurement outcomes have probabilities associated with them, and then, in addition, a different sort of randomness, which is *classical*, brought about by the inability to prepare the quantum state in a deterministic, repeatable manner. How can we describe these by a single theory? And how can we tell the two sorts of randomness apart?

We need a description that will include the case of only quantum randomness, so it is described by α and β, and that will allow us to randomize α and β to include the classical randomness as well. Recall Eq. (2.87), where we have a formula for the average over the quantum fluctuations of some quantity $\hat{\Omega}$ for a deterministic quantum state given by

$$\overline{\Omega} = \mathrm{Tr}\{\hat{\Pi}_\varphi \hat{\Omega}\}.$$

Notice how, here, the description of $|\varphi\rangle$ is taken to be its outer product (or projector matrix) $\hat{\Pi}_\varphi = |\varphi\rangle\langle\varphi|$. Moreover, it is easy to recover the ket representation from the outer product representation, and vice versa in this case, where we only have quantum randomness. On the other hand, if we are considering the case where the state $\hat{\Pi}_\varphi = |\varphi\rangle\langle\varphi|$ is non-deterministic, we must include the classical randomness in the description and write the state as a linear combination of the possible i states we discussed earlier; i.e., in terms of $\hat{\Pi}_i = |\varphi_i\rangle\langle\varphi_i|$. When writing out the linear combination, the *weight* of each $|\varphi_i\rangle$ will be determined by its corresponding p_i so that now the description is given by the matrix $\sum_i p_i |\varphi_i\rangle\langle\varphi_i|$.

This new description has it all: information about both the quantum fluctuations (which we cannot avoid) and the non-determinism of the state preparation (which we might be able to do something about, if we improve our experiment). It is sufficiently useful that it is given a name and a symbol all of its own: the **density matrix**,

$$\hat{\rho} = \sum_i p_i |\varphi_i\rangle\langle\varphi_i|. \tag{2.88}$$

Each of the collection of possible random states (usually called an *ensemble*) is specified by its random probability amplitudes, i.e., $|\varphi_i\rangle = \alpha_i|0\rangle + \beta_i|1\rangle$. Thus, the density matrix for the one-qubit case can be written as follows:

$$\hat{\rho} = \begin{bmatrix} \sum_i p_i |\alpha_i|^2 & \sum_i p_i \alpha_i \beta_i^* \\ \sum_i p_i \alpha_i^* \beta_i & \sum_i p_i |\beta_i|^2 \end{bmatrix}. \tag{2.89}$$

✪ Mathematics Tips 2.10

Density Matrix Representation from $\overline{\Omega}$. Another way to infer the density matrix description, which includes both types of randomness (classical and quantum), comes out naturally if we consider the definition of the average of a measurement operator, represented by $\hat{\Omega}$, that we saw in Eq. (2.87). The latter tells us what $\overline{\Omega}$ is when we only have one ket, $|\varphi\rangle$, to deal with. We can extend the definition for the case in which we have multiple kets with associated probabilities. Then we will get the same density matrix description $\hat{\rho} = \sum_i p_i |\varphi_i\rangle\langle\varphi_i|$ popping up its head at us.

Thus, if we have a non-deterministic state, the average over both types of randomness will be

$$\begin{aligned} \overline{\Omega} &= \sum_i \mathrm{Tr}\left\{ p_i \hat{\Pi}_i \hat{\Omega} \right\} \\ &= \mathrm{Tr}\left\{ \sum_i p_i \hat{\Pi}_i \hat{\Omega} \right\} \\ &= \mathrm{Tr}\left\{ \left[\sum_i p_i |\varphi_i\rangle\langle\varphi_i| \right] \hat{\Omega} \right\}. \end{aligned} \tag{2.90}$$

The quantity in the big square brackets is the density matrix representation.

Here are some useful facts about the density matrix which are true for states of any system, not just the single qubits we have considered here.

- **The density matrix is Hermitian**: $\hat{\rho}^\dagger = \hat{\rho}$. This comes from the definition Eq. (2.88), and the fact that the p_i are all real, and that a projector $\hat{\Pi}_i$ is also Hermitian. Remember: Hermitian matrices have real-valued eigenvalues, and eigenvectors that are orthogonal; we will denote these eigenvalues by λ_n, and the associated eigenstates by $|\phi_n\rangle$.
- **The trace of the density matrix is always unity**. Mathematically: $\mathrm{Tr}\{\hat{\rho}\} = 1$.
- **The density matrix is positive**. This is NOT the same thing as saying all the elements of $\hat{\rho}$ are positive numbers: What is does mean is that, for *any* state vector $|\phi\rangle$, the matrix element $\langle\phi|\hat{\rho}|\phi\rangle \geq 0$. That this is so can been seen pretty easily from the definition Eq. (2.88) and the fact that both the probabilities p_i and the absolute value of any complex number are both real and positive (and remember, $\langle\psi|\varphi\rangle$ is just a complex number):

$$\begin{aligned} \langle\phi|\hat{\rho}|\phi\rangle &= \sum_i p_i \langle\phi|\varphi_i\rangle\langle\varphi_i|\phi\rangle \\ &= \sum_i p_i |\langle\phi|\varphi_i\rangle|^2 \geq 0. \end{aligned} \tag{2.91}$$

> 🐚 **Interesting Facts 2.11**
>
> **What Has a Density Matrix Got to Do with "Density"?** While the use of operators like this to describe the most general state of a quantum system was pioneered by John von Neumann in 1927, he does not seem to have given it any particular name. In 1932 Eugene Wigner used quantum mechanics to look at statistical mechanics (i.e., the study of systems with a large number of particles, whose dynamics is best described in terms of probability and statistics). He used a quantity like this to describe such systems, and the name stuck when it was adapted to study random states of systems with a small number of particles.

The density matrix, just like any other Hermitian matrix, may be re-written as a combination of its eigenstates. For a single qubit, this gives the following, also known as the **spectral decomposition**:

$$\hat{\rho} = \sum_{n=1}^{2} \lambda_n |\phi_n\rangle\langle\phi_n|. \tag{2.92}$$

This may look rather similar to Eq. (2.88). However, now the sum involves the set of orthonormal eigenstates $\{|\phi_1\rangle, |\phi_2\rangle\}$, while the states $\{|\varphi_i\rangle\}$ in Eq. (2.88) are, in general, not orthonormal, and they represent what we happen to obtain (for the qubit of interest) in a random fashion; there is no implication that they are a linearly independent set. Further, the random state decomposition, also called a **convex hull decomposition**, given by Eq. (2.88) is not unique: a different set of states with different probabilities could add up to the same density matrix, making it impossible to tell the two ensembles apart; hence, analyzing the properties of the state using the probabilities p_i and the $|\varphi_i\rangle$ is rather fraught with difficulties. However, the eigenvalues λ_n represent a sort of minimalist, irreducible set of probabilities for the state of the density operator which are unique.

In quantum information, we really want to have a situation in which all of these other types of randomness have been wrung out of the system by one means or another, resulting in a situation in which only one of the eigenvalues, λ_1 say, is equal to 1, and consequently the other(s) are zero, since the eigenvalues have to add up to 1 (although we have only dealt with a single qubit so far, this is true for systems with more levels as well). States for which only one $\lambda_i = 1$ are called **pure states**, and we are back to the situation where we do not have classical randomness, only quantum. Since the analysis of quantum circuits is much easier when dealing with pure states, we will often make the assumption that our state is pure. There will be occasions requiring dealing with random states for which more than one eigenvalue is not vanishingly small; generically, these are called **mixed states**.

The **maximally mixed state** is one in which all possible states of the system are equally probable, so that, for a single qubit, $\lambda_1 = \lambda_2 = 1/2$. The density matrix for such a state can be written as

$$\begin{aligned} \hat{\rho}_{\text{max-mix}} &= \frac{1}{2}|0\rangle\langle 0| + \frac{1}{2}|1\rangle\langle 1| \\ &= \frac{1}{2}\hat{I}. \end{aligned} \tag{2.93}$$

How can we assess the degree of this "other" (classical) randomness, making it distinct from the fundamentally quantum randomness? One simple way is to think of what happens when one multiplies the density operator by itself:

$$\begin{aligned} \hat{\rho}^2 &= (\lambda_1|\phi_1\rangle\langle\phi_1| + \lambda_2|\phi_2\rangle\langle\phi_2|)\,(\lambda_1|\phi_1\rangle\langle\phi_1| + \lambda_2|\phi_2\rangle\langle\phi_2|) \\ &= \lambda_1^2|\phi_1\rangle\langle\phi_1|\phi_1\rangle\langle\phi_1| + \lambda_1\lambda_2|\phi_1\rangle\langle\phi_1|\phi_2\rangle\langle\phi_2| + \lambda_2\lambda_1|\phi_2\rangle\langle\phi_2|\phi_1\rangle\langle\phi_1| + \lambda_2^2|\phi_2\rangle\langle\phi_2|\phi_2\rangle\langle\phi_2| \\ &= \lambda_1^2|\phi_1\rangle\langle\phi_1| + \lambda_2^2|\phi_2\rangle\langle\phi_2|, \end{aligned} \tag{2.94}$$

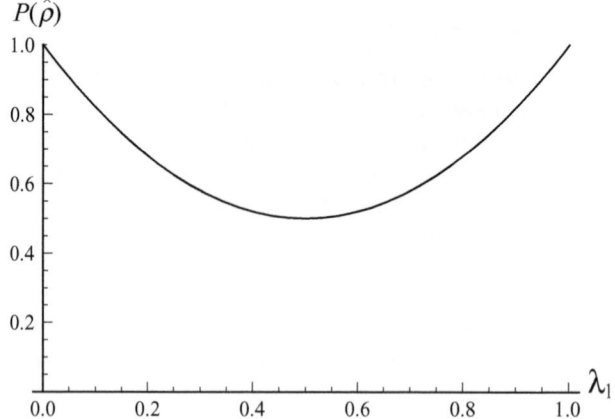

Figure 2.2   Purity of a single qubit. This is a plot of the purity, $P(\hat{\rho})$, for a single qubit versus one of the eigenvalues of the density operator, λ_1. It demonstrates that purity takes its maximum value of 1 for pure states ($\lambda_1 = 1$ or $\lambda_2 = 1$) and its minimum value when $\lambda_1 = \lambda_2 = 1/2$.

where we have used $\langle \phi_1 | \phi_2 \rangle = \langle \phi_2 | \phi_1 \rangle = 0$ and $\langle \phi_1 | \phi_1 \rangle = \langle \phi_2 | \phi_2 \rangle = 1$. If we calculate the trace of this matrix, we obtain the **purity**:

$$
\begin{aligned}
P(\hat{\rho}) &= \text{Tr}\{\hat{\rho}^2\} \\
&= \lambda_1^2 + \lambda_2^2 \\
&= (\lambda_1 + \lambda_2)^2 - 2\lambda_1\lambda_2 \\
&= 1 - 2\text{Det}\{\hat{\rho}\},
\end{aligned}
\tag{2.95}
$$

where we have used $\lambda_1 + \lambda_2 = 1$ and Eq. (2.29) in the last step. We can plot this quantity. It has the maximum value when either $\lambda_1 = 1$ and $\lambda_2 = 0$ or $\lambda_1 = 0$ and $\lambda_2 = 1$: i.e., $P(\rho)$ takes its maximum value if (and only if) the state is pure. Conversely, $P(\hat{\rho})$ has its minimum value of $1/2$ if $\lambda_1 = \lambda_2 = 1/2$, i.e., the state is maximally mixed. Values in between these two extremes represent *partially* mixed states.

🗼 **Looking a Little Bit Further 2.12**

Other Methods to Quantify the "Mixedness" of a Quantum State. There are more sophisticated ways to assess the amount of mixture in a quantum state, for example the **von Neumann entropy**:

$$
\begin{aligned}
S(\hat{\rho}) &= -\text{Tr}\{\hat{\rho} \log_2 \hat{\rho}\} \\
&= -\sum_{n=1}^{N} \lambda_n \log_2 \lambda_n,
\end{aligned}
\tag{2.96}
$$

where we have assumed the system has N possible states; $N = 2$ for a qubit. The Rényi entropy of order α is a generalization of the von Neumann entropy defined by the following expression:

$$
\mathscr{S}_\alpha(\hat{\rho}) = \frac{1}{1-\alpha} \log_2 \text{Tr}\{\hat{\rho}^\alpha\},
\tag{2.97}
$$

where α is a positive real number. Note that $\mathscr{S}_1(\hat{\rho}) = S(\hat{\rho})$ (you need to use l'Hôpital's rule to show this), and $\mathscr{S}_2(\hat{\rho}) = -\log_2 P(\hat{\rho})$. For pure states, both of these quantities are zero (to see this for the von Neumann entropy you have to show in the limit $x \to 0$ that $x \log_2 x = 0$), and both take their maximum value $\log_2 N$ for maximally mixed states.

These quantities have the advantage that they have a specific interpretation for the information content of mixed states, and play an important role in understanding things like the degree of entanglement of mixed states or the capacity of quantum channels; however, they have the drawback that they are rather more tricky to handle mathematically than the purity, $P(\hat{\rho})$, and the latter quantity is sufficient for the applications of mixed states covered in this book.

Exercises 2.6.1

1. When discussing projectors, we saw that a state $|\psi\rangle$ can be written as a matrix by taking its outer product with itself: $|\psi\rangle\langle\psi|$. This is the *density matrix representation* that we discussed in this section. Find the density matrix representation for the following states:

   (a) $|\psi_1\rangle = |0\rangle$,
   (b) $|\psi_2\rangle = \frac{1}{\sqrt{2}}\{|0\rangle + i|1\rangle\}$,
   (c) $|\psi_3\rangle = \alpha|0\rangle + \beta|1\rangle$.

2. For the following density matrices, calculate the purity, $P(\hat{\rho})$, and find the ket representation, if it exists.

   (a) $\hat{\rho} = \begin{bmatrix} 0 & 0 \\ 0 & 1 \end{bmatrix}$.

   (b) $\hat{\rho} = \frac{1}{3}\begin{bmatrix} 1 & \sqrt{2} \\ \sqrt{2} & 1 \end{bmatrix}$.

   (c) $\hat{\rho} = \begin{bmatrix} 1/4 & 0 \\ 0 & 3/4 \end{bmatrix}$.

   (d) $\hat{\rho} = \frac{1}{4}\begin{bmatrix} 3 & -i\sqrt{3} \\ i\sqrt{3} & 1 \end{bmatrix}$.

   (e) $\hat{\rho} = \frac{1}{8}\begin{bmatrix} 4 & -3 \\ -3 & 4 \end{bmatrix}$.

   (f) $\hat{\rho} = \frac{1}{2}\begin{bmatrix} 1 & 0 \\ 0 & 1 \end{bmatrix}$.

3. Find a pure state decomposition for each of the density matrices in Exercise 2. Which density matrices have more than one decomposition?

4. For each of the density matrices in Exercise 2, what type of randomness (quantum randomness, classical randomness) is in the system?

5. Find $\overline{X}, \overline{Y}$, and $\overline{Z}$ for the following states:
   (a) $|+\rangle = \frac{1}{\sqrt{2}}\{|0\rangle + |1\rangle\}$,

(b) $\hat{\rho} = \frac{1}{2} \begin{bmatrix} 1 & 0 \\ 0 & 1 \end{bmatrix}$,

(c) $\hat{\rho} = \frac{1}{4} \begin{bmatrix} 2 & i \\ -i & 2 \end{bmatrix}$.

2.7 CONCEPT CHECK AND EXERCISES

Concept Check

- Unitary transformations can be thought of as rotations of quantum states.
- If the quantum system is closed, then its dynamics can be described by a unitary matrix.
- The unitary evolution of a quantum state can be obtained through the unitary matrix representing it or through solving the Schrödinger equation; the two approaches are equivalent.
- Operations on qubits are unitary, such as the bit-flip gate $\hat{X}$ and the phase-flip gate $\hat{Z}$.
- Observables in quantum systems are represented by Hermitian matrices. Their eigenvalues, which correspond to quantities in real life, are real. Their eigenvectors, which correspond to independent states/results, are orthogonal.
- Measurements are non-unitary; their outcomes are irreversible. When a state is measured, it gets projected onto one of possible states and is no longer available for use in quantum computing operations.
- When the state of a quantum system cannot be known with certainty; i.e., at best, it can be thought of as a mixture of pure states, it does not have a ket representation and must be represented by a density matrix.

Chapter Exercises

1. Find a unitary matrix that transforms $\frac{1}{5}(4|0\rangle + 3i|1\rangle)$ into $\frac{1}{\sqrt{2}}(|0\rangle - |1\rangle)$.

2. Find the unitary matrix that transforms a basis made of the eigenvectors of $\hat{Z}$ to that made of the eigenvectors of $\hat{X}$.

3. Consider the quantum gate:

$$\hat{T} = \begin{bmatrix} 1 & 0 \\ 0 & \exp(i\pi/4) \end{bmatrix}.$$

   (a) What effect does it have on a qubit?
   (b) Factor the phase $\exp(i\pi/8)$ out of $\hat{T}$. Why is this gate sometimes called the $\pi/8$ gate?
   (c) Express $\hat{T}$ as a sum of terms involving the identity and the three Pauli matrices.

4. Why do the probabilities associated with each of the basis vectors describing the quantum state have to add up to 1?

5. Which of the following quantum states are considered equivalent in quantum mechanics?
   (a) $|\psi_1\rangle = \alpha e^{i\phi}|0\rangle + \beta e^{i\phi}|1\rangle$.
   (b) $|\psi_2\rangle = \alpha|0\rangle + \beta e^{i(\theta-\phi)}|1\rangle$.
   (c) $|\psi_3\rangle = \alpha e^{i\theta}|0\rangle + \beta e^{i\theta}|1\rangle$.
   (d) $|\psi_4\rangle = \alpha e^{i\phi}|0\rangle + \beta e^{i\theta}|1\rangle$.
   (e) $|\psi_5\rangle = \alpha e^{i(\phi-\theta)}|0\rangle + \beta|1\rangle$.

6. Given the operator product $\hat{A}\hat{B}\hat{C}$ is not Hermitian, construct a Hermitian operator involving all three of these operators.

7. The solution for the Schrödinger equation, provided the Hamiltonian is time independent, is $|\psi(t)\rangle = \exp(-i\hat{\mathcal{H}}t/\hbar)|\psi(0)\rangle$. Find the unitary operator that represents the transformation for the following Hamiltonian:

$$\hat{\mathcal{H}} = \begin{bmatrix} -\hbar\omega/2 & \hbar\Omega/2 \\ \hbar\Omega/2 & \hbar\omega/2 \end{bmatrix}.$$

8. Given the state $|\psi\rangle = |0\rangle$. If a measurement is made in the $\hat{X}$ basis, what are the possible states that $|\psi\rangle$ will be projected onto? What are the corresponding probabilities?

9. Which of the following matrices are not acceptable as density matrices? Why?

   (a) $\hat{\rho}_1 = \begin{bmatrix} 1 & 0 \\ 0 & 0 \end{bmatrix}$.

   (b) $\hat{\rho}_2 = \frac{1}{2}\begin{bmatrix} 1 & i \\ i & 1 \end{bmatrix}$.

   (c) $\hat{\rho}_3 = \frac{1}{2}\begin{bmatrix} 1 & 1 \\ 1 & -1 \end{bmatrix}$.

   (d) $\hat{\rho}_4 = \frac{1}{8}\begin{bmatrix} 5 & 4i \\ -4i & 3 \end{bmatrix}$.

10. If we start with a qubit in state $|0\rangle$ and then apply the Hadamard gate,

$$\hat{H} = \frac{1}{\sqrt{2}}\begin{bmatrix} 1 & 1 \\ 1 & -1 \end{bmatrix},$$

what state does the qubit transform to? If we measure the state of the qubit, do the probabilities to obtain $|0\rangle$ and $|1\rangle$ change after this operator is applied? What are they before and after?

11. In the scenario described in Exercise 10, assuming that the measurement is done after $\hat{H}$ is applied, draw a quantum circuit that describes it.

12. For the matrix

$$\hat{M} = \begin{bmatrix} 4 & 1 \\ 3 & 2 \end{bmatrix}:$$

   (a) Find the eigenvalues λ_i. What is the difference between the right and left eigenvalues?
   (b) Find the right eigenvectors, $|r_i\rangle$, and the left eigenvectors, $\langle l_i|$?,
   (c) Find the normalization constants for the eigenvectors so that $\langle l_i|r_j\rangle = \delta_{ij}$.
   (d) Work out $\sum_i |r_i\rangle\langle l_i|$. What does it equal?
   (e) Work out $\sum_i \lambda_i|r_i\rangle\langle l_i|$. What does it equal?

13. For the matrix

$$\hat{M} = \begin{bmatrix} 5 & 1 \\ 2 & 4 \end{bmatrix},$$

   calculate $f(\hat{M}) = \sin(\frac{\pi}{3}\hat{M})$.

14. (a) Write down the solution for the Schrödinger equation in (2.85) when the Hamiltonian has the following *time-varying* form:

$$\hat{\mathcal{H}} = \begin{bmatrix} E_0 - \hbar\omega/2 & \hbar\Omega\exp(i\omega_1 t)/2 \\ \hbar\Omega\exp(-i\omega_1 t)/2 & E_0 + \hbar\omega/2 \end{bmatrix}.$$

Hint: You cannot get the correct solution by putting the time-varying Hamiltonian in an exponent. Instead, try to find a way that the equation can be modified so that the Hamiltonian is not time varying.

(b) Suggest values of the parameters in part (a) that would allow you to perform a bit-flip gate.

15. Using Eqs. (2.85) and (2.88), show that

$$\frac{d}{dt}\hat{\rho} = \frac{1}{i\hbar}\left[\hat{\mathcal{H}}, \hat{\rho}\right],$$

where the **commutator bracket** is defined as $\left[\hat{\mathcal{H}}, \hat{\rho}\right] = \hat{\mathcal{H}}\hat{\rho} - \hat{\rho}\hat{\mathcal{H}}$.

Hint: Assume the probabilities p_i do not vary in time.

16. Suppose a qubit is in the state $|\varphi\rangle = \alpha|0\rangle + \beta|1\rangle$.

(a) Find $u = \langle\varphi|\hat{X}|\varphi\rangle$, $v = \langle\varphi|\hat{Y}|\varphi\rangle$, and $w = \langle\varphi|\hat{Z}|\varphi\rangle$ in terms of α and β. These three quantities can be thought of as a vector $\mathbf{s} = \{u, v, w\}$ in a real 3D space, analogous to (but <u>not</u> the same as) the three dimensions of everyday life. The vector is called the **Bloch vector**, and is just as valid a description of a quantum state as the state ket $|\varphi\rangle$.

(b) Show that $\mathbf{s} \cdot \mathbf{s} = 1$.

(c) Find the components of the Bloch vector for the following states:
  i. $|0\rangle$,
  ii. $|1\rangle$,
  iii. $(|0\rangle + |1\rangle)/\sqrt{2}$,
  iv. $(|0\rangle + i|1\rangle)/\sqrt{2}$.

(d) Show that the projector $\hat{\Pi} = |\varphi\rangle\langle\varphi|$ may be written as

$$\hat{\Pi} = \frac{1}{2}\left(\hat{I} + u\hat{X} + v\hat{Y} + w\hat{Z}\right) = \frac{1}{2}\left(\hat{I} + \mathbf{s} \cdot \hat{\boldsymbol{\sigma}}\right),$$

where $\hat{\boldsymbol{\sigma}} = \{\hat{X}, \hat{Y}, \hat{Z}\}$.

(e) If we write the Hamiltonian as $\hat{\mathcal{H}} = E_0\hat{I} + (\hbar\Omega/2)\mathbf{n} \cdot \hat{\boldsymbol{\sigma}}$ where $\mathbf{n} \cdot \mathbf{n} = 1$ (n.b. this expression is perfectly general for a two-level system), show that

$$\frac{d}{dt}\mathbf{s} = \Omega\,\mathbf{n} \times \mathbf{s},$$

where $\times$ denotes the vector product of two 3D vectors.

17. Show that

$$\begin{bmatrix}\hat{X} \\ \hat{Y} \\ \hat{Z}\end{bmatrix}\begin{bmatrix}\hat{X} & \hat{Y} & \hat{Z}\end{bmatrix} = \begin{bmatrix}\hat{I} & i\hat{Z} & -i\hat{Y} \\ -i\hat{Z} & \hat{I} & i\hat{X} \\ i\hat{Y} & -i\hat{X} & \hat{I}\end{bmatrix}.$$

18. (a) If $\mathbf{a} = (a_x, a_y, a_z)$ and $\mathbf{b} = (b_x, b_y, b_z)$ are any two vectors, prove that

$$(\mathbf{a} \cdot \hat{\boldsymbol{\sigma}})(\mathbf{b} \cdot \hat{\boldsymbol{\sigma}}) = \mathbf{a} \cdot \mathbf{b}\,\hat{I} + i(\mathbf{a} \times \mathbf{b}) \cdot \hat{\boldsymbol{\sigma}}.$$

(b) Prove Eq. (2.72).

3 Two Qubits and Entanglement

In this chapter, we will move on from one to two qubits. Such systems are the simplest possible in which quantum entanglement comes into play; in the opinion of many influential scientists, starting with Erwin Schrödinger in 1935, entanglement is *the* quality that defines the difference between classical and quantum systems. We will further discuss operations that can be performed on two qubits, including ones that create entanglement between them. Following that we will discuss consequences of entanglement, discussing whether hidden variables (to complete quantum theory) could possibly exist, as well as the application of teleporting quantum states.

Learning Objectives

- Learn how to extend the ket representation to multiple qubits.
- Learn about useful matrix operations: tensor product and partial trace.
- Learn about useful theorems in quantum information: Schmidt decomposition theorem and the fundamental theorem of entanglement.
- Understand the role of randomness in defining quantum entanglement and quantifying it for the pure two-qubit case.
- Learn about local and global two-qubit operations.
- Learn about quantum circuit arrangements that create entanglement.
- Follow Bell's mathematical argument that proves that the peculiarities of quantum theory cannot be dealt with using a hidden variable description.
- Understand the mathematics of quantum teleportation.

3.1 Two-Qubit Quantum States and Entanglement

As we have seen in Chapter 1 in Eq. (1.19) (but using the computational basis instead), any pure two-qubit system, made of subsystems A and B, will be in a state which can be written as

$$|\psi\rangle_{AB} = \alpha|00\rangle + \beta|01\rangle + \gamma|10\rangle + \delta|11\rangle, \tag{3.1}$$

where we have used the short-hand notation $|00\rangle = |0\rangle_A \otimes |0\rangle_B$. The *tensor product* ($\otimes$) is simply telling us "this is the state in which A is in $|0\rangle$ and B is also in $|0\rangle$". Because we now have four possible states, the space of configurations has grown by a factor of two; adding a third qubit makes eight distinct configurations, and so on.

⊕ Mathematics Tips 3.1

Tensor Products. In this book, the symbol we will use for the tensor product is ⊗. These are products between vectors or matrices. The effect of a tensor product is to make a vector space bigger. When we multiply two vectors together, if one vector has size m and the other size n, then the resulting vector will be of size $m \times n$. On the other hand, if we multiply a vector by a number, the vector will get longer or shorter, depending on the number. However, the dimension will not change. Let us look at an example: if we have one qubit, in the computational basis, its state can be written in the basis $\{|0\rangle, |1\rangle\}$. If we multiply the basis by a number, a, that will just multiply the state by a. On the other hand, if we bring a second qubit into the picture, we will increase the dimension of the system. That is because, say we also express the second qubit in the basis $\{|0\rangle, |1\rangle\}$, then the new basis will have four elements, as opposed to the initial two, given by basis $\{|0\rangle \otimes |0\rangle, |0\rangle \otimes |1\rangle, |1\rangle \otimes |0\rangle, |1\rangle \otimes |1\rangle\}$, which can be expressed more compactly as $\{|00\rangle, |01\rangle, |10\rangle, |11\rangle\}$. Do you not notice an analogy between how the basis of n qubits (obtained from the tensor product of the individual bases) and the possible numbers that can be expressed using n bits look the same? Yes, in fact the tensor product of three qubits gives the basis $\{|000\rangle, |001\rangle, |010\rangle, |100\rangle, |110\rangle, |101\rangle, |011\rangle, |111\rangle\}$.

3.1.1 Entanglement

Let us consider some of the implications of the state in Eq. (3.1) by re-writing it in the following form:

$$|\psi\rangle_{AB} = \alpha|00\rangle + \beta|01\rangle + \gamma|10\rangle + \delta|11\rangle$$
$$= (\alpha|0\rangle_A + \gamma|1\rangle_A) \otimes |0\rangle_B + (\beta|0\rangle_A + \delta|1\rangle_A) \otimes |1\rangle_B$$
$$= \sqrt{p_{A,0}}|\varphi_0\rangle_A|0\rangle_B + \sqrt{p_{A,1}}|\varphi_1\rangle_A|1\rangle_B, \tag{3.2}$$

where $p_{A,0} = |\alpha|^2 + |\gamma|^2$, $p_{A,1} = |\beta|^2 + |\delta|^2$, $|\varphi_0\rangle = (\alpha|0\rangle_A + \gamma|1\rangle_A)/\sqrt{p_{A,0}}$ and $|\varphi_1\rangle = (\beta|0\rangle_A + \delta|1\rangle_A)/\sqrt{p_{A,1}}$. Remember: all we have done is re-write $|\psi\rangle_{AB}$ in a different notation; $p_{A,0}$ and $p_{A,1}$ have the properties of probabilities, and, since $|\psi\rangle_{AB}$ is normalized, $p_{A,0} + p_{A,1} = 1$; and $|\varphi_0\rangle$ and $|\varphi_1\rangle$ are both correctly normalized state vectors of qubit A.

Recall in Section 2.6, the equations (2.88) and (2.92). They look like linear combinations of possibilities. The state as re-written in the last line of (3.2) also looks like a linear combination of possibilities. However, there is a very important difference between the two cases. In the first case; i.e., in (2.88) and (2.92), the individual states involved in the linear combination are *outer products* of kets, while the ones in (3.2) are just kets. Keep in mind that if you are dealing with *just* kets (whether of single qubits or multiple qubits), you are dealing with *pure states*. In this case, there is only one type of randomness: the inherent *quantum randomness* (specified by probability amplitudes $\sqrt{p_{A,0}}$ and $\sqrt{p_{A,1}}$). On the other hand, with a linear combination of outer products, you are dealing, in general, with mixed states, where there are two types of randomness: the inherent *quantum randomness* and the *classical randomness* of the state due to some external circumstances that rendered things non-deterministic.

As we will soon see, besides finding yourself with a mixed state because of classical randomness as we discussed in Chapter 2, you can also find yourself dealing with a mixed state if you want to focus on only one of the qubits in your system. To elaborate, if you have a pure two-qubit state ket description and decide (or have no choice) to ignore one of the qubits (by mathematically tracing it out of the ket description), then you will, in general, end up with a non-pure density matrix (no more kets!). This concept will be the basis for the quantification of entanglement that we will introduce later in this section.

Back to focusing on our two-qubit pure state in (3.2). Suppose now qubit B got lost, so that qubit A is on its own. No matter what happens to qubit B – some unitary evolution happens to it, so that $|0\rangle_B$ evolves into state $|0'\rangle_B = \hat{U}|0\rangle_B$ and $|1\rangle_B$ into $|1'\rangle_B = \hat{U}|1\rangle_B$; or someone comes along and measures it, so that its state is projected onto one of the two basis states of the measurement operator; or indeed something even more complicated. But whatever happens to B, poor old A is left entirely ignorant of the situation. In these circumstances, the state of A is random (the source of the randomness being quantum mechanical randomness over the state of B), and must thus be described by a density matrix:

$$\hat{\rho}_A = p_{A,0}|\varphi_0\rangle\langle\varphi_0| + p_{A,1}|\varphi_1\rangle\langle\varphi_1|$$

$$= \begin{bmatrix} |\alpha|^2 + |\beta|^2 & \alpha\gamma^* + \beta\delta^* \\ \alpha^*\gamma + \beta^*\delta & |\gamma|^2 + |\delta|^2 \end{bmatrix}. \tag{3.3}$$

If the roles were reversed, and it was qubit A that strayed off into the wild blue yonder, then the corresponding state of qubit B would be

$$\hat{\rho}_B = \begin{bmatrix} |\alpha|^2 + |\gamma|^2 & \alpha\beta^* + \gamma\delta^* \\ \alpha^*\beta + \gamma^*\delta & |\beta|^2 + |\delta|^2 \end{bmatrix}. \tag{3.4}$$

These two density matrices are known as the **reduced density matrices**.

They are related to the two-qubit pure state $|\psi\rangle_{AB}$ as follows. First, one forms the projector for the two-qubit state to get its density matrix:

$$\hat{\rho}_{AB} = \hat{\Pi}_{AB} = |\psi\rangle\langle\psi|_{AB} = \begin{bmatrix} |\alpha|^2 & \alpha\beta^* & \alpha\gamma^* & \alpha\delta^* \\ \alpha^*\beta & |\beta|^2 & \beta\gamma^* & \beta\delta^* \\ \alpha^*\gamma & \beta^*\gamma & |\gamma|^2 & \gamma\delta^* \\ \alpha^*\delta & \beta^*\delta & \gamma^*\delta & |\delta|^2 \end{bmatrix}. \tag{3.5}$$

Next, a **partial trace** is carried out. Like a full-blown trace, this involves a sum over certain terms of the matrix, while discarding others:

$$\mathrm{Tr}_B\{\hat{\rho}_{AB}\} = \begin{bmatrix} |\alpha|^2 & \alpha\beta^* & \alpha\gamma^* & \alpha\delta^* \\ \alpha^*\beta & |\beta|^2 & \beta\gamma^* & \beta\delta^* \\ \alpha^*\gamma & \beta^*\gamma & |\gamma|^2 & \gamma\delta^* \\ \alpha^*\delta & \beta^*\delta & \gamma^*\delta & |\delta|^2 \end{bmatrix} \quad \rightarrow \quad \hat{\rho}_A = \begin{bmatrix} |\alpha|^2 + |\beta|^2 & \alpha\gamma^* + \beta\delta^* \\ \alpha^*\gamma + \beta^*\delta & |\gamma|^2 + |\delta|^2 \end{bmatrix},$$

$$\mathrm{Tr}_A\{\hat{\rho}_{AB}\} = \begin{bmatrix} |\alpha|^2 & \alpha\beta^* & \alpha\gamma^* & \alpha\delta^* \\ \alpha^*\beta & |\beta|^2 & \beta\gamma^* & \beta\delta^* \\ \alpha^*\gamma & \beta^*\gamma & |\gamma|^2 & \gamma\delta^* \\ \alpha^*\delta & \beta^*\delta & \gamma^*\delta & |\delta|^2 \end{bmatrix} \quad \rightarrow \quad \hat{\rho}_B = \begin{bmatrix} |\alpha|^2 + |\gamma|^2 & \alpha\beta^* + \gamma\delta^* \\ \alpha^*\beta + \gamma^*\delta & |\beta|^2 + |\delta|^2 \end{bmatrix}. \tag{3.6}$$

⊕ **Mathematics Tips 3.2**

Partial Traces. As we have seen, the trace of a matrix $\hat{M}$ is the sum of all the elements along the diagonal, i.e.,

$$\mathrm{Tr}\{\hat{M}\} = \sum_{i=0}^{1} M_{ii},$$

(where we are using the convention that for a 2×2 matrix the indices denoting the elements run from 0 to 1). So, nothing wrong with that: the trace is an important scalar quantity (i.e., it is just a number, not a vector or a matrix). But how come a partial trace produces a matrix; and further, how come it involves off-diagonal elements?

To see this, we have to write the two-qubit state Eq. (3.1) in a different form: $|\psi\rangle_{AB} = \sum_{i,j=0}^{1} c_{ij}|i\rangle_A \otimes |j\rangle_B$. If we write $c_{00} = \alpha$, $c_{01} = \beta$, $c_{10} = \gamma$, and $c_{11} = \delta$, then this is identical to Eq. (3.1). In this notation the projector $\hat{\Pi}_{AB}$ is written as follows:

$$\hat{\Pi}_{AB} = \sum_{i,i',j,j'=0}^{1} c_{ij}c_{i'j'}^{*}|i\rangle\langle i'|_A \otimes |j\rangle\langle j'|_B.$$

Now let us trace the projector over qubit B only, leaving qubit A alone. We can use the result $\text{Tr}\{|j\rangle\langle j'|\} = \langle j|j'\rangle = \delta_{jj'}$, so that

$$\text{Tr}_B\{\hat{\Pi}_{AB}\} = \sum_{i,i'=0}^{1} \left(\sum_{j=0}^{1} c_{ij}c_{i'j}^{*} \right) |i\rangle\langle i'|_A.$$

This is just the same expression as Eq. (3.3), only using the new notation. For example,

$$\sum_{j=0}^{1} c_{0j}c_{1j}^{*} = c_{00}c_{10}^{*} + c_{01}c_{11}^{*} = \alpha\gamma^{*} + \beta\delta^{*} = \langle 0|\hat{\rho}_A|1\rangle,$$

as required. A similar derivation can be done when we do the partial trace over A instead of B.

This ability of one of the qubits to render the other qubit into a random state, because, initially, they were in the joint state Eq. (3.1), is a consequence of a particular property of that state called *entanglement*. Succinctly, entanglement means that one cannot reliably define a quantum state for either of the two qubits as separate entities; rather, the two qubits A and B must be considered jointly. If you consider one of the qubit in isolation, and pretend the other one does not exist, its state will be randomized, since connection to the other qubit cannot be ignored.

3.1.2 Concurrence

It is reasonable that the degree of randomness of the reduced density matrix of one of the qubits will be a good way to assess the degree of entanglement of the joint state. (We are confining ourselves to *pure* two-qubit states; there are ways of extending this argument to mixed states, but none are wholly satisfactory.) Thus it is a good idea to ask: What is the purity of $\hat{\rho}_A$? Using Eq. (2.95) the purity of $\hat{\rho}_A$ is $1 - 2\text{Det}\{\hat{\rho}_A\}$. The determinant of $\hat{\rho}_A$ can be calculated as follows:

$$\begin{aligned}
\text{Det}\{\hat{\rho}_A\} &= \left(|\alpha|^2 + |\beta|^2 \right) \left(|\gamma|^2 + |\delta|^2 \right) - \left(\alpha\gamma^{*} + \beta\delta^{*} \right) \left(\alpha^{*}\gamma + \beta^{*}\delta \right) \\
&= \alpha\alpha^{*}\gamma\gamma^{*} + \alpha\alpha^{*}\delta\delta^{*} + \beta\beta^{*}\gamma\gamma^{*} + \beta\beta^{*}\delta\delta^{*} \\
&\quad - \alpha\alpha^{*}\gamma\gamma^{*} - \beta\beta^{*}\delta\delta^{*} - \alpha\beta^{*}\gamma^{*}\delta - \alpha^{*}\beta\gamma\delta^{*} \\
&= |\alpha\delta - \beta\gamma|^2.
\end{aligned} \tag{3.7}$$

Thus, the purity of $\hat{\rho}_A$ is

$$\begin{aligned}
P(\hat{\rho}_A) &= 1 - 2|\alpha\delta - \beta\gamma|^2 \\
&= 1 - \mathcal{C}^2/2,
\end{aligned} \tag{3.8}$$

where the **concurrence** of the state is defined by

$$\boxed{\mathcal{C}(\psi_{AB}) = 2|\alpha\delta - \beta\gamma|.} \tag{3.9}$$

The purity of $\hat{\rho}_B$ is the same as the purity of $\hat{\rho}_A$, so it does not matter which qubit is chosen. Further, because the purity of the reduced density matrix is a simple function of concurrence, concurrence itself is a pretty good way to tell how entangled a *pure* two-qubit state is.

One other property of the concurrence: the eigenvalues of $\hat{\rho}_A$ (and of $\hat{\rho}_B$, since the two matrices have the same eigenvalues), which we will denote λ_0 and λ_1, are given by Eq. (2.27): $(T \pm R)/2$. Here $T = 1$ and $R = \sqrt{1 - 4|\alpha\delta + \beta\gamma|^2}$; thus,

$$\lambda_0 = \left(1 - \sqrt{1 - \mathcal{C}^2}\right)/2, \quad \lambda_1 = \left(1 + \sqrt{1 + \mathcal{C}^2}\right)/2. \tag{3.10}$$

Since concurrence is defined as the absolute value of an expression; it is a real number whose minimum value is 0. Its maximum value can be figured out from Eq. (3.7): $\mathcal{C}^2 = 4\text{Det}\{\hat{\rho}_A\} = 4\lambda_1\lambda_2$ where λ_1 and λ_2 are the eigenvalues of $\hat{\rho}_A$. Since λ_1 and λ_2 are real and positive, and $\lambda_1 + \lambda_2 = 1$, the maximum value of $\lambda_1\lambda_2$ is $1/4$ when $\lambda_1 = \lambda_2 = 1/2$; thus $\mathcal{C}$ has the maximum possible value of 1. What do these extreme values of $\mathcal{C}$ imply about the states $|\psi_{AB}\rangle$?

First, a non-entangled or **separable state** is a special case of the general two-qubit state in which the two qubits, A and B, are completely isolated from each other, either directly or via a third party, and thus have no chance to interact or become correlated. In this case each qubit can have a well-defined pure state of its own, and the combined state of this *separable* system will be

$$\begin{aligned}
|\psi\rangle_{AB}^{(\text{separable})} &= |\varphi\rangle_A \otimes |\varepsilon\rangle_B \\
&= \left(\alpha_A|0\rangle_A + \beta_A|1\rangle_A\right) \otimes \left(\alpha_B|0\rangle_B + \beta_B|1\rangle_B\right) \\
&= (\alpha_A\alpha_B)\,|00\rangle_{AB} + (\alpha_A\beta_B)\,|01\rangle_{AB} + \\
&\quad (\beta_A\alpha_B)\,|10\rangle_{AB} + (\beta_A\beta_B)\,|11\rangle_{AB},
\end{aligned} \tag{3.11}$$

where α_A and β_A are the probability amplitudes for the state of qubit A, and α_B and β_B are those for qubit B. Thus, for separable systems, the two-qubit probability amplitudes appearing in Eq. (3.1) are $\alpha = \alpha_A\alpha_B$, $\beta = \alpha_A\beta_B$, $\gamma = \beta_A\alpha_B$ and $\delta = \beta_A\beta_B$. If we substitute these expressions into Eq. (3.9), we find $\mathcal{C}(\psi_{AB}^{(\text{separable})}) = 0$.

The states for which $\mathcal{C} = 1$ correspond to the reduced density matrices of both qubits being maximally mixed (see Eq. (2.93)), in which the individual qubit states are completely random; the quantum state is entirely joint between the two qubits: these are called **maximally entangled states**, and we will have more to say about them shortly.

One important property of concurrence is that this measure does not change when we apply a unitary operator to either qubit *individually*. This is essential for a measure of entanglement to be reliable. We want to know the global correlation between the two qubits which does not depend on **local** operations that might be performed on either qubit. This property can be seen by applying a unitary operator to either qubit; e.g. for qubit A:

$$\begin{aligned}
(\hat{U} \otimes \hat{I})|\varphi\rangle_{AB} &= \alpha|0'0\rangle + \beta|0'1\rangle + \gamma|1'0\rangle + \delta|1'1\rangle \\
&= \alpha\{a|00\rangle - b^*|10\rangle\} + \beta\{a|01\rangle - b^*|11\rangle\} + \gamma\{b|00\rangle + a^*|10\rangle\} + \delta\{b|01\rangle + a^*|11\rangle\} \\
&= \{a\alpha + b\gamma\}|00\rangle + \{a\beta + b\delta\}|01\rangle + \{a^*\gamma - b^*\alpha\}|10\rangle + \{a^*\delta - b^*\beta\}|11\rangle,
\end{aligned}$$

$$\tag{3.12}$$

where $|0'\rangle_A$ and $|1'\rangle_A$ are defined in Eq. (2.65). Thus,

$$
\begin{aligned}
\alpha'\delta' - \beta'\gamma' &= \{a\alpha + b\gamma\}\{a^*\delta - b^*\beta\} - \{a\beta + b\delta\}\{a^*\gamma - b^*\alpha\} \\
&= |a|^2\alpha\delta - ab^*\alpha\beta + a^*b\gamma\delta - |b|^2\beta\gamma - |a|^2\beta\gamma + ab^*\alpha\beta - a^*b\gamma\delta + |b|^2\alpha\delta \\
&= \alpha\delta - \beta\gamma,
\end{aligned}
\tag{3.13}
$$

where we have used Eq. (2.61). Notice that in (3.12) the expression $(\hat{U}_A \otimes \hat{I})$, where a local operator acts on qubit A only, the fact that no (effective) operator is applied on qubit B is shown through the identity operator, $\hat{I}$, being applied on it. A similar proof can be used to show that the concurrence does not change when a unitary is applied to qubit B.

✦ Mathematics Tips 3.3

Tensor Product of 2×2 Matrices. Given two 2×2 matrices $\hat{A}$ and $\hat{B}$, if the matrix representation of $\hat{A}$ is given by

$$
\hat{A} = \begin{bmatrix} a_1 & a_2 \\ a_3 & a_4 \end{bmatrix},
\tag{3.14}
$$

and that of $\hat{B}$ is given by

$$
\hat{B} = \begin{bmatrix} b_1 & b_2 \\ b_3 & b_4 \end{bmatrix}
\tag{3.15}
$$

then the 4×4 matrix representing the tensor product $\hat{A} \otimes \hat{B}$ is obtained by multiplying each of the four a_i separately by the matrix of $\hat{B}$ constructing the final matrix to look like this (in block form):

$$
\begin{bmatrix} a_1\hat{B} & a_2\hat{B} \\ a_3\hat{B} & a_4\hat{B} \end{bmatrix} = \begin{bmatrix} a_1b_1 & a_1b_2 & a_2b_1 & a_2b_2 \\ a_1b_3 & a_1b_4 & a_2b_3 & a_2b_4 \\ a_3b_1 & a_3b_2 & a_4b_1 & a_4b_2 \\ a_3b_3 & a_3b_4 & a_4b_3 & a_4b_4 \end{bmatrix}.
\tag{3.16}
$$

3.1.3 Schmidt Decomposition

The reduced density matrices $\hat{\rho}_A$ and $\hat{\rho}_B$ are, like all density matrices, Hermitian. From Eq. (2.34) we know that there are unitary matrices, constructed from the eigenvectors, with the properties

$$
\hat{U}_A^\dagger \hat{\rho}_A \hat{U}_A = \hat{U}_B^\dagger \hat{\rho}_B \hat{U}_B = \hat{\Lambda},
\tag{3.17}
$$

where $\hat{\Lambda}$ is the diagonal matrix of eigenvalues of $\hat{\rho}_A$ and $\hat{\rho}_B$, i.e.,

$$
\hat{\Lambda} = \begin{bmatrix} \lambda_0 & 0 \\ 0 & \lambda_1 \end{bmatrix}.
\tag{3.18}
$$

If we apply the tensor product of $\hat{U}_A$ and $\hat{U}_B$ to $\hat{\Pi}_{AB}$ (Eq. (3.5)) we get the following rather sparse matrix:

$$\left(\hat{U}_A^\dagger \otimes \hat{U}_B^\dagger\right) \hat{\Pi}_{AB} \left(\hat{U}_A \otimes \hat{U}_B\right) = \begin{bmatrix} \lambda_0 & 0 & 0 & \sqrt{\lambda_0 \lambda_1} \\ 0 & 0 & 0 & 0 \\ 0 & 0 & 0 & 0 \\ \sqrt{\lambda_0 \lambda_1} & 0 & 0 & \lambda_1 \end{bmatrix}. \tag{3.19}$$

Equivalently, this implies the **Schmidt decomposition theorem**, which states that any two-qubit pure state can be written as

$$|\psi\rangle_{AB} = \hat{U}_A \otimes \hat{U}_B \left(\sqrt{\lambda_0}|00\rangle + \sqrt{\lambda_1}|11\rangle\right). \tag{3.20}$$

Add this to your quantum toolbox as a tool that can help you simplify solving problems. Its proof is rather involved, and uses a result from linear algebra called the **singular value decomposition** (see Appendix B).

Exercises 3.1.4

1. Write the following N-qubit states in a compact form (for example, $|0\rangle \otimes |0\rangle = |00\rangle$):
   (a) $|1\rangle \otimes |1\rangle \otimes |1\rangle \otimes |1\rangle$,
   (b) $\frac{1}{\sqrt{3}} (|0\rangle \otimes |0\rangle \otimes |1\rangle + |0\rangle \otimes |1\rangle \otimes |0\rangle + |1\rangle \otimes |0\rangle \otimes |0\rangle)$,
   (c) $|0\rangle \otimes |1\rangle \otimes |1\rangle \otimes |0\rangle \otimes |1\rangle \otimes |1\rangle \otimes |0\rangle$.
2. Prove that the general two-qubit (ket) state $|\psi\rangle_{AB} = \alpha|00\rangle + \beta|01\rangle + \gamma|10\rangle + \delta|11\rangle$ is pure.
3. Show, using an example, that a two-qubit (density matrix) state can be mixed; i.e., it is not in general pure.
4. Given qubits A and B with the corresponding states:

$$|\psi\rangle_A = a_0|0\rangle + a_1|1\rangle,$$
$$|\psi\rangle_B = b_0|0\rangle + b_1|1\rangle.$$

   (a) Write out the tensor product of the two states to obtain $|\psi\rangle_{AB}$, the state of the two qubits.
   (b) Find the reduced density matrix of the two qubits; i.e., $\hat{\rho}_A$ and $\hat{\rho}_B$.
   (c) Find the purity of the reduced density matrices.
   (d) What can you conclude about the entanglement in the system? Does it matter to a given qubit when the other qubit is traced out?
5. Given the general two-qubit pure state $|\psi\rangle_{AB} = \alpha|00\rangle + \beta|01\rangle + \gamma|10\rangle + \delta|11\rangle$.
   (a) Find the reduced density matrix of the two qubits; i.e., $\hat{\rho}_A$ and $\hat{\rho}_B$.
   (b) Calculate the (von Neumann) entropy of the reduced density matrices, which is defined in terms of their eigenvalues λ_0 and λ_1, as follows: $S(\hat{\rho}) = -\sum_{n=0}^{1} \lambda_n \log_2 \lambda_n$.
   (c) Write the expression for the entropy you found (also known as the entanglement of formation) in terms of the expression for concurrence of pure states that you learned about in this section.
6. For a pure two-qubit system, the purity of the reduced density matrix, which is a monotonic function of the concurrence of the system, is what tells us whether the system is entangled or not. If it is 0, then we know the system is not entangled. Otherwise, we know it is. However, it is important to stress that this simple technique does not work if the two-qubit system cannot be written as a ket; i.e., is mixed and can only be represented by a density matrix. Find the purity of the reduced density matrices of the following states to find out why it is not appropriate to use as a measure of entanglement for mixed states:

(a) The maximally entangled state; i.e., it has $\mathcal{C} = 1 : |\Psi^+\rangle = \frac{1}{\sqrt{2}}\{|01\rangle + |01\rangle\}$.

(b) The maximally mixed state, which is known to have no entanglement, given by the density matrix: $\hat{\rho} = \frac{1}{4}\{|00\rangle\langle00| + |01\rangle\langle01| + |10\rangle\langle10| + |11\rangle\langle11|\}$.

What did you find out?

7. Find the concurrence for the following states. State whether each is separable, partially entangled, or maximally entangled.

(a) $|\psi_1\rangle = \frac{1}{\sqrt{6}}\left(|00\rangle + |01\rangle + \sqrt{2}|10\rangle + \sqrt{2}|11\rangle\right)$.

(b) $|\psi_2\rangle = \frac{1}{\sqrt{2}}\left(|00\rangle - |11\rangle\right)$.

(c) $|\psi_3\rangle = \frac{1}{\sqrt{5}}\left(|00\rangle + 2|11\rangle\right)$.

(d) $|\psi_4\rangle = |10\rangle$.

(e) $|\psi_5\rangle = \frac{1}{2}\left(|00\rangle + |01\rangle + |10\rangle + |11\rangle\right)$.

8. For a general pure two-qubit state (made of qubit A and qubit B), show that applying a local unitary on qubit B does not change the concurrence of the state.

9. Find the concurrence of the state in (3.20) in terms of λ_0 and λ_1.

10. Calculate all possible tensor products of the form $\hat{\sigma}_i \otimes \hat{\sigma}_j$ in matrix form. *Hint: Use Eq. (2.48).*

11. In Eq. (3.17) we used the unitaries $\hat{U}_A$ and $\hat{U}_B$ to diagonalize the reduced density matrices. Show that these operators are not unique: one can multiply either of them by a phase matrix

$$\begin{bmatrix} \exp i\chi_0 & 0 \\ 0 & \exp i\chi_1 \end{bmatrix},$$

where χ_0 and χ_1 are arbitrary phases, without altering their diagonalization property. It may be necessary to perform such a multiplication, with careful choice of phases in order to obtain real-valued off-diagonal elements in Eq. (3.19).

3.2   Operations on Two Qubits

Next we turn our attention to the kind of operations that can be applied to two-qubit systems, including local and global operations. We also discuss what kind of operations can create entanglement, an important resource in quantum computation.

3.2.1   Local Unitary Operators

As we saw earlier in this chapter, local unitary operators, or local unitaries, are operators that are applied on only one of the qubits. Recall the general form of a unitary operator acting on one qubit that we found in the previous chapter (Eq. (2.60)). We can apply it on one of the qubits in (3.1). For example, if we want to apply this unitary on the first qubit, then this is the form of the unitary on the whole system:

$$\hat{U} \otimes \hat{I}. \tag{3.21}$$

Notice how in this case the identity $\hat{I}$ is applied to the second qubit to leave it undisturbed locally. Likewise, applying $\hat{U}$ on the second qubit only requires the following tensor product of operators to be applied on (3.1):

$$\hat{I} \otimes \hat{U}. \tag{3.22}$$

In order to find the matrix representations of these local operators, see Mathematics Tips 3.3.

As you will find out in Exercise 2, when you apply local operations, you do not change the fact that a state can be written as a product of an exclusive part that corresponds to one qubit and another part that corresponds to the other qubit:

$$\left(\hat{U} \otimes \hat{V}\right)|0\rangle|0\rangle = \left(\hat{U}|0\rangle\right) \otimes \left(\hat{V}|0\rangle\right). \tag{3.23}$$

Therefore, **local operations can never increase entanglement.** That is why you will notice that in studies of the evolution of quantum systems in which all that is of concern is entanglement, if applying local operations simplifies solving the problem, it is acceptable to do so, even though one is actually *changing* the states of the individual qubits.

3.2.2 Creating Entanglement

Here we discuss how to create a general entangled two-qubit state. What we want is to transfer an easily created separable state such as $|0\rangle|0\rangle$ to the general entangled form in (3.1). At this stage we are equipped with most of the tools we need to achieve that. One more thing we will need in this mission is a global unitary operation that will change the state of the second qubit, depending on what the state of the first qubit is. It will be introduced very shortly.

In the first step, we apply a local operator on qubit A. Next, we apply the global operator (we just hinted at) on the whole system, and finally we use the Schmidt decomposition theorem to achieve the desired state.

Step 1

Take the initial state $|\psi\rangle - |0\rangle|0\rangle$ and apply the following unitary on the first qubit:

$$\hat{U}_1 = \sqrt{\lambda_0}\hat{I} - i\sqrt{\lambda_1}\hat{Y} = \begin{bmatrix} \sqrt{\lambda_0} & -\sqrt{\lambda_1} \\ \sqrt{\lambda_1} & \sqrt{\lambda_0} \end{bmatrix}. \tag{3.24}$$

This transforms the initial state as follows:

$$|\psi\rangle \longrightarrow |\psi_1\rangle = \sqrt{\lambda_0}|00\rangle + \sqrt{\lambda_1}|10\rangle. \tag{3.25}$$

In terms of quantum circuit diagrams, this transformation looks like this.

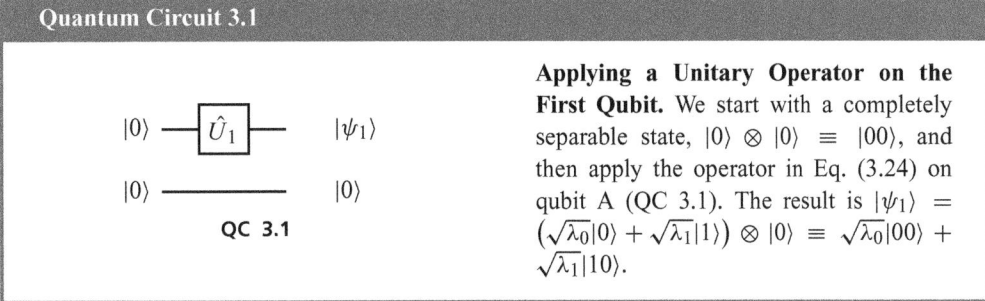

Quantum Circuit 3.1

$|0\rangle \ \boxed{\hat{U}_1} \ |\psi_1\rangle$

$|0\rangle \ \text{———} \ |0\rangle$

QC 3.1

Applying a Unitary Operator on the First Qubit. We start with a completely separable state, $|0\rangle \otimes |0\rangle \equiv |00\rangle$, and then apply the operator in Eq. (3.24) on qubit A (QC 3.1). The result is $|\psi_1\rangle = \left(\sqrt{\lambda_0}|0\rangle + \sqrt{\lambda_1}|1\rangle\right) \otimes |0\rangle \equiv \sqrt{\lambda_0}|00\rangle + \sqrt{\lambda_1}|10\rangle$.

Would it not be nice if the 0 in $|10\rangle$ could be made into 1? In this case, we would almost (since both λ_0 and λ_1 would also have to equal $1/2$) have a maximally entangled state. We can do that with a special global operator called a *CNOT gate*.

Step 2

A $\text{CNOT}_{A,B}$ gate bit-flips qubit B if and only if qubit A is in $|1\rangle$ and leaves qubit B untouched if qubit A is in $|0\rangle$. It leaves qubit A untouched. We use the subscripts to denote which two qubits are involved: the first, A, is generically called the **control qubit**, while the second, B, is the **target qubit**. Thus $\text{CNOT}_{2,3}$ is a CNOT gate in which qubit #3 is bit-flipped conditional on the state of qubit #2. This is accomplished *reversibly*; i.e., without measuring qubit A. The operator that achieves this is constructed as follows:

$$\text{CNOT}_{2,1} = \hat{\Pi}_0 \otimes \hat{I} + \hat{\Pi}_1 \otimes \hat{X} = \begin{bmatrix} 1 & 0 & 0 & 0 \\ 0 & 1 & 0 & 0 \\ 0 & 0 & 0 & 1 \\ 0 & 0 & 1 & 0 \end{bmatrix}. \tag{3.26}$$

Note that we have departed from our usual practice of denoting unitary operators with single upper-case letters with hats on them. In diagrams we dispense with the subscripts, since it is clear which qubit is the target and which the control; when writing out the operators algebraically, the subscripts are needed to avoid confusion. Qubits will conventionally be numbered from right to left in formulas and from bottom to top in circuits.

Recall $\hat{\Pi}_0$ and $\hat{\Pi}_1$ are the projectors we discussed earlier in Eq. (2.86). In quantum circuit diagrams, there are two different ways the CNOT gate can be represented.

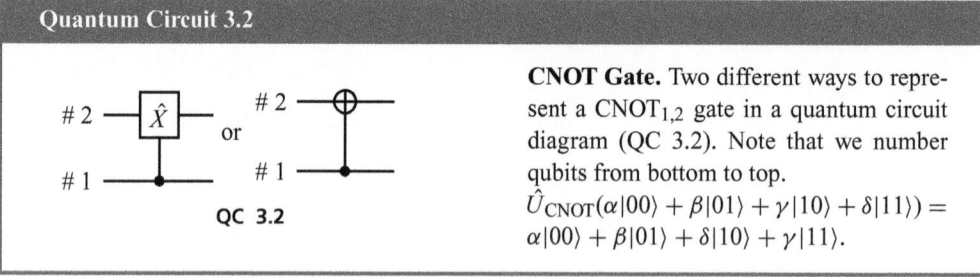

Quantum Circuit 3.2

QC 3.2

CNOT Gate. Two different ways to represent a $\text{CNOT}_{1,2}$ gate in a quantum circuit diagram (QC 3.2). Note that we number qubits from bottom to top.
$\hat{U}_{\text{CNOT}}(\alpha|00\rangle + \beta|01\rangle + \gamma|10\rangle + \delta|11\rangle) = \alpha|00\rangle + \beta|01\rangle + \delta|10\rangle + \gamma|11\rangle.$

This will transform the state we have into

$$|\psi_1\rangle \longrightarrow |\psi_2\rangle = \sqrt{\lambda_0}|00\rangle + \sqrt{\lambda_1}|11\rangle. \tag{3.27}$$

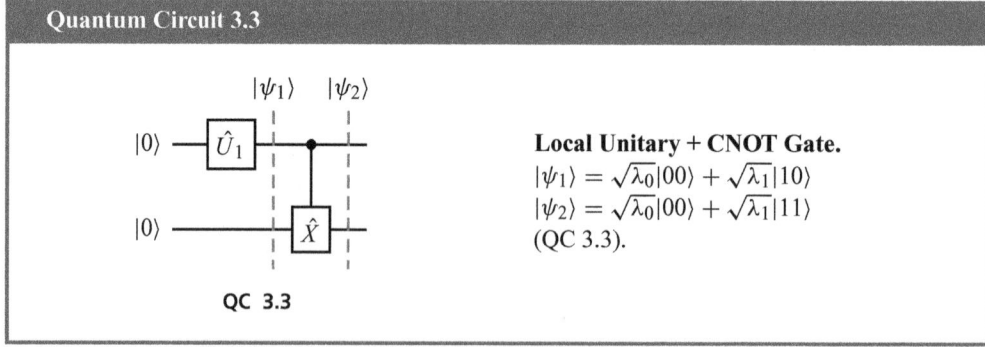

Quantum Circuit 3.3

QC 3.3

Local Unitary + CNOT Gate.
$|\psi_1\rangle = \sqrt{\lambda_0}|00\rangle + \sqrt{\lambda_1}|10\rangle$
$|\psi_2\rangle = \sqrt{\lambda_0}|00\rangle + \sqrt{\lambda_1}|11\rangle$
(QC 3.3).

This has produced an entangled state, albeit in a particular basis. We want an entangled state in the most general two-qubit basis.

Step 3

Recall the Schmidt decomposition theorem represented in (3.20). It literally implies that local unitary operators applied to both qubits A and B can make the state in (3.27) into any two-qubit state we want, which will depend on the choice of the operators. Therefore, the state can be transformed further as follows:

$$\begin{aligned} |\psi_3\rangle &= \hat{U}_2 \otimes \hat{U}_3 \left(\sqrt{\lambda_0}|00\rangle + \sqrt{\lambda_1}|11\rangle \right) \\ &= \alpha|00\rangle + \beta|01\rangle + \gamma|10\rangle + \delta|11\rangle, \end{aligned} \qquad (3.28)$$

where $\hat{U}_2$ and $\hat{U}_3$ are the necessary unitary operators required to achieve the final state. The following gives the complete quantum circuit diagram that describes all previous steps to create entanglement in the system.

Quantum Circuit 3.4

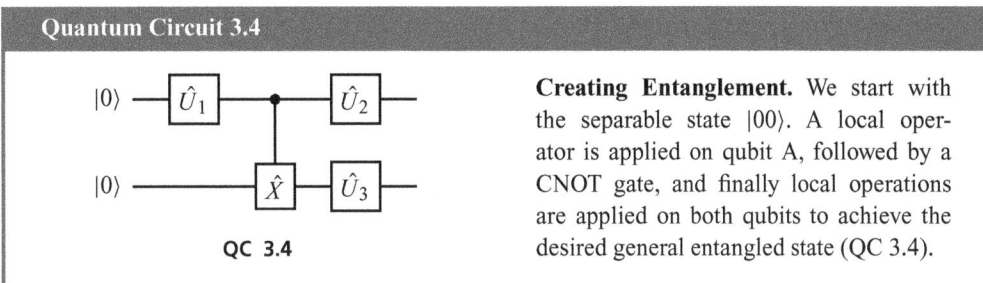

QC 3.4

Creating Entanglement. We start with the separable state $|00\rangle$. A local operator is applied on qubit A, followed by a CNOT gate, and finally local operations are applied on both qubits to achieve the desired general entangled state (QC 3.4).

3.2.3 Universal Two-Qubit Gates

The CNOT gate is a crucial global operator in the creation of entanglement. As we saw, it operates by changing the state of one qubit, depending on what the state of the other qubit is. Needing only one operator to achieve this simplifies quantum engineering tremendously. That is why the CNOT gate warrants being called a *universal* gate.

There are other **universal gates** that can also be used, in conjunction with local operators, instead of the CNOT gate to create *any* entangled state. In fact, these can be obtained from the CNOT by applying specific local unitaries. For example, the controlled-Z (CZ) gate is constructed very similarly to the CNOT gate, except for the fact that $\hat{Z}$ is used instead of the $\hat{X}$ in (3.26):

$$CZ_{2,1} = \hat{\Pi}_0 \otimes \hat{I} + \hat{\Pi}_1 \otimes \hat{Z} = \begin{bmatrix} 1 & 0 & 0 & 0 \\ 0 & 1 & 0 & 0 \\ 0 & 0 & 1 & 0 \\ 0 & 0 & 0 & -1 \end{bmatrix}. \qquad (3.29)$$

Note for the CZ gate the property that $CZ_{1,2} = CZ_{2,1}$.

This is how the CZ is represented in quantum circuit diagrams.

Quantum Circuit 3.5

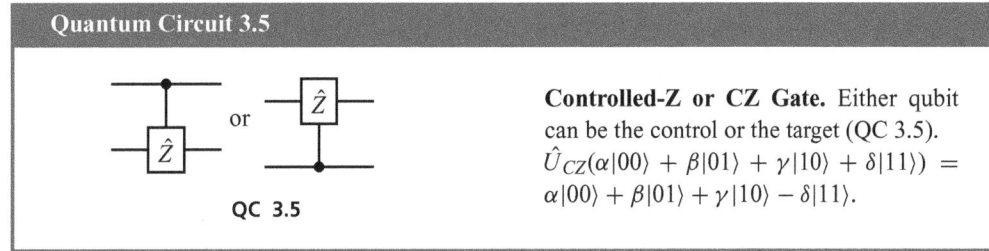

QC 3.5

Controlled-Z or CZ Gate. Either qubit can be the control or the target (QC 3.5). $\hat{U}_{CZ}(\alpha|00\rangle + \beta|01\rangle + \gamma|10\rangle + \delta|11\rangle) = \alpha|00\rangle + \beta|01\rangle + \gamma|10\rangle - \delta|11\rangle$.

The $\hat{X}$ and $\hat{Z}$ operators are related to each other in terms of the Hadamard gate, $\hat{H}$, as follows:

$$\hat{X} = \hat{H}\hat{Z}\hat{H}, \tag{3.30}$$

where

$$\hat{H} = \frac{1}{\sqrt{2}}\begin{bmatrix} 1 & 1 \\ 1 & -1 \end{bmatrix}. \tag{3.31}$$

That is why, to obtain the CNOT gate from the CZ gate, one can apply the local operators necessary as described in (3.30).

Quantum Circuit 3.6

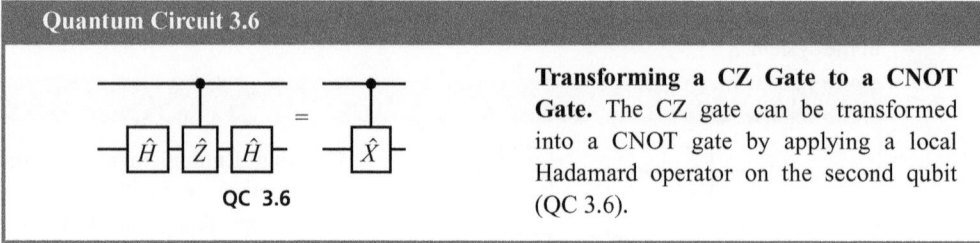

QC 3.6

Transforming a CZ Gate to a CNOT Gate. The CZ gate can be transformed into a CNOT gate by applying a local Hadamard operator on the second qubit (QC 3.6).

Another example of a universal gate is the ROOT-SWAP gate, which is obtained from the non-universal gate known as the SWAP gate. The latter swaps the states of the two qubits so that

$$
\begin{aligned}
|00\rangle &\longrightarrow |00\rangle, \\
|01\rangle &\longrightarrow |10\rangle, \\
|10\rangle &\longrightarrow |01\rangle, \\
|11\rangle &\longrightarrow |11\rangle.
\end{aligned} \tag{3.32}
$$

Notice that the SWAP gate only has an effect when the two qubits are different, and since it is not a universal gate, entanglement *cannot* be created by swapping. The following are the matrix and quantum circuit diagram representations of this gate:

$$\text{SWAP}_{1,2} = \frac{1}{2}\left(\hat{I}\otimes\hat{I} + \hat{X}\otimes\hat{X} + \hat{Y}\otimes\hat{Y} + \hat{Z}\otimes\hat{Z}\right) = \begin{bmatrix} 1 & 0 & 0 & 0 \\ 0 & 0 & 1 & 0 \\ 0 & 1 & 0 & 0 \\ 0 & 0 & 0 & 1 \end{bmatrix}. \tag{3.33}$$

Like the CZ gate, SWAP is symmetric in the interchange of qubits: $\text{SWAP}_{1,2} = \text{SWAP}_{2,1}$.

Quantum Circuit 3.7

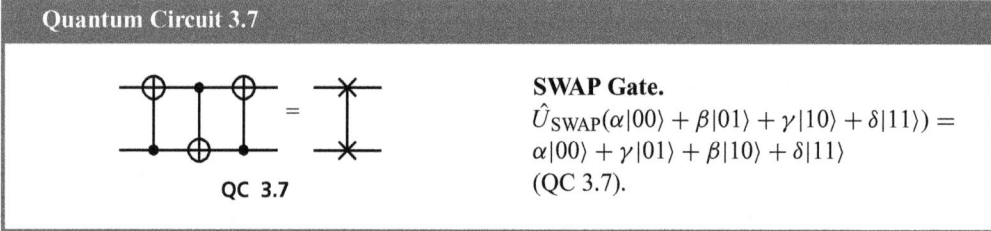

QC 3.7

SWAP Gate.
$\hat{U}_{\text{SWAP}}(\alpha|00\rangle + \beta|01\rangle + \gamma|10\rangle + \delta|11\rangle) = \alpha|00\rangle + \gamma|01\rangle + \beta|10\rangle + \delta|11\rangle$
(QC 3.7).

✛ **Mathematics Tips 3.4**

The Function of a Matrix. Given the matrix $\hat{M}$, when asked to calculate a function of this matrix $f(\hat{M})$ (e.g., $f(x) = x^2, f(x) = \sin x$, etc.), a common mistake is to take the function of the matrix elements of the matrix. This is NOT what we mean in this book. Here, the way to evaluate the function of the matrix is to:

1. Find the eigenvalues, λ_i, and the eigenvectors, $|v_i\rangle$, of $\hat{M}$.
2. Evaluate the function for each eigenvalue: $f(\lambda_i)$.
3. Construct the outer product of each eigenvector: $|v_i\rangle\langle v_i|$.
4. To obtain $f(\hat{M})$, multiply each function (of an eigenvalue) with its corresponding eigenvector outer product, then add them all up. In other words, $f(\hat{M}) = \sum_i f(\lambda_i)|v_i\rangle\langle v_i|$.

If the square root of the SWAP gate is taken, then the resulting operator is universal and can, therefore, be used to create entanglement. This gate is represented by $\sqrt{\text{SWAP}}$. The way to find the matrix representation of this operator from the matrix representation of the SWAP operator, as seen in (3.33), is to find its eigenvalues and eigenvectors, and then to create a sum made of terms such that each term consists of the square root of an eigenvalue multiplied by the outer product of the corresponding eigenvector. Working such a sum out should give the following matrix representation of the $\sqrt{\text{SWAP}}$ operator:

$$\sqrt{\text{SWAP}}_{1,2} = \begin{bmatrix} 1 & 0 & 0 & 0 \\ 0 & (1-i)/2 & (1+i)/2 & 0 \\ 0 & (1+i)/2 & (1-i)/2 & 0 \\ 0 & 0 & 0 & 1 \end{bmatrix}. \tag{3.34}$$

One chooses a two-qubit gate to make things as easy as possible for the chosen technology.

3.2.4 Maximally Entangled States

If entanglement is *the* resource that is being exploited in quantum technologies, it is not surprising that the states that maximize the amount of entanglement in the system have a special role.

Since we learned earlier how to create any entangled state, this means that we can easily create a maximally entangled state; i.e., a state with concurrence equalling unity.

If we start with the state $|00\rangle$, then apply a Hadamard gate on the first qubit, followed by the CNOT gate on both qubits, we end up with the following maximally entangled state (see the next quantum circuit):

$$|\beta_0\rangle = \frac{1}{\sqrt{2}}(|00\rangle + |11\rangle). \tag{3.35}$$

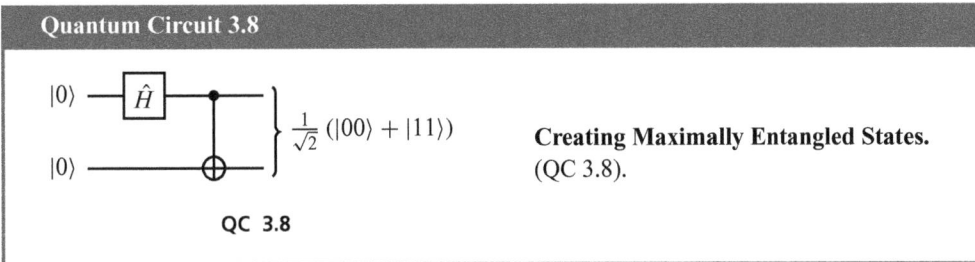

Quantum Circuit 3.8

$|0\rangle$ — $\hat{H}$ — $\left.\begin{matrix} \\ \end{matrix}\right\}\frac{1}{\sqrt{2}}(|00\rangle + |11\rangle)$

$|0\rangle$ —

Creating Maximally Entangled States. (QC 3.8).

QC 3.8

Recall that local unitaries can change a state, but have no effect on the entanglement of the system. We can use this fact to generate other maximally entangled states from the state in (3.35). While there are several ways to do that, we will achieve it as follows:

$$|\beta_k\rangle = i\hat{I} \otimes \hat{\sigma}_k |\beta_0\rangle. \tag{3.36}$$

In Exercise 15, you will find that

$$|\beta_1\rangle = \frac{i}{\sqrt{2}} (|01\rangle + |10\rangle),$$

$$|\beta_2\rangle = \frac{-1}{\sqrt{2}} (|01\rangle - |10\rangle), \tag{3.37}$$

$$|\beta_3\rangle = \frac{i}{\sqrt{2}} (|00\rangle - |11\rangle).$$

The orthonormal set $\{|\beta_0\rangle, |\beta_1\rangle, |\beta_2\rangle, |\beta_3\rangle\}$ is complete, and forms the **Bell basis** (named after John S. Bell), which can be used as a basis to describe any two-qubit state as follows:

$$|\psi\rangle = A|\beta_0\rangle + B|\beta_1\rangle + C|\beta_2\rangle + D|\beta_3\rangle. \tag{3.38}$$

Effect of Local Operations on Entangled States: The Fundamental Theorem of Entanglement
Let us now consider what happens when we apply some arbitrary operation $\hat{M}$ to one of the two qubits in a maximally entangled state, for example $|\beta_2\rangle$. Note that we do not assume anything about $\hat{M}$ other than it is a 2×2 matrix (e.g. Eq. (2.3)); the effect on the qubit states is $|0'\rangle = \hat{M}|0\rangle = a|0\rangle + c|1\rangle$ and $|1'\rangle = \hat{M}|1\rangle = b|0\rangle + d|1\rangle$.

$$\begin{aligned}
\left(\hat{I} \otimes \hat{M}\right)|\beta_2\rangle &= -\frac{1}{\sqrt{2}} \left(|0\rangle|1'\rangle - |1\rangle|0'\rangle\right) \\
&= -\frac{1}{\sqrt{2}} \left(b|00\rangle + d|01\rangle - a|10\rangle - c|11\rangle\right) \\
&= -\frac{1}{\sqrt{2}} \left(\{d|0\rangle - c|1\rangle\}|1\rangle - \{-b|0\rangle + a|1\rangle\}|0\rangle\right) \\
&= \mathrm{Det}\{\hat{M}\}\left(\hat{M}^{-1} \otimes \hat{I}\right)|\beta_2\rangle, \tag{3.39}
\end{aligned}$$

where, in the last step, we have used the formula for the inverse matrix, Eq. (2.11). Thus, applying any linear operation to one qubit of a maximally entangled pair is equivalent to applying a related operation (in this case, the inverse) to the other qubit. This seeming automatic and instantaneous transmission of an action from one qubit to the other is central to both the fundamental implications and the technological applications of quantum entanglement, and encapsulates its essential "spookiness"; thus it seems not inappropriate to dub this result the *fundamental theorem of entanglement*. How the $\hat{M}$ is transformed when it is "transmitted" from one qubit to the other is dependent on the exact form of the maximally entangled state being used.

Exercises 3.2.5

1. Using Eq. (2.60) write $\hat{U} \otimes \hat{I}$ and $\hat{I} \otimes \hat{U}$ in matrix form.
2. Show that entanglement cannot be created by applying local unitaries alone on the qubits. *Hint: Start with a completely separable state.*
3. What happens to the concurrence if two different local operators, none of which is the identity, are applied to each of the two qubits at the same time?

4. Prove the identity for QC 3.9:

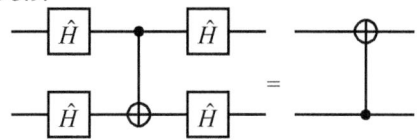

QC 3.9

5. Given the state $|\psi\rangle = \frac{1}{2}(|00\rangle + |01\rangle + |10\rangle + |11\rangle)$, apply a Hadamard gate on each qubit and simplify. Did these local unitaries change the state to be easier to work with or make it more complicated? How did the concurrence change?

6. How can $\hat{U}_1$ in (3.24) be redefined in terms of $\hat{X}$ so that the creation of the state in (3.25) is still achieved?

7. What effect does a $CNOT_{1,2}$ gate have on the state $a|00\rangle + b|01\rangle$? (Recall: the convention is $CNOT_{control,target}$, and qubits are numbered from right to left in kets and from bottom to top in circuit diagrams.)

8. Show that the CZ gate is invariant under swapping of qubit roles; i.e., $\hat{\Pi}_0 \otimes \hat{I} + \hat{\Pi}_1 \otimes \hat{Z} \equiv \hat{I} \otimes \hat{\Pi}_0 + \hat{Z} \otimes \hat{\Pi}_1$.

9. Using QC 3.4 and starting with the state $|00\rangle$, what values for the parameters λ_0 and λ_1, and what local unitaries $\hat{U}_2$ and $\hat{U}_3$ should be chosen to create the maximally entangled state: $\frac{1}{\sqrt{2}}(|00\rangle + |11\rangle)$?

10. Draw a quantum circuit to create a state of concurrence equal to 120/169. Include the value of the parameters of any relevant unitaries.

11. Given $f(x) = x^2$, calculate $f(\hat{H})$, where $\hat{H}$ is the Hadamard gate given in (3.31).

12. Assuming $\hat{M}$ is Hermitian and traceless, find $\hat{M}$ so that

$$\sin(\pi\hat{M}) = \sin(5\pi)\begin{bmatrix} 4/5 & 3/5 \\ 3/5 & -4/5 \end{bmatrix}.$$

13. Use the SWAP gate representation in (3.33) to obtain the ROOT-SWAP gate representation in (3.34).

14. Using the definition of the ROOT-SWAP gate (Eq. (3.34)), show that the circuit in QC 3.10 is equivalent to the CZ gate:

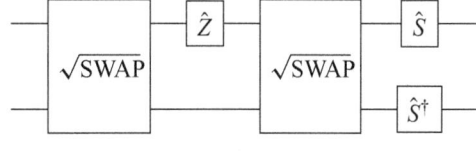

QC 3.10

where

$$\hat{S} = \begin{bmatrix} 1 & 0 \\ 0 & i \end{bmatrix}. \tag{3.40}$$

15. Use (3.36) to find $|\beta_1\rangle$, $|\beta_2\rangle$, and $|\beta_3\rangle$ in (3.37).

16. Show that the set $\{|\beta_0\rangle, |\beta_1\rangle, |\beta_2\rangle, |\beta_3\rangle\}$, known as the Bell basis, is orthonormal, and, therefore, forms a complete set of vectors that can be used as a basis in a two-qubit space.

17. Work out the expression for concurrence in terms of the coefficients A, B, C, D that define the quantum state in the Bell basis (3.38). Find the condition on the Bell basis probability amplitudes for the state to be maximally entangled.

18. Repeat the derivation of the fundamental theorem of entanglement for the other three Bell states, finding how the operation $\hat{M}$ is transformed on transmission.

3.3   Bell's Inequalities and Hidden Variable Theories

In this section, we will investigate one of the first, and still one of the most fascinating applications of entanglement. The apparent instantaneous transmission of the outcome of a measurement, regardless of the basis in which the measurement is made, naturally leads to speculation that there must be some predetermination of that outcome, somehow encoded in a *hidden variable* that describes the qubits' true configuration more deeply than quantum mechanics. In a hidden variable theory, the randomness of quantum mechanics is nothing more than an admission of our ignorance of those hidden variables.

A consequence of two qubits being in an entangled state is that, if we separate the two qubits by some arbitrary distance, even light years, a measurement of one qubit instantaneously affects the state of the other qubit. Special relativity requires that information cannot travel faster than the speed of light, so there is no way that somehow a signal was sent by the first qubit to the second. For example, if the quantum state before measurement was $\alpha|00\rangle + \beta|11\rangle$, and measurement of the first qubit yields the result "0", there is no way it has enough time to send this information to the second qubit, telling it that it is now in the state $|0\rangle$.

We are speaking rather loosely about what sort of "information" cannot be communicated faster than the speed of light. The individual with that first qubit cannot choose the outcome of the measurement, in the way that they can choose to wear black or white socks that morning. Rather it is the qubit that has "chosen" to be "0" or "1"; but qubits are not people capable of making choices of their own free will. If we limit the notion of information to decisions made by people, like "sell my Enron stock" or "invade on 6th June", then the outcome of the measurement of the first qubit is not included in this definition, and thus entanglement is not violating the *letter* of the law of special relativity; but still, it does seem to violate the *spirit* of the law.

If the possibility of sending information between entangled qubits is not a plausible explanation for the predictions of quantum mechanics, then what is a more feasible explanation? Another possibility is that perhaps quantum theory is not the complete description of the system, and there is a deeper notion of a system's configuration, not easily accessible to the puny efforts of modern day experimental physics, in which the outcome of any measurement is *predetermined*: With this complete description, we will have a definite answer to what the results will be upon measurement. The whole argument about the possibility of information travelling faster than the speed of light would be rendered irrelevant in this case. In other words, the quantum randomness we discussed in Chapter 1 is turned into an example of the classical randomness discussed in Section 2.6. However, John S. Bell devised an experimental test of the conjecture that there is a hidden variable theory that can reproduce the predictions of quantum mechanics; and the not-so-puny efforts of modern day experimental physics show that hidden variable theories fail the test.

The way Bell went about proving this is by first assuming the validity of a hidden variable description. Then he considered a simple experimental arrangement, in which two qubits, prepared in an entangled state, are measured in different ways, and the correlations between the two measurement outcomes are recorded. For example, if every time the two observers have exactly the same outcome from a measurement then these two measurements are highly correlated; if the two observers see the same outcome half the time, and different outcomes the rest of the time, the outcomes are not correlated at all. Now the initial assumption of a hidden variable description implies that there are certain inequalities involving these correlations. Thus, if these correlations are measured and the inequalities are violated, then the system cannot be described by a hidden variable theory.

In Bell's original work, he used the singlet state (a state related by a global phase to one of the Bell states we discussed earlier in Eq. (3.37)), and we will use it here. It is given by

$$|\beta_2\rangle = \frac{1}{\sqrt{2}} (|01\rangle - |10\rangle).$$ (3.41)

When we measure a qubit, the measurement outcome is an eigenvalue that corresponds to the eigenvector into which the state has been rendered as a result of this measurement. If this measurement is meant to determine whether the state is projected into $|0\rangle$ or into $|1\rangle$, for example, then the operator that describes this can be written as

$$\hat{\Omega} = \begin{bmatrix} 0 & 0 \\ 0 & 1 \end{bmatrix}.$$ (3.42)

Notice that the eigenvalues 0 and 1 are on the diagonal of the matrix, and their corresponding eigenvectors are $|0\rangle$ and $|1\rangle$. When discussing the violation of Bell's inequalities in two qubits, the measurement outcomes most commonly used are -1 and 1, and we will shift our eigenvalues to accommodate this: simply, we are recording the measurement event in which a qubit was found in state $|0\rangle$ as having the value -1. To achieve that, we transform the matrix $\hat{\Omega}$ to $\hat{Z}$ as follows:

$$\hat{I} - 2\hat{\Omega} = \hat{Z}.$$ (3.43)

The eigenvectors of both $\hat{\Omega}$ and $\hat{Z}$ are the same. It is only the corresponding eigenvalue of $|0\rangle$ that has changed from 0 to -1. For the analysis discussed here, it really does not matter whether we are using $\hat{\Omega}$ or $\hat{Z}$; they are just translations of each other. The important thing is to work with a **dichotomic observable**; i.e., an observable that has two possible outcomes upon measurement. Moreover, the results of the local measurement that we are making on each qubit should be invariant under rotation. For example, the result of measuring qubit A in the singlet (3.41) should always give the opposite result for qubit B, whether we are performing the local measurement in the computational basis or any other basis obtained by rotating the computational one using (2.71), for example, so that the new measurement operator is now $\hat{U}^\dagger \hat{Z} \hat{U}$.

The Bell inequality that we will set up will involve four different measurement operators: two for qubit A (labelled $\hat{a}$ and $\hat{b}$) and two for qubit B (labelled $\hat{c}$ and $\hat{d}$). All of these operators are dichotomic and are chosen so that measuring them yields -1 or 1. Matrices with the latter property can be expressed in terms of the Pauli matrices. That is why, for example, $\hat{a}$ can be written as

$$\hat{a} = \mathbf{a} \cdot \hat{\boldsymbol{\sigma}},$$ (3.44)

where $\hat{\boldsymbol{\sigma}}$ is the vector made of the three Pauli operators $(\hat{\sigma}_x, \hat{\sigma}_y, \hat{\sigma}_z)$.

The operator $\hat{a}$, as well as the other three operators $(\hat{b}, \hat{c}, \hat{d})$ under consideration are quantum operators. When measuring them, depending on the state of the system, using quantum theory, we can find the probabilities associated with each outcome, the average value of the operator, etc. In the early days of quantum theory, it was suggested that perhaps it does not give the full description of the system, and if it did, then this probabilistic nature of answers would go away. Hence the existence of the hidden variables was proposed. Here we will represent these missing or hidden variables by λ, which we will claim predetermines the outcome of any quantum measurement. This means that the value of the measurement outcome, a, of $\hat{a}$ is a function of λ:

$$a = a(\lambda). \tag{3.45}$$

Similarly, $b = b(\lambda), c = c(\lambda), d = d(\lambda)$. If the probability distribution of λ is $P(\lambda)$, then the average of a is given by

$$\bar{a} = \int a(\lambda)P(\lambda)d\lambda. \tag{3.46}$$

The quantum mechanical average of an observable $\hat{a}$, $\langle \hat{a} \rangle$, which is confirmed by experimental observations, should agree with the average in (3.46), if a hidden variable theory is possible here. In other words,

$$\bar{a} = \langle \hat{a} \rangle. \tag{3.47}$$

To calculate the quantum mechanical average of an operator, the operator is sandwiched between the bra and ket representing the state, and the resulting expression is evaluated. Here, the state of the system is the singlet in (3.41), so in order to find the average of $\hat{a}$, the following expression has to be evaluated:

$$\langle \hat{a} \rangle = \langle \beta_2 | \hat{a} | \beta_2 \rangle. \tag{3.48}$$

Let us first set up the Bell's inequality that we will analyze here. First define the following algebraic expression:

$$f(\lambda) = \{a(\lambda) + b(\lambda)\}c(\lambda) + \{a(\lambda) - b(\lambda)\}d(\lambda). \tag{3.49}$$

⊕ Mathematics Tips 3.5

Absolute Value Relation. We know from mathematics that

$$\frac{1}{n}\sum_{i}^{n}|x_i| \geq \left|\frac{1}{n}\sum_{i}^{n}x_i\right|, \tag{3.50}$$

$$\overline{|x|} \geq |\bar{x}|,$$

where n represents the number of terms in the sum, and x_i is a number. What (3.50) tells us is that if we take the absolute value of the numbers, which can be negative, before averaging, then the average will in general be greater than if we averaged the numbers first before taking the absolute value of the resulting average.

Remember that all the variables, $a(\lambda), b(\lambda), c(\lambda)$, and $d(\lambda)$, take two possible values: ± 1. Knowing this, we can calculate the possible values for $f(\lambda)$, by considering the possible cases, as follows:

1. If $a = b$, then $f(\lambda) = 2ac = \pm 2$.
2. If $a = -b$, then $f(\lambda) = 2ad = \pm 2$.

Therefore, $f(\lambda) = \pm 2$. Combining this and taking the absolute value of (3.49), we get

$$\left| a(\lambda)c(\lambda) + b(\lambda)c(\lambda) + a(\lambda)d(\lambda) - b(\lambda)d(\lambda) \right| = 2. \tag{3.51}$$

Applying (3.50) to (3.51), we obtain the following Bell's inequality:

$$S = \left| \overline{ac} + \overline{bc} + \overline{ad} - \overline{bd} \right| \leq 2. \tag{3.52}$$

We dropped showing the λ dependence in (3.52) to simplify the presentation in the following discussion.

If a hidden variable theory is capable of predicting the outcomes of quantum theory, then each of the averages in (3.52) should agree with the corresponding quantum mechanical average. For example, $\overline{ac}$ should equal $\langle \hat{a} \otimes \hat{c} \rangle$, $\overline{bc}$ should equal $\langle \hat{b} \otimes \hat{c} \rangle$, and so on, so that (3.52) is satisfied. Let us see if that is the case.

By using the fundamental theorem of entanglement as well as the definitions of dot and cross products, one can show that

$$\langle \hat{a} \otimes \hat{c} \rangle = \sum_{i,j=1}^{3} a_i c_j \langle \beta_2 | \hat{\sigma}_i \otimes \hat{\sigma}_j | \beta_2 \rangle = -\mathbf{a} \cdot \mathbf{c} \tag{3.53}$$

and similarly for other correlations.

Now, let us choose some particular values for a, b, c, and d:

$$\mathbf{a} = [1,0,0], \quad \mathbf{b} = [0,0,1], \quad \mathbf{c} = \frac{-1}{\sqrt{2}}[1,0,1], \text{ and } \mathbf{d} = \frac{1}{\sqrt{2}}[-1,0,1]. \tag{3.54}$$

These are not chosen at random, but rather by the fact that they have the particular property that S takes its maximum value. Substituting (3.54) into (3.53) and then into (3.52), we get

$$S = \left| \frac{1}{\sqrt{2}} + \frac{1}{\sqrt{2}} + \frac{1}{\sqrt{2}} - \frac{1}{\sqrt{2}} \right| = 2\sqrt{2}. \tag{3.55}$$

But $2\sqrt{2}$ is larger than 2! Thus, quantum mechanics *violates the Bell's inequality* in (3.52). This means that the inequality (3.52) derived assuming that quantum systems can be described by a hidden variable theory is not consistent with the predictions of quantum mechanics, which we are confident about because they are, so far, agreeing with experimental observations. Therefore, there is no hidden variable description of quantum systems, and we do not know why quantum mechanics *knows* the results of the distant measurements instantaneously.

Bell's inequalities are not unique. One can set up various ones pertaining to the experimental setup and state to demonstrate the violation we are discussing. The one we considered here was set up by Clauser and his colleagues in 1969 (**CHSH inequality**), before he and Freedman demonstrated the violation of Bell's inequalities for the first time in an experiment in 1972. The violation was demonstrated again in improved experimental setups such as the one done by Aspect *et al.* in 1982. In 2015, three international groups demonstrated the violations of Bell's inequalities with large spatial distances and highly efficient detectors, removing some of the stronger objections (or "loopholes") that had been raised about earlier experiments.

Exercises 3.3.1

1. Work out the following state: $(\hat{X} \otimes \hat{X})|\beta_2\rangle$.
2. Use the fundamental theorem of entanglement to show that $(\hat{U} \otimes \hat{U})|\beta_2\rangle = |\beta_2\rangle$, where $\hat{U}$ is a unitary matrix of determinant 1. What is the concurrence before and after the operator is applied?

3. Show that for the singlet state $|\beta_2\rangle$

$$\langle \hat{a} \otimes \hat{c} \rangle = -\mathbf{a}.\mathbf{c}.$$

Hint: The orthonormality of the Bell basis Eq. (3.37), the fundamental theorem of entanglement Eq. (3.39), and the products of Pauli matrices (Chapter 2, Exercise 17) might come in handy.

4. John Stewart Bell, the discoverer of Bell's inequalities, had a friend called Bertlmann, who would always wear mis-matched socks, one of which would be pink. Thus, as soon as one of his feet entered the room, and you saw what colour sock it had on, instantaneously one would know whether or not the other sock was pink. Is this an example of quantum entanglement? How would you describe the situation quantum mechanically? Explain the difference between qubits and socks.

5. There is more than one Bell's inequality. Suppose a_1, b_1, b_2, and c_2 are four variables, all of which can have values ± 1, with subscripts corresponding to which of the two qubits is being measured. Show that when there is perfect anti-correlation between the values of variables measured for the two qubits (so that $b_1 = -b_2$ always)

$$a_2(b_1 - c_1) = \pm(1 + b_2 c_1),$$

and hence demonstrate that a local hidden variable theory implies

$$|\langle \hat{a} \otimes \hat{b} \rangle - \langle \hat{a} \otimes \hat{c} \rangle| - \langle \hat{b} \otimes \hat{c} \rangle \leq 1.$$

Find a set of measurements that violates this inequality.

3.4   Quantum Teleportation

An interesting process that we can do with entanglement is **quantum teleportation**. This is a common operation used to test a new quantum technology as it matures, since it requires only two CZ gates and three qubits. Before we delve into this application of entanglement, we need to make it clear that we are *not* physically teleporting or transferring the qubit, as in a science fiction TV show like *Star Trek*. What we are doing is *transferring the state* of a qubit to another qubit. As we will soon see, in order to achieve that, we will need to use an entangled pair of qubits as well as classical communication.

 Interesting Facts 3.6

The Literary History of Teleportation. The idea of instantaneous transportation in both space and time was not lost on William Shakespeare. In the Prologue of *Henry V* he decrees that we should carry his characters "here and there, jumping o'er times"; incidentally, this was a few years before he predicted the indeterminacy of reality in *Hamlet* ("To be or not to be..."). More recently, the idea of teleportation by mechanical (rather than magical or mystical) means has been a staple of science fiction stories for nearly 150 years. Among our favorite examples are the "Travelling-car of the Sutenrāa, an ingenious mechanical device that your intelligence could not comprehend" envisioned by Fred T. Jane in 1897 in his book *To Venus in Five Seconds*; and the "disintegrator-reintegrator" of *The Fly* by George Langelaan (1957), a device that had an unfortunate side-effect of combining the DNA of the teleportee with that of a house-fly (it was made into a couple of pretty scary movies). But pride of place has to be the TV show *Star Trek*, first broadcast in 1966, in which "transporter beams" took characters between the orbiting spaceship and the surface of alien planets. Interestingly enough, the producers of the show originally envisioned landings on alien planets by a detachable saucer-shaped section of the starship, but the special effects budget was limited and that would have cost too much; necessity is the mother of invention.

Quantum Circuit 3.11 illustrates the three steps involved in teleporting a quantum state, which we will shortly describe in detail.

Quantum Circuit 3.11

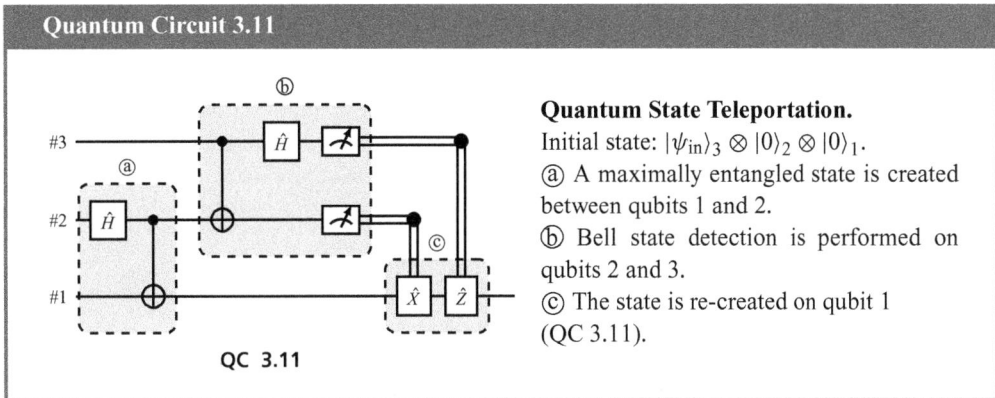

QC 3.11

Quantum State Teleportation.
Initial state: $|\psi_{\text{in}}\rangle_3 \otimes |0\rangle_2 \otimes |0\rangle_1$.
ⓐ A maximally entangled state is created between qubits 1 and 2.
ⓑ Bell state detection is performed on qubits 2 and 3.
ⓒ The state is re-created on qubit 1 (QC 3.11).

Step 1

We begin with the input state where qubit #3 is in an unknown superposition state, $|\psi_{\text{in}}\rangle$, and qubits #1 and #2 each start in state $|0\rangle$:

$$|\psi_1\rangle = \underbrace{(\alpha|0\rangle + \beta|1\rangle)}_{|\psi_{\text{in}}\rangle} \otimes (|0\rangle|0\rangle). \tag{3.56}$$

The first step is to create maximal entanglement between qubits #1 and #2 by applying a Hadamard gate on one qubit followed by a CNOT gate on both qubits:

$$|\psi_{\text{in}}\rangle \otimes |00\rangle \xrightarrow{\hat{I} \otimes \hat{H} \otimes \hat{I}} |\psi_{\text{in}}\rangle \otimes \frac{1}{\sqrt{2}}(|00\rangle + |10\rangle) \xrightarrow{\text{CNOT}_{2,1}} \frac{1}{\sqrt{2}}|\psi_{\text{in}}\rangle \otimes (|00\rangle + |11\rangle) = |\psi_2\rangle. \tag{3.57}$$

Step 2

The next step in quantum teleportation is the *Bell state detection*. First, a CNOT gate is applied to qubits #3 and #2:

$$\left(\text{CNOT}_{3,2} \otimes \hat{I}\right)|\psi_2\rangle = \frac{1}{\sqrt{2}}(\alpha|000\rangle + \alpha|011\rangle + \beta|110\rangle + \beta|101\rangle). \tag{3.58}$$

Next, a Hadamard gate is applied to qubit #3:

$$\begin{aligned}
|\psi_3\rangle &= \left(\hat{H}_A \otimes \hat{I} \otimes \hat{I}\right)\left(\text{CNOT}_{3,1} \otimes \hat{I}\right)|\psi_2\rangle \\
&= \frac{1}{2}\left[\alpha\left(|000\rangle + |100\rangle + |011\rangle + |111\rangle\right) + \beta\left(|010\rangle - |110\rangle + |001\rangle - |101\rangle\right)\right] \\
&= \frac{1}{2}\left[|00\rangle(\alpha|0\rangle + \beta|1\rangle) + |01\rangle(\beta|0\rangle + \alpha|1\rangle) + |10\rangle(\alpha|0\rangle - \beta|1\rangle) + |11\rangle(-\beta|0\rangle + \alpha|1\rangle)\right] \\
&= \frac{1}{2}\left[|00\rangle\left(\hat{I}|\psi_{\text{in}}\rangle\right) + |01\rangle\left(\hat{X}|\psi_{\text{in}}\rangle\right) + |10\rangle\left(\hat{Z}|\psi_{\text{in}}\rangle\right) + |11\rangle\left(\hat{X}\hat{Z}|\psi_{\text{in}}\rangle\right)\right].
\end{aligned} \tag{3.59}$$

Now, both #2 and #3 are measured, projecting the state of qubit #1 to the following:

$$|\psi_{\text{proj}}\rangle = \hat{X}^b \hat{Z}^a |\psi_{\text{in}}\rangle, \tag{3.60}$$

where $\{a, b\} \in \{0, 1\}$. The set $\{0, 1\}$ corresponds to the possible measurement outcomes for qubits #2 and #3.

Step 3
To reconstruct the state $|\psi_{\text{in}}\rangle$ in qubit #1 we need to apply the appropriate gates on (3.60), as follows:

$$\boxed{|\psi_{\text{out}}\rangle_c = \hat{Z}^a \hat{X}^b |\psi_{\text{proj}}\rangle_c = |\psi_{\text{in}}\rangle.} \tag{3.61}$$

Whoever is manipulating qubit #1 is given the information about the results of measuring qubits #2 and #3 through classical communication to know the appropriate a and b to use in (3.61). In other words, *the initial state of qubit #3 is teleported to qubit #1.*

Exercises 3.4.1

1. Circuit-Based Proof for Teleportation

Quantum Circuit 3.12

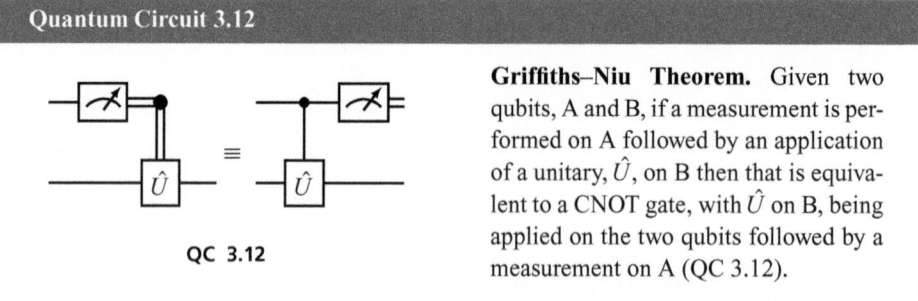

QC 3.12

Griffiths–Niu Theorem. Given two qubits, A and B, if a measurement is performed on A followed by an application of a unitary, $\hat{U}$, on B then that is equivalent to a CNOT gate, with $\hat{U}$ on B, being applied on the two qubits followed by a measurement on A (QC 3.12).

In this exercise, we will use the Griffiths–Niu theorem. (See Quantum Circuit 3.12 to work out a circuit-based proof for teleportation step by step.) Recall the teleportation circuit as seen in Quantum Circuit 3.11.

(a) Use the Griffiths–Niu theorem to show that the teleportation circuit in Quantum Circuit 3.11 has the same effect as QC 3.13. Use the latter in the next step.

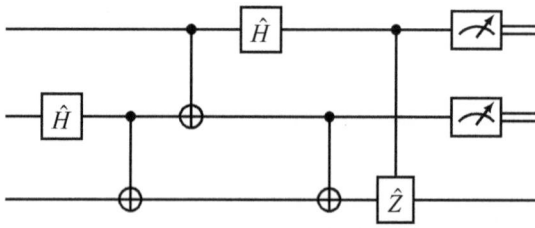

QC 3.13 The teleportation circuit re-expressed.

(b) Use the identity $\hat{H}\hat{X}\hat{H} = \hat{Z}$, as seen in QC 3.14, to re-write the appropriate part of the circuit in QC 3.13.

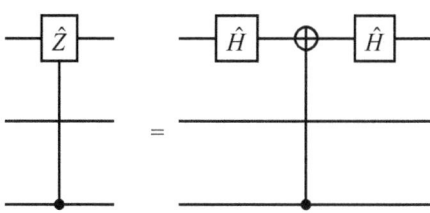

QC 3.14 The $\hat{H}\hat{X}\hat{H} = \hat{Z}$ identity.

(c) Use $\hat{H}\hat{H} = \hat{I}$ and the fact that $\hat{H}|0\rangle$ is an eigenstate of $\hat{X}$ to simplify the circuit further.

(d) Work out what appropriate single CNOT gate replaces the CNOT gate combination in QC 3.15.

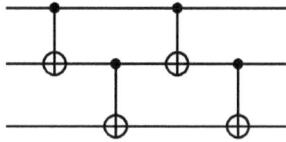

QC 3.15 A CNOT gate combination.

(e) Apply what you found in (d) on the teleportation circuit to simplify it further.

(f) You will find that qubit #2 will no longer be involved in any multi-qubit gates. In other words, it has become a *spectator* qubit. Move the Hadamard gate acting on this qubit all the way to the right just before measurement.

(g) Make a SWAP gate by creating a combination of a CNOT gate (see the previous section), acting on qubits #3 and #1, by inserting a CNOT gate with qubit #1 being the control qubit. Since qubit C is in state $|0\rangle$, that will have no effect. You should end up with the circuit shown in QC 3.16.

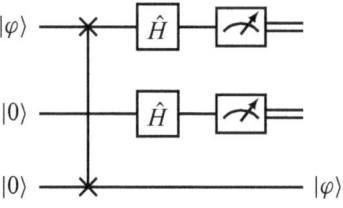

QC 3.16 Teleporting a state from qubit A to qubit C.

This very simplified version of the circuit compactly expresses that teleportation happens. However, the more involved, but equivalent, circuit we discussed in Quantum Circuit 3.11 is more practical in real life.

2. **The No-Cloning Theorem**. In this exercise, you will prove the no-cloning theorem, which states that you cannot clone or reproduce an arbitrary quantum state. That, of course, includes the qubit states we are focusing on in this book.

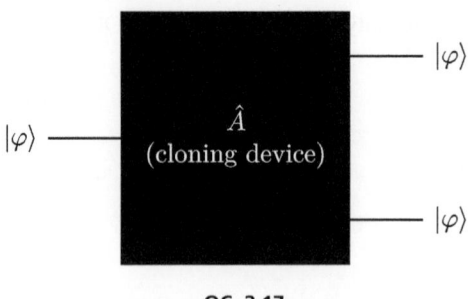

QC 3.17

The operator $\hat{A}$ represents a conjectural device that can clone qubits (QC 3.17); i.e., if you input a single qubit in an arbitrary quantum state, the output will be two qubits both in that same state, such as the following:

$$\hat{A}|0\rangle \longrightarrow |00\rangle,$$
$$\hat{A}|1\rangle \longrightarrow |11\rangle. \tag{3.62}$$

Using (3.62), find $\hat{A}|\psi\rangle$, where $|\psi\rangle = \frac{1}{\sqrt{2}}(|0\rangle + |1\rangle)$. Does the answer you get equal $|\psi\rangle \otimes |\psi\rangle$? Comment on how that proves or disproves the no-cloning theorem.

3. Here is a scheme for sending signals faster than the speed of light using entanglement (QC 3.18).

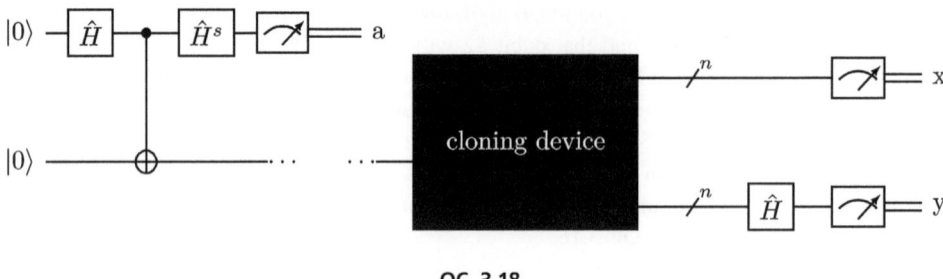

QC 3.18

Alice and Bob each have one qubit of an entangled qubit pair. Alice chooses to apply a Hadamard or not to her qubit before measurement. She then performs the measurement as shown. Then, the signal that she is actually sending is encoded in whether or not she applied the Hadamard; i.e., the classical bit, s, which can be 0 or 1. Bob, who has the other half of the entangled qubit pair can be in a far distant location. He has a device that amplifies the individual qubit into 2^n qubits. On half of these, he performs a measurement, in the computational basis. On the other half, he applies a Hadamard before he measures in the computational basis. If everything is working as advertised, the signal, s, can be deduced from the outputs of the measurements. If Alice did not apply a Hadamard, i.e. $s = 0$, then in Bob's second set of measurements (when the Hadamard was applied), there will be equal likelihood of 0 and 1, while the first set of measurements should all be either 0 or 1. Conversely, if $s = 1$, the first set of Bob's measurements will produce random output, and the second set will always produce the same result. Note that, in principle, this communication happens instantly. What is wrong with this argument?

3.5 CONCEPT CHECK AND EXERCISES

Concept Check

- The smallest quantum system in which entanglement can exist is a two-qubit system.
- Entanglement mathematically refers to the inability to write a state of a system as a product of the states of its subsystems.
- There are two types of randomness that can exist in a quantum system: the inevitable *quantum randomness* and the *classical randomness* that is due to our ignorance. In the former case, the state can be described by a ket, while in the latter case, it can only be described by a density matrix.
- Concurrence is a useful tool to quantify entanglement. It is based on the concept that if there is entanglement in the system of two qubits, then choosing to be ignorant of one will introduce (classical) randomness into the system. This randomness can be used to quantify how entangled the state is.
- A universal gate, such as a CNOT gate, is essential for creating entanglement.
- Bell gave guidelines to set up inequalities (pertaining to specific experiments) to test whether a local hidden variable theory can reproduce the predictions of quantum mechanics.
- In general, quantum systems violate Bell's inequalities. This means that they cannot be described by a hidden variable theory.
- Quantum teleportation involves transferring a quantum *state* from one quantum system, such as an atomic qubit, to another.

Chapter Exercises

1. Find the 4×4 unitary matrix that transforms the two-qubit computational basis $\{|00\rangle, |01\rangle, |10\rangle, |11\rangle\}$ into the Bell basis Eq. (3.37). Draw the quantum circuit corresponding to this transformation.
2. Recall the fair coin exercise (11) from the end-of-chapter exercises in Chapter 1, where you worked out the probabilities of obtaining various numbers of Heads for situations with separable quantum coins. Now, let us add entanglement into it. Consider the following maximally entangled states (0 takes the place of H, and 1 takes the place of T):

   (a) $\frac{1}{\sqrt{2}} (|00\rangle + |11\rangle)$,

   (b) $\frac{1}{\sqrt{2}} (|01\rangle + |10\rangle)$.

   Does the probability distribution of the outcomes change? If so, how?
3. Consider QC 3.19:

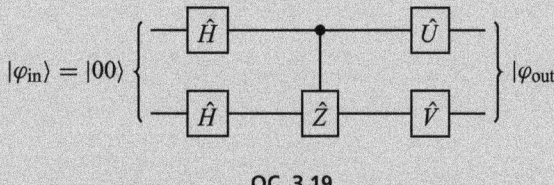

$$|\varphi_{in}\rangle = |00\rangle \qquad |\varphi_{out}\rangle$$

QC 3.19

where

$$\hat{U} = \frac{1}{\sqrt{2}} \begin{bmatrix} \cos(\theta) & -i\sin(\theta) \\ i\sin(\theta) & -\cos(\theta) \end{bmatrix}$$

and

$$\hat{V} = \frac{1}{\sqrt{2}} \begin{bmatrix} \cos(\phi) & -i\sin(\phi) \\ i\sin(\phi) & -\cos(\phi) \end{bmatrix},$$

with $\theta = \pi\sqrt{5/13}$ and $\theta = \pi\sqrt{12/13}$.

Find the concurrence of the input and of the output states.

4. Simplify QC 3.20:

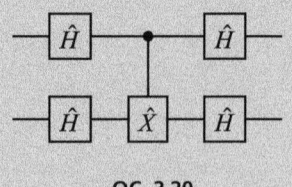

QC 3.20

5. Consider QC 3.21:

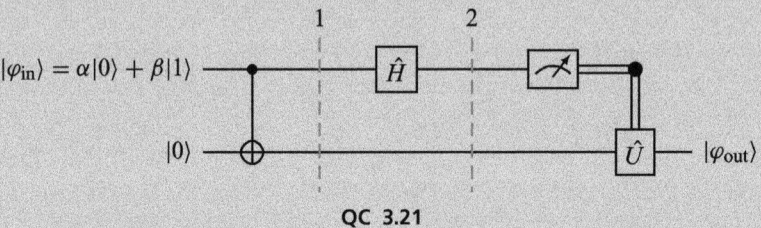

QC 3.21

The double wires indicate classical control, i.e., the unitary $\hat{U}$ is applied if the measurement outcome was "1", and is not applied if the measurement outcome was "0".

(a) Write down the quantum states at points 1 and 2 as shown on the diagram.

(b) What is the controlled-unitary $\hat{U}$ that is required for $|\varphi_{in}\rangle = |\varphi_{out}\rangle$?

6. Consider QC 3.22. The state $|\varphi\rangle = \alpha|0\rangle + \beta|1\rangle$ ($|\alpha|^2 + |\beta|^2 = 1$) is an arbitrary single-qubit state we wish to teleport to another qubit (i.e., the bottom row of the circuit), utilizing two entangled qubits in the state $|\chi\rangle = \gamma|00\rangle + \delta|11\rangle$ (both γ and δ are real; $\gamma^2 + \delta^2 = 1$).

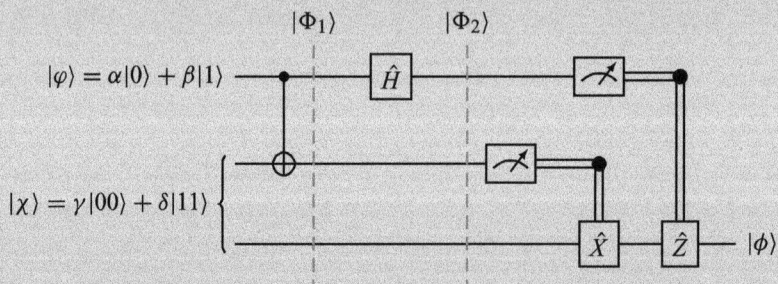

QC 3.22

(a) Write down the three-qubit state $|\Phi_1\rangle$.

(b) Write down the three-qubit state $|\Phi_2\rangle$.

(c) Assuming $\gamma = \delta = 1/\sqrt{2}$, find the output state $|\phi\rangle$.

(d) If $\gamma \neq \delta$, find the possible output states $|\phi\rangle$ and their associated probabilities.

(e) Find the averaged **fidelity** of the output state with the input state $\bar{F} = \overline{|\langle\varphi|\phi\rangle|^2}$, where the overbars indicate an average over the possible output states, and show that $\bar{F} \geq (1 + C)/2$, where C is the concurrence of $|\chi\rangle$.

7. **(a)** Show that any single-qubit unitary operator can be written as the product of two distinct Hermitian matrices, each with eigenvalues ± 1.

   (b) Prove the following quantum circuit identity for QC 3.23:

   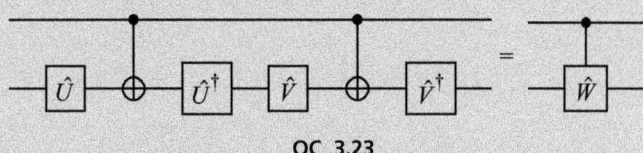

   QC 3.23

   where $\hat{U}$, $\hat{V}$, and $\hat{W} = \hat{V}^\dagger \hat{X} \hat{V} \hat{U}^\dagger \hat{X} \hat{U}$ are unitary operators.

   (c) Prove that any unitary operator can be expressed in the form $\hat{V}^\dagger \hat{X} \hat{V} \hat{U}^\dagger \hat{X} \hat{U}$ (thus this quantum circuit allows *any* controlled unitary to be performed using local unitaries and CNOTs). *Hint: Recall Eq. (2.72).*

4 Quantum Algorithms

This is the chapter where we start examining how the peculiar properties of quantum mechanics can be exploited to improve computers. We have seen how quantum mechanics diverges from classical mechanics in the important aspect that a quantum system can be in multiple configurations at the same time, and seen some interesting implications of this when more than one quantum system is involved, such as Bell's inequalities and teleportation. Now we are going to study how quantum systems can be used to perform calculations, and how some tasks can be performed with quantum technologies with potentially far greater efficiency than their classical counterparts.

> **Learning Objectives**
>
> - Learn how binary numbers are stored in quantum systems.
> - Become familiar with more quantum circuit diagram conventions.
> - Learn about the acceptable way to evaluate and represent a function in quantum systems.
> - Learn about the general setup for quantum algorithms.
> - Understand the quantum circuits of various algorithms and what each accomplishes.

4.1 Storing Numbers in Quantum Systems

Not surprisingly, any sort of computation using quantum systems necessarily involves the storage and manipulation of numbers. How can we do this with a bunch of two-level quantum systems? First, we will confine ourselves to the *whole* numbers, $\{0, 1, 2, \ldots\}$, just as conventional computers do in their most basic level. These numbers are conveniently expressed in the base 2 form: $0_{10} = 0_2$, $1_{10} = 1_2$, $2_{10} = 10_2$, $3_{10} = 11_2$, $4_{10} = 100_2$, and so on, where the subscript denotes what base we have used. (Note that we can drop the leading zeros, from the left, only starting with the first largest non-zero digit.) The individual digits of the binary number are called *bits*, short for binary digits. The larger the number, the more bits are required; thus $4_{10} = 100_2$ is a three-bit number, $27_{10} = 11011_2$ is a five-bit number and $346_{10} = 101011010_2$ is a nine-bit number and so on. Suppose you want to encode a generic number x using binary digits: how can you find the number $L(x)$ of bits you would need? The answer, assuming $x > 0$, can be found using the logarithm function in base 2: $L(x) = \lfloor \log_2 x \rfloor + 1$, where $\lfloor y \rfloor$ is the "floor" of y; i.e., the largest whole number smaller than or equal to y.

> 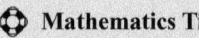 **Mathematics Tips 4.1**
>
> **Nomenclature: *L*-Bit Integers.** Given a variable x, we say that x is an *L-bit integer* if it can be written as a binary number with L bits. In this case, the number of integers that x represents is 2^L,

and the range of its values is $0 \leq x \leq 2^L - 1$. For example, if x is a three-bit integer, this means that $L = 3$, and the $2^3 = 8$ possible values that x can take, 0_{10}, 1_{10}, 2_{10}, 3_{10}, 4_{10}, 5_{10}, 6_{10}, and 7_{10}, are in binary notation given by 000_2, 001_2, 010_2, 011_2, 100_2, 101_2, 110_2, and 111_2, where X_n is the number representation in base n.

Using this idea of binary encoding of numbers (which is how classical computer logic works at its most basic level), we can also encode numbers in quantum states. Suppose we have an L-bit number x, whose binary digits are $\{x_{L-1}, x_{L-2}, \ldots, x_0\}$; or in other words:

$$x = x_{L-1}2^{L-1} + \cdots + x_2 2^2 + x_1 2^1 + x_0 2^0 = \sum_{m=0}^{L-1} x_m 2^m. \tag{4.1}$$

(A word of caution here: some authors use a different convention, writing x_0 as the most significant bit value, and numbering them from left to right.)

Remember that the binary digits $\{x_{L-1}, x_{L-2}, \ldots, x_0\}$ all have only two possible values, 0 or 1. Then the L-qubit state representing the number x is

$$|x\rangle = |x_{L-1}\rangle \otimes |x_{L-2}\rangle \otimes \cdots \otimes |x_2\rangle \otimes |x_1\rangle \otimes |x_0\rangle = \bigotimes_{m=0}^{L-1} |x_m\rangle. \tag{4.2}$$

The set of all 2^L states $\{|0\rangle, |1\rangle, |2\rangle, \ldots, |2^L - 1\rangle\}$, or equivalently $\{|x\rangle \mid x \in \mathbb{W}, x < 2^L\}$, where we have used set notation ($\mid$ means "such that", and $\mathbb{W}$ denotes the whole numbers $\{0, 1, 2, \ldots\}$), forms the *computational basis* for 2^L qubits; each element of the basis is made up of a tensor product of L single-qubit kets.

As an example, suppose we want to store the number 13 in a set of qubits. First, $\log_2 13 = 3.70044$, so that $L(13) = \lfloor 3.70044 \rfloor + 1 = 3 + 1 = 4$; i.e., 13 is a four-bit number. Now let's convert 13 into base 2: we get $13_{10} = 1101_2$, confirming that indeed we do need four bits to represent the number. The quantum state representing this number is therefore $|13_{10}\rangle = |1\rangle \otimes |1\rangle \otimes |0\rangle \otimes |1\rangle = |1101\rangle$. It will be much more common that we use the binary representation of numbers inside the ket, and this will be the default; we will use the subscript when using decimal, so $|13_{10}\rangle = |1101\rangle$. The circuit used to store 13 is then given by QC 4.1.

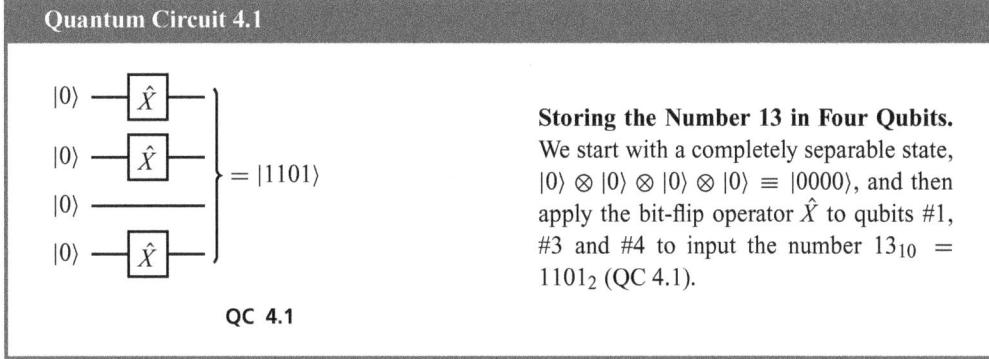

Quantum Circuit 4.1

$= |1101\rangle$

QC 4.1

Storing the Number 13 in Four Qubits. We start with a completely separable state, $|0\rangle \otimes |0\rangle \otimes |0\rangle \otimes |0\rangle \equiv |0000\rangle$, and then apply the bit-flip operator $\hat{X}$ to qubits #1, #3 and #4 to input the number $13_{10} = 1101_2$ (QC 4.1).

In quantum circuit diagrams, we will adopt the convention that the lowest qubit corresponds to x_0, working upwards, with $|x_{L-1}\rangle$ the highest (see Quantum Circuit 4.1).

Circuit diagrams with a large number of qubits, each storing an individual bit of a number x, can easily start to be very cumbersome. It is often more convenient to draw them as a *qubit bundle*, indicated by the diagonal line and a symbol to indicate the number of qubits in the bundle (L, in this case).

Quantum Circuits 4.2 and 4.3

QC 4.2

QC 4.3

Conventions for Qubit Bundles

- If x is an L-bit number, x_0 is the least significant bit, x_{L-1} is the most significant bit (QC 4.2), i.e.,

$$x = x_{L-1}2^{L-1} + \cdots + x_2 2^2 + x_1 2^1 + x_0 2^0.$$

- The left-most qubit stores the most significant bit, the right-most qubit the least significant bit:

$$|x\rangle \equiv \bigotimes_{m=0}^{L-1} |x_m\rangle = |x_{L-1}\rangle \otimes \cdots \otimes |x_1\rangle \otimes |x_0\rangle.$$

- Multiple tensor products of operators:

$$\bigotimes_{k=0}^{k=L-1} \hat{U}_k \equiv \hat{U}_{L-1} \otimes \hat{U}_{L-2} \otimes \cdots \otimes \hat{U}_1 \otimes \hat{U}_0.$$

- Tensor exponentials of operators (QC 4.3):

$$\hat{U}^{\otimes L} = \hat{U} \otimes \hat{U} \otimes \cdots \otimes \hat{U}.$$

Similarly when we have the *same* operation applied to all the qubits of a bundle, it is convenient to use the "tensor exponential" notation $\hat{U}^{\otimes L}$, which is shorthand for $\hat{U} \otimes \hat{U} \otimes \cdots \otimes \hat{U}$. When *different* operations are applied to each qubit, the following notation is used:

$$\bigotimes_{k=0}^{k=L-1} \hat{U}_k \equiv \hat{U}_{L-1} \otimes \hat{U}_{L-2} \otimes \cdots \otimes \hat{U}_1 \otimes \hat{U}_0. \tag{4.3}$$

This notation is summarized in Quantum Circuits 4.2 and 4.3.

Exercises 4.1.1

1. Suppose y is a four-bit integer, what are the possible values that it can take? How are these values represented in binary notation?
2. Find the binary representation of the following base-10 numbers: (a) 3; (b) 23; (c) 42; (d) 128.
3. *In this exercise, you will be creating a table with multiple columns to understand the formula for finding the number of bits required to represent a number x, in base 10. Budget enough space for six columns to organize your findings.*
   (a) In Column 1, list the numbers $x = 0$ to $x = 16$; the numbers 1, 2, 4, 8, and 16 should also be written as a power of 2, e.g., "$8 = 2^3$".
   (b) In Column 2 write corresponding binary (base 2) forms of these numbers.
   (c) In Column 3 write the number of bits appearing in the corresponding entry in Column 2 (excluding leading zeros).

Notice that every time x reaches the value 2^n (in base 10), where n is a whole number, the number of bits required to represent it is $n + 1$. This means that all values of x less than 2^n can be represented using (at most!) n bits depending on the value. In this case, x is an *n-bit number*. When the number to be represented is between 2^{n-1} (inclusive) and 2^n (exclusive), then this means that n bits, but not fewer, are enough to represent it.

(d) To get information about the number of bits, we have to deal with exponential functions like 2^n. The inverse of an exponential is a logarithm. Thus, to *extract* from the exponent 2^n the information about the number of bits needed to represent the number we take the logarithm, base 2.

In the table, write out three more columns: Column 4 lists $\log_2 x$ (exclude $x = 0$ as it is undefined for logarithmic functions); Column 5 will have $\lfloor \log_2 x \rfloor$ ($\lfloor a \rfloor$ is the largest whole number smaller than or equal to a); and finally in Column 6 write $\lfloor \log_2 x \rfloor + 1$. You should find that the entries in Column 6 are equal to those in Column 3.

4. Using the formula $L(x) = \lfloor \log_2 x \rfloor + 1$, find the number of bits needed to represent the following base 10 numbers in binary: (a) $x = 7$; (b) $x = 15$; (c) $x = 16$; (d) $x = 157$; (e) $x = 550$; (f) $x = 3000$.

5. Write the following binary numbers in base 10: (a) 111; (b)1111; (c) 0110; (d) 1000000; (e) 11001101; (f) 111111111. *Hint: Equation (4.1).*

6. (a) Represent the following base 10 numbers as L-qubit states; i.e., use the ket representation: (a) $x = 7$; (b) $x = 15$; (c) $x = 20$.
(b) Draw the quantum circuit representation for each of the states in (a), including the quantum bundle representation equivalence as well.

7. Prove the orthogonality of two L-qubit states, i.e., $\langle x|y \rangle = \delta_{x,y}$.

8. Is the set of states (in base 10) $\{|0_{10}\rangle, |1_{10}\rangle, |2_{10}\rangle, |3_{10}\rangle\}$ a basis? For what space?

9. Given the following state (in binary): $|1100101\rangle$ with the operator $\hat{Z}$ applied on each qubit. Draw the quantum circuit that describes this scenario.

10. Draw the quantum circuit to create the following states (all numbers are base 10):

(a) $|4_{10}\rangle$,

(b) $|7_{10}\rangle$,

(c) $(|5_{10}\rangle + |13_{10}\rangle)/\sqrt{2}$,

(d) $(|0_{10}\rangle - |7_{10}\rangle)/\sqrt{2}$.

4.2 Manipulating Numbers Stored in a Quantum System

If we can store a number, the next question is: can we perform a calculation with it, such as finding the value of some function $f(x)$? This is a pretty straightforward thing to do on a conventional computer, or pocket calculator: we type in the value of a number (or *argument*), x, then press a key on our calculator, or run a few lines of code on our computer, and our device spits out the value of $f(x)$. If we cannot do something similar on a quantum device, it would not really be much of a computer.

Let us take the simplest approach: our first guess would be to apply some operator on $|x\rangle$ which transforms the number stored in the qubits to $|f(x)\rangle$, just like the screen of a pocket calculator replaces the input value with the output value; i.e.,

$$|x\rangle \longrightarrow |f(x)\rangle. \tag{4.4}$$

As we previously learned, whatever operation we apply on our qubits, it should be reversible, or unitary in the language of quantum mechanics. In other words, there must be an inverse operation of the form

$$|x\rangle \longrightarrow |f^{-1}(x)\rangle. \tag{4.5}$$

Unfortunately this is not possible, as in general not all functions have an inverse. For example, suppose $f(x) = \sin^2(\pi x/2)$: this is a perfectly good function, and is easy to calculate on a pocket calculator; we find that when x is a natural number, $f(x)$ only takes the possible values 0 or 1, for even and odd values of x, respectively. Thus, the inverse of the function cannot possibly exist: given the value of $f(x)$, the best we can do is say whether x was even or odd, which is equivalent to figuring out only the value of its last bit, x_0. Thus, for functions without an inverse, a unitary operation to perform the transformation given in Eq. (4.4) just cannot exist! It looks like a quantum computer cannot even do simple things we can do on a pocket calculator!

Well, perhaps we should not be so pessimistic: instead of replacing the value of x by the value of $f(x)$, let us try a different tack. We will have one set of qubits to store the value of the argument x, and a distinct separate set of qubits to store the value of the function $f(x)$. We will call these the **argument register** and the **function register**, respectively. Let the input state of the argument register be $|x\rangle$ and the input state of the function register be $|y\rangle$.

⊕ Mathematics Tips 4.2

Bitwise Addition Modulo 2. A useful and easy way to implement arithmetic operations on bits is **addition modulo 2**, which is defined as the remainder when the sum of two bits is divided by 2; it is denoted $\oplus$ and, for two one-bit numbers is defined as follows:

$$\begin{aligned}
0 \oplus 0 &= 0, \\
0 \oplus 1 &= 1, \\
1 \oplus 0 &= 1, \\
1 \oplus 1 &= 0.
\end{aligned} \tag{4.6}$$

Alternatively, when dealing with logic (where "true" and "false" replace 1 and 0), the equivalent operation is called the "XOR" or exclusive-or; thus a XOR b has the value "true" if either of the variables a or b is true, but not both.

For two L-bit numbers x and y, $z = x \oplus y$ is defined to be the L-bit number whose bits are the sum modulo 2 of the corresponding bits of x and y (see Eq. (4.1)), i.e., $z_m = x_m \oplus y_m$, for $m = 0, 1, \ldots, L - 1$.

For example, let $x = 7_{10} = 111_2$ and $y = 4_{10} = 100_2$; then $z = 011_2 = 3_{10}$.

Addition modulo 2 (without the "bitwise") is just the remainder when $a + b$ is divided by 2, and is always a one-bit number, regardless of the size of a and b. (We don't give this operation a special symbol.)

Now consider the following operation:

$$|x\rangle|y\rangle \longrightarrow \hat{U}_f|x\rangle|y\rangle = |x\rangle|y \oplus f(x)\rangle, \tag{4.7}$$

where $y \oplus z$ is bitwise addition modulo 2 (as explained in Mathematics Tips 4.2). This operation is reversible: In fact, since $y \oplus f(x) \oplus f(x) = y$, then $\hat{U}_f\hat{U}_f|x\rangle|y\rangle = |x\rangle|y\rangle$. This is how a unitary quantum device can evaluate functions. In particular, if $y = 0$ then

$$\hat{U}_f|x\rangle|0\rangle = |x\rangle|f(x)\rangle, \tag{4.8}$$

and then measurement of the qubits in the function register allows us to find the value of $f(x)$. In other words, our pessimism was misplaced: a quantum device, despite being constrained to evolve via unitary operations, can emulate the action of pocket calculators. Using the compact qubit bundle notation, the circuit for function evaluation on a quantum computer looks like QC 4.4.

Quantum Circuit 4.4

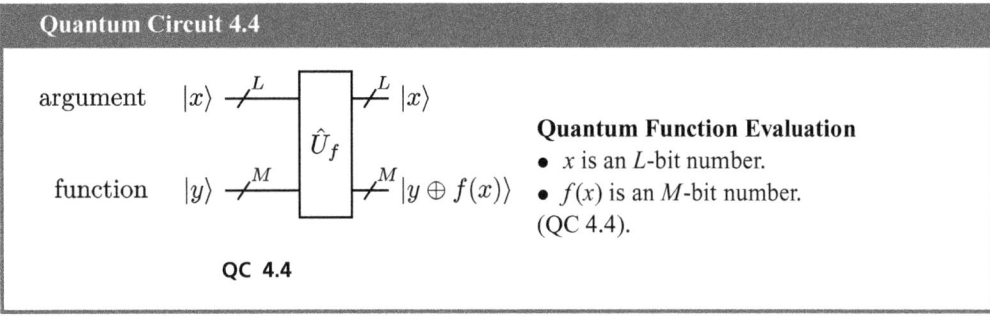

Quantum Function Evaluation
- x is an L-bit number.
- $f(x)$ is an M-bit number.
(QC 4.4).

QC 4.4

⊕ **Mathematics Tips 4.3**

Ket Representation of the Argument Registers and the Function Registers. In a quantum computational problem, if we are working with an L-bit integer *argument* x, and we intend to evaluate the M-bit *function* $f(x)$, then when using the ket notation, each of $|x\rangle$ and $|f(x)\rangle$ is represented by L kets: $\underbrace{|..\rangle|..\rangle \cdots |..\rangle}_{L}$ and M kets: $\underbrace{|..\rangle|..\rangle \cdots |..\rangle}_{M}$, respectively. See Eqs. (4.1) and (4.2).

For example, if $L = 3$ and $M = 2$, then each $|x\rangle$ will be represented by three kets: $|..\rangle|..\rangle|..\rangle$, and each $|f(x)\rangle$ will be represented by two kets: $|..\rangle|..\rangle$, respectively. The number inside each individual ket is chosen so that the overall ket gives the binary representation for $|x\rangle$ and $|f(x)\rangle$. For example, for $|x = 0\rangle$, the ket representation is $|0\rangle|0\rangle|0\rangle = |000\rangle$, for $|x = 1\rangle$, the ket representation is $|001\rangle$, for $|x = 2\rangle$, the ket representation is $|010\rangle$, and so on until one obtains all $2^L = 2^3 = 8$ representations of the possible values of $|x\rangle$.

In this case, if $f(5) = 0$, i.e., the function evaluated at $x = 5$ gives 0, then the ket representation for this evaluation is $|101\rangle|00\rangle$, where $|101\rangle$ represents the number 5 and $|00\rangle$ the number 0.

Exercises 4.2.1

1. Work out the following bitwise addition modulo 2 expressions:
   (a) $(101)_2 \oplus (111)_2$.
   (b) $(1111)_2 \oplus (0101)_2$.
   (c) $(01001)_2 \oplus (10010)_2$.
2. Find (a) $5 \oplus 7$; (b) $10 \oplus 8$; (c) $10 \oplus 2$; (d) $16 \oplus 15$. All numbers are in base 10.
3. *More practice with bitwise addition modulo 2.*
   (a) Suppose x, y, and z are one-bit numbers. Show that bitwise addition modulo 2 is associative, i.e., $(x \oplus y) \oplus z = x \oplus (y \oplus z)$.
   (b) Show that bitwise addition modulo 2 is associative for numbers of length L.
   (c) Prove $y \oplus y \oplus z = z$.
4. For $f(x) = \sin^2(\pi x/2)$ write down the matrix representation of $\hat{U}_f$ for the two-qubit space with $L = M = 1$ (i.e., x and y both take values 0 and 1, so the space is spanned by the basis $\{|00\rangle, |01\rangle, |10\rangle, |11\rangle\}$). Show that this matrix is unitary.
5. Same as Exercise 4 only with $L = 2$ and $M = 1$.
6. In (4.7), how would the value of $f(x)$ be retrieved if the initial function register was set to $|0\rangle$? What if it was set to $|1\rangle$ instead?

4.3   Quantum Parallelism and Quantum Algorithms

So far we have seen how a quantum computer might be used as a rather elaborate and expensive pocket calculator, able in principle to evaluate a function and have the result read out. However, a quantum computer is in fact far more powerful than that. With a pocket calculator, if you want to know all the values of $f(x)$, you are forced to evaluate it over and over again, once for each of the 2^L possible values of x; this is rather tiresome and repetitive, especially if L is a large number. However, in a quantum computer, you can prepare the argument register in a *superposition state of all possible values of x*.

A single Hadamard gate applied to a single qubit in state $|0\rangle$ results in the state $(1/\sqrt{2})(|0\rangle + |1\rangle)$; i.e., an equal superposition of the states corresponding to the possible values of a single bit. For two qubits we find

$$\hat{H} \otimes \hat{H}|00\rangle = \left(\frac{|0\rangle + |1\rangle}{\sqrt{2}}\right) \otimes \left(\frac{|0\rangle + |1\rangle}{\sqrt{2}}\right)$$

$$= \frac{1}{2}\left(|00\rangle + |01\rangle + |10\rangle + |11\rangle\right),$$

$$= \frac{1}{2}\left(|0_{10}\rangle + |1_{10}\rangle + |2_{10}\rangle + |3_{10}\rangle\right),$$

$$= \frac{1}{2}\sum_{x=0}^{3}|x\rangle; \tag{4.9}$$

i.e., an equal superposition of all the states corresponding to the possible values of two bits $00_2 = 0_{10}$, $01_2 = 1_{10}$, $10_2 = 2_{10}$, and $11_2 = 3_{10}$. Similarly, we find for three qubits:

$$\hat{H} \otimes \hat{H} \otimes \hat{H}|000\rangle = \frac{1}{2\sqrt{2}}\sum_{x=0}^{7}|x\rangle. \tag{4.10}$$

You can start to see a pattern evolving here: in the general case, involving L qubits, we get

$$\hat{H}^{\otimes L}|0\rangle^{\otimes L} = \frac{1}{\sqrt{2^L}}\sum_{x=0}^{2^L-1}|x\rangle = |\text{all}_L\rangle; \tag{4.11}$$

i.e., an equal superposition of all possible number states of the L-qubit bundle. This state occurs so often we have given it a special notation $|\text{all}_L\rangle$.

Quantum Circuit 4.5

$|0\rangle^{\otimes L}$ —L— $\boxed{\hat{H}^{\otimes L}}$ —L— $\frac{1}{\sqrt{2^L}}\sum_{x=0}^{2^L-1}|x\rangle$

QC 4.5

Multiple Hadamards in Parallel (I)
- Start with L qubits all in state $|0\rangle$.
- Result is an equal superposition of all possible computational basis states (QC 4.5):

$$|\text{all}_L\rangle = \frac{1}{\sqrt{2^L}}\sum_{x=0}^{2^L-1}|x\rangle.$$

 Interesting Facts 4.4

The Origin of the Word "Algorithm". One of the great mathematicians of the Islamic Golden Age was Muhammad bin Mūsā Al-Khawārizmi – Muhammad, son of Moses, from Khawārizm (a prosperous city on the Silk Road, now called Khiva in modern-day Uzbekistan). During the first half of the ninth century CE he worked at the famous centre of scholarship *The House of Wisdom* in Baghdad, then the capital city of the Abbasid Caliphate. One of his most influential works was *Kitāb al-Jabr wa-l-Muqābala* (The book on restoration and balancing); *al-Jabr*, meaning "the restoration" (i.e., when you change one side of an equation, you must restore the balance by doing the same thing on the other side) is the origin of the word **algebra**. In European scholarship, the concept of the **algorithm**, a series of mathematical and logical steps in the proper order designed to solve a problem, grew from concepts that originated in the works of Al-Khawārizmi, and was named after him (with due allowance for their somewhat shaky transliteration of Arabic).

Taking this superposition state and applying the function evaluation operator $\hat{U}_f$, we find that we have evaluated $f(x)$ for all possible values of x in parallel:

$$\hat{U}_f |\text{all}_L\rangle \otimes |0\rangle = \frac{1}{\sqrt{2^L}} \sum_{x=0}^{2^L-1} |x\rangle |f(x)\rangle. \tag{4.12}$$

This is *quantum parallelism*: the 2^L evaluations of the function that were needed for a classical device have been replaced by a *single* evaluation. This is the beating heart that gives quantum computers their power.

However, at second glance, this does not seem to have helped much: we have a state that is rather complicated, and is dependent on all the values $\{f(0), f(1), f(2), \ldots, f(2^L - 1)\}$, but it is not obvious it presents any particular advantage. In particular, if you measured the argument register, all you would achieve is to project the superposition onto one particular value of x, and collapse the function register into the corresponding $|f(x)\rangle$ state, which, on measurement, will yield the value of the function. This is just the same as the previous case, with the added uncertainty of which value of x you end up evaluating. This should hardly be surprising: the maximum number of binary digits one can access by measuring the argument and function registers is limited to the total number of qubits.

4.3.1 Quantum Algorithms

Suppose we are not actually that interested in all the possible values of the function $f(x)$, but instead we want to know some over-arching property of the function. Examples of this might be: we want to know if $f(x)$ is just a constant function or not; or, if $f(x)$ is zero except for one particular value of x for which $f(x) = 1$, what is that special value of x? Or if $f(x)$ is periodic, what is its period? In conventional computation, finding answers to these problems requires evaluating the function for a large number of values of the argument, even though the answer we seek does not actually require a large number of binary digits. But in a quantum computer, we have already done the hard work of multiple evaluations of $f(x)$ with a single application of $\hat{U}_f$; can we extract the desired over-arching property from the state Eq. (4.12)?

There are two things we can do to help with this: first, there is no special reason why the initial state of the function register has to be $|0\rangle$ prior to using $\hat{U}_f$: it may be a useful strategy, depending on what question we are asking, to prepare it in some other state. Second, after evaluating the function, but before measuring the argument register, we can perform some post-processing operation. We will see explicit examples of both function register preparation and argument register post-processing in what follows.

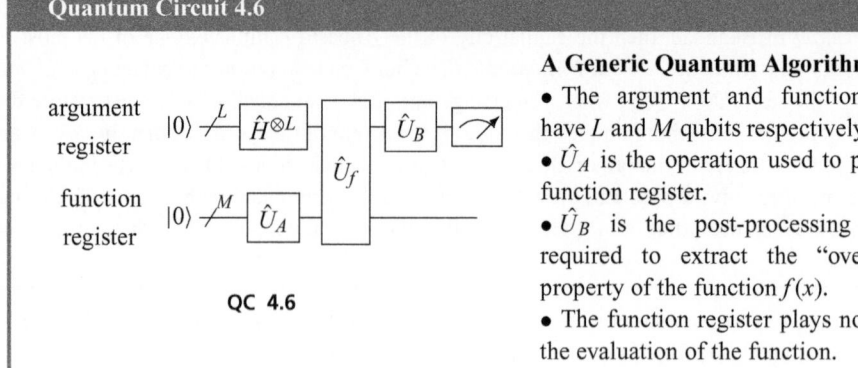

Quantum Circuit 4.6

argument register $|0\rangle$ $\hat{H}^{\otimes L}$ $\hat{U}_B$

function register $|0\rangle$ $\hat{U}_A$ $\hat{U}_f$

QC 4.6

A Generic Quantum Algorithm
- The argument and function registers have L and M qubits respectively (QC 4.6).
- $\hat{U}_A$ is the operation used to prepare the function register.
- $\hat{U}_B$ is the post-processing operation required to extract the "over-arching" property of the function $f(x)$.
- The function register plays no role after the evaluation of the function.

Exercises 4.3.2

1. How many terms are there in the following three-qubit states if you write them out in full using the computational basis:
   (a) $|0\rangle \otimes |0\rangle \otimes |0\rangle$,
   (b) $(\hat{I} \otimes \hat{H} \otimes \hat{I})|0\rangle \otimes |0\rangle \otimes |0\rangle$,
   (c) $(\hat{H} \otimes \hat{H} \otimes \hat{I})|0\rangle \otimes |0\rangle \otimes |0\rangle$,
   (d) $(\hat{H} \otimes \hat{H} \otimes \hat{H})|0\rangle \otimes |0\rangle \otimes |0\rangle$.
2. How many terms appear in $|\text{all}_L\rangle$?
3. Consider the function $f(x) = \sin^2(\pi x/2)$ we encountered in the previous Section 4.2.1, Exercise 4: write out the state given by Eq. (4.12) for the case $L = M = 1$. What information can be extracted by measuring this state?
4. Same as Exercise 3 only with $L = 2$ and $M = 1$.

4.4 Some Simple Examples

4.4.1 Deutsch's Problem

We will start with a very simple problem, indeed arguably the simplest possible problem imaginable. As we discussed above, in our context a function is some mathematical operation that maps one natural number, the argument x, onto another natural number, the function's value, $f(x)$. Ignoring the rather trivial example that x has only one possible value, the simplest case is that x has two possible values $\{0, 1\}$, and similarly $f(x)$ can take two different values (again from the set $\{0, 1\}$). With these constraints, there are only four possible functions (see Fig. 4.1). As we can see, there are two different types of one-bit functions: those that are *constant* (i.e., $f(0) = f(1)$) and those that are *balanced* (i.e., $f(0) \neq f(1)$).

Suppose we have a device that, when we put in a number (the argument x) the value of an unknown function $f(x)$ will be the output. Both input and output have one bit: they can both be value either 0 or 1. The device that evaluates the function is called an **oracle**. This term, widely used in the theory of computers, comes from the ancient Greek high priestess who would deign to answer a question posed by each supplicant. The problem is simple: we do not know which of the four possible functions the oracle is evaluating, but can we, with only a single consultation of the oracle, determine if the unknown function $f(x)$ is *constant* or *balanced*?

	$f(0)$	$f(1)$		
(a)	0	0		"constant"
(b)	0	1		"balanced"
(c)	1	0		"balanced"
(d)	1	1		"constant"

Figure 4.1   Constant and balanced functions.

All we can do is try: if we put in an argument value $x = 1$ say, and receive the response $f(1) = 1$, we can tell that we are dealing with either case (b) or case (d), one of which is balanced, the other constant. Instead, putting in $x = 0$ likewise yields ambiguous results, regardless of the oracle's response. Only by asking the oracle for *both* of the values of $f(0)$ *and* $f(1)$ can we determine whether the function is balanced or constant. (The oracle only responds to natural number inputs; asking the oracle for the value of $f(1/2)$ results in something dire happening, for which there are many precedents in ancient mythology.) In order to determine that the function is constant or balanced, without being picky about which of the constant and balanced cases one is dealing with, one would still have to evaluate the function *twice* to accomplish that. This is the limitation we have to face in the classical case. **Deutsch's algorithm**, named for its discoverer David Deutsch, allows us to achieve that with only *one* computation.

As we previously learned, whatever operation we apply on our qubits has to be unitary. This must also be true of the quantum oracle. Thus, our quantum oracle performs the following operation on the input qubit (see Eq. (4.7)):

$$\hat{U}_f|x\rangle|y\rangle = |x\rangle|y \oplus f(x)\rangle. \tag{4.13}$$

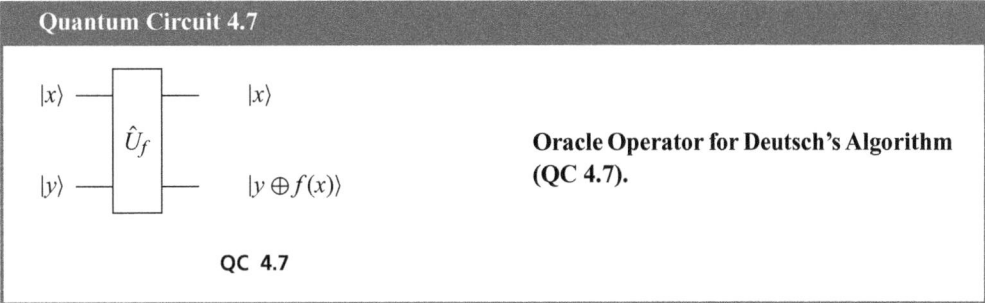

Quantum Circuit 4.7

$|x\rangle$ ——— $|x\rangle$

$\hat{U}_f$

$|y\rangle$ ——— $|y \oplus f(x)\rangle$

Oracle Operator for Deutsch's Algorithm (QC 4.7).

QC 4.7

As before, inputting the state $|\varphi_{\text{in}}\rangle = |x, 0\rangle$ yields $|\varphi_{\text{out}}\rangle = |x, f(x)\rangle$, and measurement then just gives you the value of the function $f(x)$; in other words, we have not made any advance over the classical case. Instead, let us try using quantum superposition of all possible values of the argument: $|\varphi_{\text{in}}\rangle = \frac{1}{\sqrt{2}}(|0, 0\rangle + |1, 0\rangle)$ (where we have put commas between the state designators inside the ket, to help keep track of the argument and function registers). The oracle then yields

$$|\varphi_{\text{out}}\rangle = \hat{U}_f|\varphi_{\text{in}}\rangle$$
$$= \frac{1}{\sqrt{2}}(|0, f(0)\rangle + |1, f(1)\rangle). \tag{4.14}$$

We have used quantum parallelism to evaluate the function for both possible arguments simultaneously. The resultant state in some sense contains the information we seek: if $f(x)$ is balanced, the output state is entangled; if $f(x)$ is constant, the state is separable. If we possessed some way to determine the degree of entanglement from a single measurement of a two-qubit state, the problem would be solved. Alas, no such measurement exists; entanglement is a correlation, which can only be measured by taking an average of repeated measurements. In this case, a measurement of the output qubits would yield either $\{0, f(0)\}$ or $\{1, f(1)\}$ (but not both), which is exactly where we started.

Is there any hope? Let us see what happens when we try a little bit of manipulation on the function register prior to using the oracle. How about a superposition of both the input qubits: $|\varphi_{in}\rangle = \frac{1}{2}(|0,0\rangle + |1,0\rangle - |0,1\rangle - |1,1\rangle)$. This state is not entangled, and can be created from the original input state by a simple circuit.

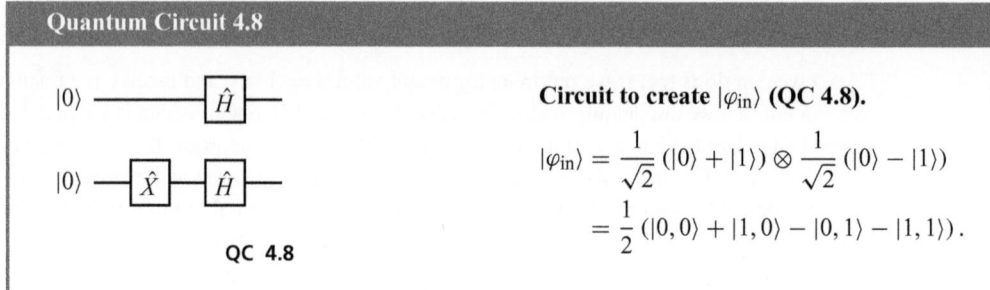

Quantum Circuit 4.8

$|0\rangle$ —— $\hat{H}$ ——

$|0\rangle$ —— $\hat{X}$ —— $\hat{H}$ ——

QC 4.8

Circuit to create $|\varphi_{in}\rangle$ (QC 4.8).

$$|\varphi_{in}\rangle = \frac{1}{\sqrt{2}}(|0\rangle + |1\rangle) \otimes \frac{1}{\sqrt{2}}(|0\rangle - |1\rangle)$$

$$= \frac{1}{2}(|0,0\rangle + |1,0\rangle - |0,1\rangle - |1,1\rangle).$$

In other words, our function register preparation operator (see Quantum Circuit 4.6) is $\hat{U}_A = \hat{H}\hat{X}$. Now let us try applying the oracle operator to this state. Here is what we get:

$$\hat{U}_f|\varphi_{in}\rangle = \frac{1}{2}(|0, f(0)\rangle + |1, f(1)\rangle - |0, 1 \oplus f(0)\rangle - |1, 1 \oplus f(1)\rangle). \tag{4.15}$$

Well, at first glance, this is not helping very much. Persevering, let us consider the constant case when $f(0) = f(1)$:

$$\hat{U}_f|\varphi_{in}\rangle = \frac{1}{2}(|0, f(0)\rangle + |1, f(0)\rangle - |0, 1 \oplus f(0)\rangle - |1, 1 \oplus f(0)\rangle)$$

$$= \frac{1}{2}(|0\rangle + |1\rangle) \otimes (|f(0)\rangle - |1 \oplus f(0)\rangle). \tag{4.16}$$

In the case $f(0) = 0$, this state reduces to $\frac{1}{\sqrt{2}}(|0\rangle + |1\rangle) \otimes \frac{1}{\sqrt{2}}(|0\rangle - |1\rangle)$; if $f(0) = 1$ we get $\frac{1}{\sqrt{2}}(|0\rangle + |1\rangle) \otimes \frac{1}{\sqrt{2}}(|1\rangle - |0\rangle)$; hence, for constant functions:

$$\hat{U}_f|\varphi_{in}\rangle = \frac{(-1)^{f(0)}}{2}(|0\rangle + |1\rangle) \otimes (|0\rangle - |1\rangle). \tag{4.17}$$

For a balanced function one can similarly show (see Exercise 2 at the end of this section) that

$$\hat{U}_f|\varphi_{in}\rangle = \frac{(-1)^{f(0)}}{2}(|0\rangle - |1\rangle) \otimes (|0\rangle - |1\rangle). \tag{4.18}$$

Notice the subtle difference between the two output states, Eq. (4.17) for the case of a constant function and Eq. (4.18) for the case of a balanced function. The only difference is a minus sign in the superposition state of the first (argument) qubit. This difference cannot be measured directly (see Exercise 3). However, if we apply a Hadamard gate to the first qubit, the state is quite different. Recall $\hat{H}(1/\sqrt{2})(|0\rangle + |1\rangle) = |0\rangle$ and $\hat{H}(1/\sqrt{2})(|0\rangle - |1\rangle) = |1\rangle$. Our post-processing operation $\hat{U}_B = \hat{H}$. Measurement of the first qubit after this terminal Hadamard yields "0" for a constant function, and

"1" for a balanced function: an unequivocal answer to the question we originally posed, and the oracle operator $\hat{U}_f$ was used only once.

Putting it all together, here is the circuit.

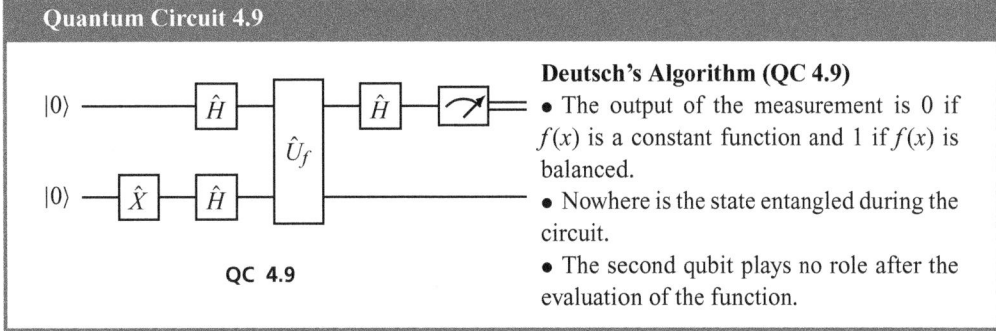

Quantum Circuit 4.9

QC 4.9

Deutsch's Algorithm (QC 4.9)
• The output of the measurement is 0 if $f(x)$ is a constant function and 1 if $f(x)$ is balanced.
• Nowhere is the state entangled during the circuit.
• The second qubit plays no role after the evaluation of the function.

Deutsch's algorithm was originally proposed by David Deutsch of the University of Oxford in 1985, and played a crucial role in the historical development of the field. For the first time it was demonstrated that, using the principle of superposition inherent in quantum mechanics – i.e., that a physical system could be in two configurations at the same instant – allows more efficient computation. By itself, it is not a particularly useful algorithm, since it has no known application. Further, asserting that it demonstrates some vaguely defined superiority over a conventional computer is problematical: the quantum oracle, built using quantum technologies, is a very different beast than its conventional counterpart (i.e., a few lines of code in a subroutine).

4.4.2 Deutsch–Jozsa Algorithm

Deutsch's algorithm considered in the last section serves as an hors d'oeuvres to whet our appetite: it demonstrates a factor of two increase in efficiency (in terms of the number of calls to an oracle) for a computer capable of performing calculations with qubits. But given the constant enhancement of conventional computational capabilities in the past 50 years, is this all that important? *Moore's law*, named for Gordon Moore, one of the founders of Intel Corporation, points out that computer chip capabilities have doubled every two years or so. This advance of conventional technology has been sustained consistently over the past five decades. Thus it begs the question: to achieve a factor of two improvement, as Deutsch's algorithm promises, why not just wait a couple of years, rather than going to the trouble of building a quantum computer?

The **Deutsch–Jozsa algorithm**, invented in 1992, provides the answer to this: again, it is a solution to a rather contrived mathematical problem, which has no known application. But it demonstrated that, rather than just a factor of two improvement, a quantum computer can be *exponentially* more efficient.

Here's the problem: just as in Deutsch's algorithm, the function $f(x)$ always has the value either 0 or 1. Now, however, the argument x is an L-bit number, so it can take any of the 2^L integer values in the range 0 to $2^L - 1$; for example, if $L = 3$, x has values in the range $0, 1, \ldots, 7$. The function $f(x)$ is either constant (i.e., $f(x) = f(0)$, for all values of x) or balanced (i.e., half the possible values of x return 0, and the other half return 1). *Any* set of 2^{L-1} values of x that give the value 0 will make the function balanced, not just the simple case of the function having alternating values. No other type of function is considered. The goal, as before, is to determine whether $f(x)$ is balanced or constant with the smallest possible calls on the oracle that evaluates the function.

Classically one can determine whether $f(x)$ is constant or balanced by at most $2^{L-1} + 1$ evaluations. If you evaluated the function for half the possible values of x and obtained the same outcome, you still could not reliably deduce whether the function is constant or balanced; only one half plus 1 will reveal the truth. Yes, you could get lucky and determine the answer by just two evaluations: e.g., if $f(0) \neq f(1)$, you can immediately deduce the function must be balanced; this however is a hit-or-miss game, and one cannot rely on getting lucky all the time. The average of the best case (2 evaluations) and the worst case ($2^{L-1}+1$ evaluations) is $(2^{L-1}+3)/2 = 2^{L-2}+3/2$ evaluations; this is a reasonable estimate of the expected number of calls on the oracle.

As we will see, quantumly we can deduce the answer with 100% accuracy with a *single* call to the oracle. Thus, we have replaced $2^{L-2} + 3/2$ oracle calls by 1: this is an exponential speed-up because $2^{L-2} + 3/2$ involves 2 raised to the power of $L - 2$, where L is the "size" of the problem (i.e., L, the number of binary digits required to specify the argument of the function involved).

Using the qubit bundle notation, the Deutsch–Jozsa algorithm looks very similar to Deutsch's algorithm.

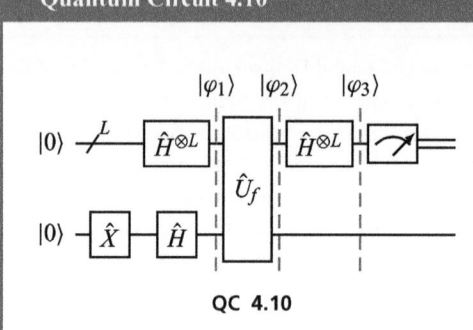

Quantum Circuit 4.10

$|\varphi_1\rangle$ $|\varphi_2\rangle$ $|\varphi_3\rangle$

QC 4.10

Deutsch–Jozsa Algorithm (QC 4.10)
- The algorithm will not work unless $f(x)$ is either constant or balanced.
- If the output of the measurement is 0, $f(x)$ is a constant function; any other result implies $f(x)$ is balanced.
- Nowhere is the state entangled during the circuit.
- The second qubit plays no role after the evaluation of the function.

Let us break it down to see how it works. The argument register is prepared in the state $|\text{all}_L\rangle$, the equal superposition of all possible values of x (see Quantum Circuit 4.5), while the function register (which only has one qubit) is prepared in just the same way as in Deutsch' algorithm; thus the $(L + 1)$-qubit state $|\varphi_1\rangle$ which is input to the oracle is given by

$$|\varphi_1\rangle = |\text{all}_L\rangle \otimes (|0\rangle - |1\rangle)/\sqrt{2}. \tag{4.19}$$

What is the output state after the oracle? Let us sneak up on this problem and take it unawares by considering what happens to just one of the component states, $|\varphi_{1,x}\rangle = |x\rangle (|0\rangle - |1\rangle)/\sqrt{2}$; applying $\hat{U}_f$ to this state we get

$$\hat{U}_f|\varphi_{1,x}\rangle = |x\rangle \otimes (|f(x)\rangle - |1 \oplus f(x)\rangle)/\sqrt{2}. \tag{4.20}$$

Now recall: the function $f(x)$ is equal to either 0 or 1, and $1 \oplus 1 = 0$, and $1 \oplus 0 = 1$. Thus, the state output by the oracle will be $|x\rangle (|0\rangle - |1\rangle)/\sqrt{2}$ if $f(x) = 0$ or $|x\rangle (|1\rangle - |0\rangle)/\sqrt{2}$ if $f(x) = 1$; combining these two results, we see the output state will be

$$\hat{U}_f|\varphi_{1,x}\rangle = (-1)^{f(x)}|x\rangle \otimes (|0\rangle - |1\rangle)/\sqrt{2}. \tag{4.21}$$

Now applying this to the entire superposition state $|\varphi_1\rangle$ we find the output state of the oracle:

$$|\varphi_2\rangle = \hat{U}_f|\varphi_1\rangle$$

$$= \left(\sum_{x=0}^{2^L-1} \frac{(-1)^{f(x)}}{\sqrt{2^L}}|x\rangle\right) \otimes (|0\rangle - |1\rangle)/\sqrt{2}. \tag{4.22}$$

Now let us find out how this state is transformed by the last set of single-qubit Hadamard gates applied in parallel to the L qubits of the argument register. Again, we keep things simple by considering the simplest case, $L = 1$, first. Recall that $\hat{H}|0\rangle = (|0\rangle + |1\rangle)/\sqrt{2}$ and $\hat{H}|1\rangle = (|0\rangle - |1\rangle)/\sqrt{2}$, which can be combined to give the following formula:

$$\hat{H}|x\rangle = \frac{1}{\sqrt{2}}\left(|0\rangle + (-1)^x|1\rangle\right)$$

$$= \frac{1}{\sqrt{2}}\sum_{y=0}^{1}(-1)^{xy}|y\rangle. \tag{4.23}$$

Now, for two qubits ($L = 2$), the two Hadamards are applied in parallel on the state $|x\rangle$. As before, we write the base-2 digits of the number x as x_1 and x_0, and we number the digits in decreasing order from left to right; see Eq. (4.2).

$$\hat{H}^{\otimes 2}|x\rangle = \left(\hat{H} \otimes \hat{H}\right)(|x_1\rangle \otimes |x_0\rangle)$$

$$= \frac{1}{\sqrt{2}}\left(|0\rangle + (-1)^{x_1}|1\rangle\right) \otimes \frac{1}{\sqrt{2}}\left(|0\rangle + (-1)^{x_0}|1\rangle\right)$$

$$= \frac{1}{\sqrt{2}}\left(\sum_{y_1=0}^{1}(-1)^{x_1y_1}|y_1\rangle\right) \otimes \frac{1}{\sqrt{2}}\left(\sum_{y_0=0}^{1}(-1)^{x_0y_0}|y_0\rangle\right)$$

$$= \frac{1}{2}\sum_{y=0}^{3}(-1)^{x_1y_1+x_0y_0}|y\rangle. \tag{4.24}$$

Notice we have in the exponent of (-1) the sum of products of the binary digits of the integer x (corresponding to the input state $|x\rangle$) and of the binary digits of the integer y (corresponding to the states $|y\rangle$ appearing in the sum which gives the output state). If we think of the binary digits, or bits, of x, i.e., $x = (x_{L-1}x_{L-2}\ldots x_0)_2$ as the components of a vector $\{x_{L-1}, x_{L-2}, \ldots, x_0\}$, and similarly the bits of the number y as the components of another vector $\{y_{L-1}, y_{L-2}, \ldots, y_0\}$, then the sum appearing in the exponent is equivalent to the scalar or dot product. Also, since this appears in the exponent of -1, it makes no difference if the addition is done modulo 2 (i.e., $\oplus$ rather than $+$): this is because $(-1)^a = 1$ if $a = 2$ or any power of 2. Thus, we can usefully introduce the compact notation for this, which applies to any pair of L-bit integers:

$$x \cdot y = x_{L-1}y_{L-1} \oplus x_{L-2}y_{L-2} \oplus \cdots \oplus x_0y_0. \tag{4.25}$$

The result we have shown for a pair of qubits can be scaled up to apply for an arbitrary number (see Chapter 4, Exercise 3). Thus, we obtain the result shown in Quantum Circuit 4.11.

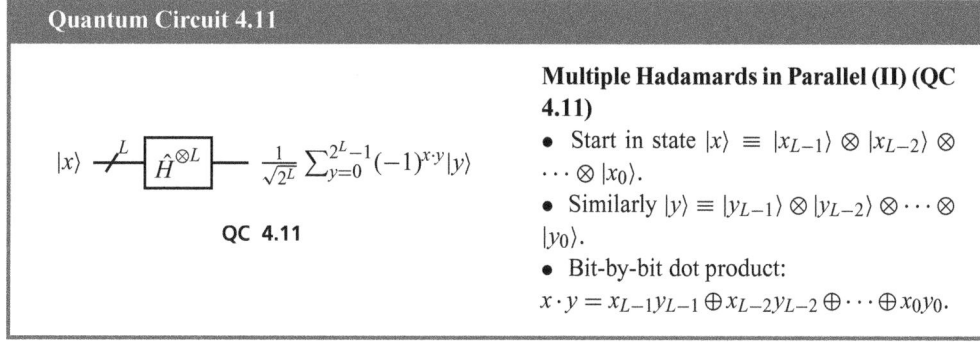

Quantum Circuit 4.11

QC 4.11

Multiple Hadamards in Parallel (II) (QC 4.11)

- Start in state $|x\rangle \equiv |x_{L-1}\rangle \otimes |x_{L-2}\rangle \otimes \cdots \otimes |x_0\rangle$.
- Similarly $|y\rangle \equiv |y_{L-1}\rangle \otimes |y_{L-2}\rangle \otimes \cdots \otimes |y_0\rangle$.
- Bit-by-bit dot product:
  $x \cdot y = x_{L-1}y_{L-1} \oplus x_{L-2}y_{L-2} \oplus \cdots \oplus x_0y_0$.

Now let us return to the Deutsch–Jozsa circuit. The final state, just prior to measurement is

$$
\begin{aligned}
|\varphi_3\rangle &= \left(\hat{H}^{\otimes L} \otimes \hat{I} \right) |\varphi_2\rangle \\
&= \left(\sum_{x,y=0}^{2^L-1} \frac{(-1)^{f(x)+x\cdot y}}{2^L} |y\rangle \right) \otimes \frac{1}{\sqrt{2}} (|0\rangle - |1\rangle) \\
&= \left(\sum_{y=0}^{2^L-1} c_y |y\rangle \right) \otimes \frac{1}{\sqrt{2}} (|0\rangle - |1\rangle),
\end{aligned}
\tag{4.26}
$$

where c_y is the probability amplitude for the argument register to be in the state $|y\rangle$; it is given by the formula

$$
c_y = \frac{1}{2^L} \sum_{x=0}^{2^L-1} (-1)^{f(x)+x\cdot y}.
\tag{4.27}
$$

Thus, when we do the final step of the algorithm and measure the L-qubits of the argument register, we will obtain an integer y (i.e., the outcome of the measurement of qubit 0 is y_0, the outcome of the measurement of qubit 1 is y_1, and so on) with probability $|c_y|^2$.

Consider the case that $f(x)$ is a constant function. The probability amplitude c_y can then be written as

$$
\begin{aligned}
c_y &= \frac{(-1)^{f(0)}}{2^L} \sum_{x=0}^{2^L-1} (-1)^{x\cdot y} \\
&= \frac{(-1)^{f(0)}}{2^L} \sum_{x_0=0}^{1} \sum_{x_1=0}^{1} \cdots \sum_{x_{L-1}=0}^{1} (-1)^{x_0 y_0 + x_1 y_1 + \cdots x_{L-1} y_{L-1}} \\
&= \frac{(-1)^{f(0)}}{2^L} \prod_{i=0}^{L-1} \sum_{x_i=0}^{1} (-1)^{x_i y_i} \\
&= \frac{(-1)^{f(0)}}{2^L} \prod_{i=0}^{L-1} \left[1 + (-1)^{y_i} \right].
\end{aligned}
\tag{4.28}
$$

If $y_0 = y_1 = \cdots = y_{L-1} = 0$, and using the fact that $(-1)^0 = 1$, each term in square brackets has the value 2; but if *any* of y_i are not equal to zero, then one of the square brackets has value 0 and so $c_y = 0$; or, in other words,

$$
c_y = \begin{cases} (-1)^{f(0)} & \text{if } y = 0 \\ 0 & \text{if } y \neq 0 \end{cases} \qquad (f(x) \text{ constant}).
\tag{4.29}
$$

If $f(x)$ is a balanced function, then it has two possible values $f(0)$ and $1 \oplus f(0)$. Thus, when evaluating the sum for c_y in the case that $y = 0$ there will be 2^{L-1} terms involving $(-1)^{f(0)}/2^L$ and the same number of terms involving $(-1)^{1\oplus f(0)}/2^L = -(-1)^{f(0)}/2^L$. These terms will cancel, and so the result is $c_0 = 0$. Thus, we have the result as stated in Quantum Circuit 4.10: with a single consultation of the oracle operator one can distinguish between a constant function and a balanced function with 100% accuracy. Note the algorithm does not work if $f(x)$ is neither balanced nor constant: other functions will in general have non-zero probability amplitudes for the state $|0\rangle$.

Tantalizing though this exponential improvement this may seem, the same reservations we expressed about Deutsch's problem apply: claiming an exponential speed-up seems hardly honest, given the technological resources required for creating a quantum oracle and a large register of qubits to interface with it are far different than writing a few extra lines of code required to perform the classical approach. Further, as we have noted above, the exponential gain itself is rather equivocal, since, classically, there is a chance that one might solve the problem with as few as two oracle calls; and finally, there is no known application for this problem.

4.4.3 Bernstein–Vazirani Problem

The Bernstein–Vazirani problem is another illustrative example of how a quantum computer achieves a speed-up over any conventional approach to a calculation, and is closely related to the Deutsch–Jozsa algorithm: only the oracle has changed. In this case, instead of generic properties of a function (i.e., is it balanced or constant), we consider a specific function $f(x) = a \cdot x$, where a is an unknown parameter and x the argument of the function as before. Both a and x are L-bit numbers, and "$\cdot$" is the bit-by-bit scalar product defined by Eq. (4.25). The question in this case is: how many calls on the oracle are required to determine a?

Classically, one can find the value of the m-th bit of a by realizing that 2^m has the binary representation $(00\ldots 1 \ldots 00)_2$, where the bit value 1 appears in the m position; see Eq. (4.1). Thus, if $x = 2^m$ the bit values of x will be 0 except for $x_m = 1$. Applying this to Eq. (4.25) we find $f(2^m) = a_m$, the m-th bit value of the unknown number a. To obtain all the bits of a we need to repeat the evaluation L times. Quantumly, we can achieve the desired result with a single call on the oracle, representing a linear enhancement of efficiency.

Using Eq. (4.22) we see that the state at the output of the Bernstein–Vazirani oracle will be

$$
\begin{aligned}
|\varphi_2^{\mathrm{BV}}\rangle &= \hat{U}_f |\varphi_1\rangle \\
&= \left(\sum_{x=0}^{2^L-1} \frac{(-1)^{a \cdot x}}{\sqrt{2^L}} |x\rangle \right) \otimes \left(|0\rangle - |1\rangle \right) / \sqrt{2}.
\end{aligned}
\tag{4.30}
$$

Comparing this to Quantum Circuit 4.11, we see that

$$
|\varphi_2^{\mathrm{BV}}\rangle = \left(\hat{H}^{\otimes L} |a\rangle \right) \otimes \left(|0\rangle - |1\rangle \right) / \sqrt{2}.
\tag{4.31}
$$

And since $\hat{H}^2 = \hat{I}$ (this is true, no matter how many qubits we apply the Hadamards to), we see the output state, immediately prior to measurement, is

$$
\begin{aligned}
|\varphi_3^{\mathrm{BV}}\rangle &= \left(\hat{H}^{\otimes L} \right) \otimes \hat{I} |\varphi_2^{\mathrm{BV}}\rangle \\
&= |a\rangle \otimes \left(|0\rangle - |1\rangle \right) / \sqrt{2}.
\end{aligned}
\tag{4.32}
$$

Thus, we see measurement of the argument register yields the value of a, with only a single call on the oracle.

Unlike the Deutsch–Jozsa algorithm, it is relatively straightforward to devise quantum circuits to implement the Bernstein–Vazirani oracle. In Quantum Circuit 4.12, for example, is the oracle for the case $a = 9_{10} = 1001_2$.

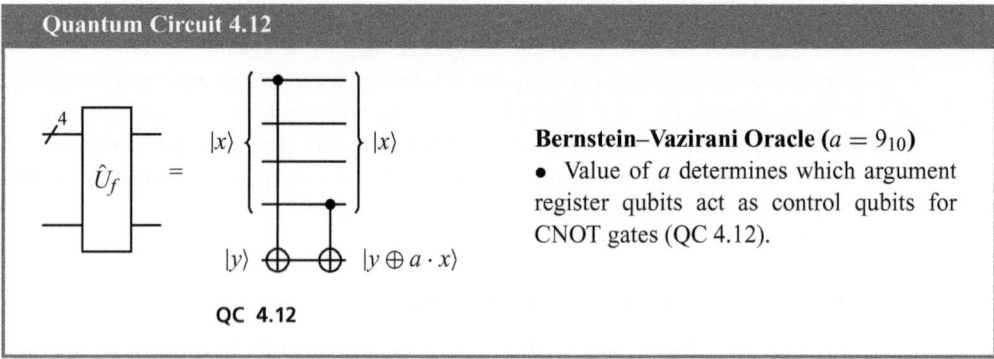

Quantum Circuit 4.12

QC 4.12

Bernstein–Vazirani Oracle ($a = 9_{10}$)
• Value of a determines which argument register qubits act as control qubits for CNOT gates (QC 4.12).

Quantum Circuit 4.12 for the oracle allows us to perform a derivation of the result using quantum circuit identities (Mermin, 2007, p. 54). Substitute the quantum oracle circuit into the full circuit diagram, including all four argument register qubits and their Hadamard gates (QC 4.13):

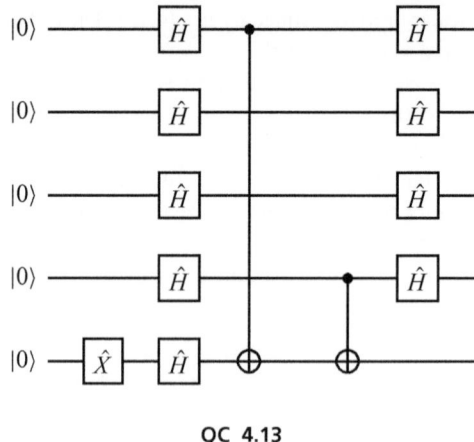

QC 4.13

Rearranging, and using the fact that $\hat{H}\hat{H} = \hat{I}$, this can be re-drawn as (QC 4.14):

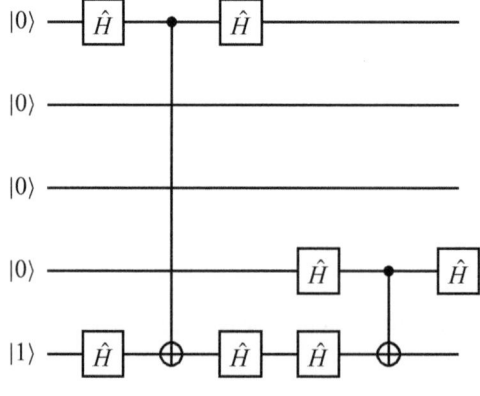

QC 4.14

Recall the identity in QC 4.15 (see Exercises 3.2.5, Exercise 4 on page 65):

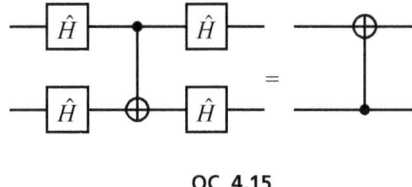

QC 4.15

Applying this to our circuit, we find QC 4.16:

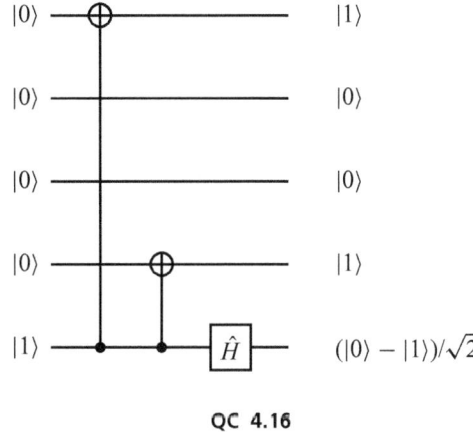

QC 4.16

In other words, the four argument register qubits are in the state $|a\rangle$, as expected.

Exercises 4.4.4

1. Write down the quantum circuits to execute the unitary operation, $\hat{U}_f$, for the four different functions given in Fig. 4.1. *Hint: You will only need to use at most a local $\hat{X}$ and a CNOT gate.*
2. Starting from Eq. (4.15) show that, for a *balanced* function (i.e., $f(1) = 1 \oplus f(0)$) that $\hat{U}_f|\varphi_{\text{in}}\rangle = \frac{(-1)^{f(0)}}{2}(|0\rangle - |1\rangle) \otimes (|0\rangle - |1\rangle)$. *Hint: Since $f(x)$ is a one-bit number, $1 \oplus (1 \oplus f(0)) = f(0)$.*
3. The output state of the oracle in Deutsch's algorithm can be summarized as

$$\hat{U}_f|\varphi_{\text{in}}\rangle = \frac{1}{2}\left((-1)^{f(0)}|0\rangle + (-1)^{f(1)}|1\rangle\right) \otimes (|0\rangle - |1\rangle).$$

   What are the possible results of a measurement of the first qubit in the computational basis (i.e., $\{|0\rangle, |1\rangle\}$)? Include possible outcomes and their associated probabilities.
4. Write out the summations in Eq. (4.23) in full for the single-qubit states $|0\rangle$ and $|1\rangle$.
5. Consider the function $f(x) = \sin^2(\pi x/2)$ being analyzed by the Deutch–Jozsa algorithm (Quantum Circuit 4.10).
   (a) Write out explicitly (i.e., no summations) the state $|\varphi_2\rangle$ for a two-qubit argument register.
   (b) Write out explicitly the state $|\varphi_3\rangle$ for a two-qubit argument register.
   (c) What are the possible outcomes of measurement of the argument register in $|\varphi_3\rangle$, and their associated probabilities?
   (d) Repeat parts (a), (b), and (c) for a three-qubit argument register.

6. Repeat Exercise 5 for $f(x) = \sin^2(\pi x)$.

7. Work out $(-1)^{\sum_{n=0}^{L-1} x_n y_n}$ and $(-1)^{x \cdot y}$, where $x \cdot y$ is defined by Eq. (4.25), for the following pairs:

(a) $x = 2, y = 3$.

(b) $x = 8, y = 3$.

(c) $x = 15, y = 11$.

4.5   Finding a Needle in a Haystack: Grover's Search Algorithm

The next example of quantum algorithms has an important distinction over what we have seen so far: it solves a problem that actually has applications. Consider a situation that many of us might have faced: a missed call on your phone from someone who has blocked their caller ID. How do you find who the caller was just from knowledge of the number alone: it could be an individual you have a strong desire to talk to, or it could be someone trying to sell you vinyl siding or air-duct cleaning. Let us suppose you do not want to just call the number (you might end up having someone talk your ear off about the not-to-be-neglected hazards of dirty air ducts), so you are reduced to leafing through an alphabetized phone directory name by name to find the person with the appropriate number.

There are 10 million possible seven (decimal) digit numbers for each area code: 000-0000 to 999-9999; some of these, of course, have special purposes, like the police or directory inquires etc., and will never be used for ordinary telephone numbers, but 10^7 is a reasonable upper limit for the sake of argument. Using the formula $L(x) = \lfloor \log_2 x \rfloor + 1$ we discussed in Section 4.1, that implies the telephone number is 24-bits long.

If you had to search through an alphabetized phone directory for the name that goes with the mystery number, 10 million is a rather inconveniently large number. You might just get lucky and find it on the first try, or you might have to wait until the very last query; on average $10^7/2$ (or 2^{L-1}) is approximately the number of look-ups you might intuitively expect to have to perform before you find the name. There is a theorem called the *no free lunch theorem* in search and optimization which states, in effect, you cannot really do much better no matter how sophisticated your classical search algorithm; but, maybe quantum computers *can* serve you up some free lunch?

A quantum search algorithm was originally proposed by Lov K. Grover in 1996. It is shown diagrammatically in Quantum Circuit 4.17, and as you can see it conforms pretty closely to the generic form of quantum algorithm shown in Quantum Circuit 4.6, with the added refinement that the function oracle and the post-processing unitary are executed multiple times.

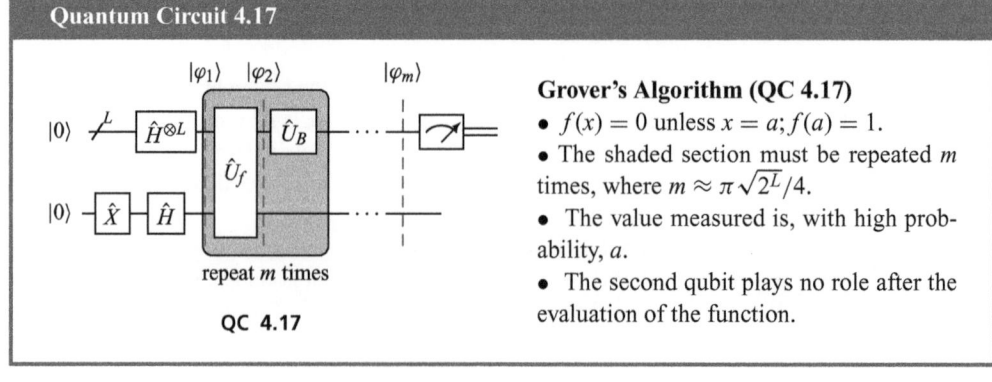

Quantum Circuit 4.17

QC 4.17

Grover's Algorithm (QC 4.17)
- $f(x) = 0$ unless $x = a$; $f(a) = 1$.
- The shaded section must be repeated m times, where $m \approx \pi \sqrt{2^L}/4$.
- The value measured is, with high probability, a.
- The second qubit plays no role after the evaluation of the function.

To understand how it works, first we simplify the search task to a mathematical problem. The function $f(x)$ is defined as follows:

$$f(x) = \begin{cases} 1 & \text{if } x = a \\ 0 & \text{if } x \neq a \end{cases} \tag{4.33}$$

where a is the unknown special value we seek. In the phone book example, suppose we have the telephone number 555-1234; each person in the alphabetized directory has a number indicating their place in the list (e.g. Aaron A. Aaronson is $x = 0$, Zygmut Zyzz is $x = 999,999$, etc.); $f(x)$ compares the telephone number of the customer at place x in the alphabetic list with the known telephone number (555-1234) for which we are searching; if the telephone number is different, then $f(x) = 0$, if the telephone number is the same as the known telephone number, then $f(x) = 1$, and the corresponding value of x is a, the place in the list we seek.

Let us suppose we have a unitary quantum oracle of the type already discussed which evaluates $f(x)$ for this search problem. Since the value of $f(x)$ is either 0 or 1, we can use Eq. (4.22) to find the state after the first application of $\hat{U}_f$:

$$|\varphi_2\rangle = \left(\sum_{x=0}^{2^L-1} \frac{(-1)^{f(x)}}{\sqrt{2^L}} |x\rangle \right) \otimes |-\rangle, \tag{4.34}$$

where $|-\rangle = (|0\rangle - |1\rangle)/\sqrt{2}$. Now let us use Eq. (4.33),

$$|\varphi_2\rangle = \left(\frac{1}{\sqrt{2^L}} \sum_{\substack{x-0 \\ x\neq a}}^{2^L-1} |x\rangle - \frac{1}{\sqrt{2^L}} |a\rangle \right) \otimes |-\rangle$$

$$= (|\text{all}_L\rangle - 2\sin(\epsilon)|a\rangle) \otimes |-\rangle, \tag{4.35}$$

where $|\text{all}_L\rangle$ is the equal superposition of all computational basis states (see Quantum Circuit 4.5) and $\sin(\epsilon) = 1/\sqrt{2^L} = \langle a|\text{all}_L\rangle$.

The argument register has L qubits, and thus its Hilbert space has a dimension of 2^L, but during Grover's algorithm only two states, $|a\rangle$ and $|\text{all}_L\rangle$, come into play, and so we can simplify things a lot by concentrating on the two-dimensional subspace spanned by these two. Since they are not orthogonal, we define the normalized perpendicular component of $|a\rangle$ by the formula

$$|a_\perp\rangle = (|a\rangle - \sin(\epsilon)|\text{all}_L\rangle)/\cos(\epsilon), \tag{4.36}$$

which implies

$$|a\rangle = \cos(\epsilon)|a_\perp\rangle + \sin(\epsilon)|\text{all}_L\rangle. \tag{4.37}$$

Since L is a large number (otherwise searching the database would be pretty easy), then ϵ is small; we can think of ϵ as the angle between the vectors $|a_\perp\rangle$ and $|a\rangle$: they are very nearly parallel.

Using $|a_\perp\rangle$ and $|\text{all}_L\rangle$ as basis states, we find that the action of the oracle on an arbitrary combination of $|\text{all}_L\rangle$ and $|a_\perp\rangle$ in the argument register gives us

$$\hat{U}_f (\alpha|\text{all}_L\rangle + \beta|a_\perp\rangle) \otimes |-\rangle = (\alpha'|\text{all}_L\rangle + \beta'|a_\perp\rangle) \otimes |-\rangle, \tag{4.38}$$

where α and β are arbitrary probability amplitudes, and α' and β' are given by the matrix formula

$$\begin{bmatrix} \alpha' \\ \beta' \end{bmatrix} = \begin{bmatrix} \cos(2\epsilon) & -\sin(2\epsilon) \\ -\sin(2\epsilon) & -\cos(2\epsilon) \end{bmatrix} \begin{bmatrix} \alpha \\ \beta \end{bmatrix}. \tag{4.39}$$

This looks very suggestive: the application of the function operator $\hat{U}_f$ has produced something that looks rather like a rotation: the output state has a larger component of $|a_\perp\rangle$ than it did to begin with, and if we can increase this component so the output is almost entirely made of $|a_\perp\rangle$, then when we measure the output state, we have a pretty good probability of getting the value of a.

But looks can be deceptive: the matrix in Eq. (4.39) is *not* a rotation. In fact, if we apply $\hat{U}_f$ twice, we get right back to where we started (try multiplying the matrix by itself). Thus blindly applying $\hat{U}_f$ in the hope of "amplifying" the component of $|a_\perp\rangle$ in the output will not have much effect.

Before we descend into the gloom of pessimism and give up on quantum computing ever being able to do anything useful, recall we have one more card to play: what about the post-processing operation $\hat{U}_B$? Can we figure out an operation that will help our cause? Let us try the following:

$$\hat{U}_B|\text{all}_L\rangle = -|\text{all}_L\rangle, \tag{4.40}$$

$$\hat{U}_B|a_\perp\rangle = |a_\perp\rangle. \tag{4.41}$$

Thus,

$$\left(\hat{U}_B \otimes \hat{I}\right)\hat{U}_f\left(\alpha|\text{all}_L\rangle + \beta|a_\perp\rangle\right) \otimes |-\rangle = \left(\alpha''|\text{all}_L\rangle + \beta''|a_\perp\rangle\right) \otimes |-\rangle, \tag{4.42}$$

where

$$\begin{bmatrix} \alpha'' \\ \beta'' \end{bmatrix} = \begin{bmatrix} -1 & 0 \\ 0 & 1 \end{bmatrix}\begin{bmatrix} \cos(2\epsilon) & -\sin(2\epsilon) \\ -\sin(2\epsilon) & -\cos(2\epsilon) \end{bmatrix}\begin{bmatrix} \alpha \\ \beta \end{bmatrix}$$

$$= -\begin{bmatrix} \cos(2\epsilon) & -\sin(2\epsilon) \\ \sin(2\epsilon) & \cos(2\epsilon) \end{bmatrix}\begin{bmatrix} \alpha \\ \beta \end{bmatrix}. \tag{4.43}$$

The matrix in Eq. (4.43) *is* a rotation. If we apply it multiple times, the arguments in the trigonometric functions add rather than cancel; in fact, multiplying the matrix by itself m times we get

$$\begin{pmatrix} \cos(2\epsilon) & -\sin(2\epsilon) \\ \sin(2\epsilon) & \cos(2\epsilon) \end{pmatrix}^m = \begin{pmatrix} \cos(2m\epsilon) & -\sin(2m\epsilon) \\ \sin(2m\epsilon) & \cos(2m\epsilon) \end{pmatrix}. \tag{4.44}$$

This can be proven using matrix multiplication, standard trigonometry formulas, and the method of induction. Hence, for the state $|\varphi_m\rangle$ in Quantum Circuit 4.17, we obtain the formula

$$|\varphi_m\rangle = \left\{\left(\hat{U}_B \otimes \hat{I}\right)\hat{U}_f\right\}^m|\varphi_1\rangle$$

$$= (-1)^m\left\{\cos(2m\epsilon)|\text{all}_L\rangle + \sin(2m\epsilon)|a_\perp\rangle\right\} \otimes |-\rangle. \tag{4.45}$$

The probability we obtain the outcome a when we measure the argument register is therefore given by

$$P(a) = |\langle a|\varphi_m\rangle|^2$$

$$= [\cos(2m\epsilon)\sin(\epsilon) + \sin(2m\epsilon)\cos(\epsilon)]^2$$

$$= \sin^2\left[(2m+1)\epsilon\right]. \tag{4.46}$$

If $(2m+1)\epsilon = \pi/2$ then we obtain the answer a with probability 1. However, since m has to be a natural number, we can only get close to the probability 1; using the approximation $\epsilon \approx 1/\sqrt{2^L}$, the optimal number of repetitions is

$$m = \text{Round}\left(\frac{\pi/2\epsilon - 1}{2}\right) \approx \text{Round}\left(\frac{\pi\sqrt{2^L} - 2}{4}\right), \tag{4.47}$$

where Round(x) is the nearest integer to the real number x. Recall that, classically, we expected to have to perform $2^L/2$ queries of the phone book in order to find our unknown number. Quantumly, this has been replaced by about $\sqrt{2^L}$ repetitions of the $\left(\left(\hat{U}_B \otimes \hat{I}\right)\hat{U}_f\right)$ sequence of quantum operations; in other words, it offers a polynomial rather than an exponential speed-up.

Exercises 4.5.1

1. Prove that $|a_\perp\rangle$ defined in Eq. (4.36) is normalized and orthogonal to $|\text{all}_L\rangle$.
2. Work out formulas for $\hat{U}_f|\text{all}_L\rangle|-\rangle$ and $\hat{U}_f|a_\perp\rangle|-\rangle$, and hence prove Eq. (4.39).
3. Prove Eq. (4.44).
4. Using Eq. (4.47) and the approximation $\epsilon \approx 1/\sqrt{2^L}$ find the optimum number of repetitions m_{opt} in Grover's algorithm for (a) $L = 3$; (b) $L = 5$; (c) $L = 10$; (d) $L = 24$.
5. Use Eq. (4.46) to find the probability of measuring the correct value of a at the output of Grover's algorithm for the four cases in Exercise 4.

4.6 CONCEPT CHECK AND EXERCISES

Concept Check

- The Deutsch algorithm distinguishes a constant one-bit function from a one-bit balanced function with one oracle call: this proved the quantum advantage.
- The Deutsch–Jozsa algorithm distinguishes a constant L-bit function from an L-bit balanced function with one oracle call: this proved the possibility of exponential speed-up.
- The Bernstein–Vazirani algorithm determines a hidden parameter with one oracle call, with linear speed-up. The oracles are easy to design.
- The Grover algorithm determines which value of an otherwise constant function is different. It gives a polynomial speed-up.

Chapter Exercises

1. **Qubit registers as classical data storage devices**
   In Section 4.1 we introduced the standard method of storing numbers in qubits, in which an L-bit number required L qubits; thus, the number $5_{10} = 101_2$ is represented by the three-qubit state $|101\rangle$. Equivalently, we could change our basis, for example by applying a Hadamard gate to each qubit, in which case the following state represents the number 5_{10}:

$$(\hat{H} \otimes \hat{H} \otimes \hat{H})|101\rangle = \frac{1}{2\sqrt{2}}\,(|000\rangle - |001\rangle + |010\rangle - |011\rangle$$
$$-|100\rangle + |101\rangle - |110\rangle + |111\rangle).$$

This new representation is an equal superposition of all possible three-qubit states (i.e., each state of the computational basis appears with equal probability of $1/8$), with the numerical bit values now being encoded in the phases (i.e., $+1$ has phase 0, -1 has phase π).

Note, however, there are eight different probability amplitudes each of which, in principle, could be either $+1$ or -1, meaning such a three-qubit state could, in principle, store a seven-bit number. This represents a different quantum computing paradigm, in which we could store numbers exponentially more efficiently.

For example, let x be a three-bit number with binary digits $\{x_0, x_1, x_2\}$, i.e., $x = \sum_{n=0}^{2} x_n 2^n$. This number can be represented as a quantum state of two qubits as follows:

$$|\psi(x)\rangle = \frac{1}{2}\left(|11\rangle + (-1)^{x_2}|10\rangle + (-1)^{x_1}|01\rangle + (-1)^{x_0}|00\rangle\right)$$

$$= \frac{1}{2^{L/2}}\left(|2^L - 1\rangle + \sum_{y=0}^{2^L-2}(-1)^{x_y}|y\rangle\right).$$

(a) Why can we not use the probability amplitude of the state $|11\rangle$ to store a fourth bit?

(b) Find $\langle\psi(x)|\psi(x')\rangle$.

(c) Devise simple quantum circuits to store the following numbers using this scheme:
   (i) $x = 7$; (ii) $x = 3$; (iii) $x = 4$.

(d) À propos the previous question, why would such an approach to data storage be of dubious utility?

2. Work through Grover's algorithm for $L = 3$ and $a = 5$. Use two repetitions of the $\hat{U}_f$ and $\hat{U}_B$ combination. Write out the states after each repetition in the computational basis. What is the probability of getting the correct result?

3. Show that

$$\hat{H}^{\otimes L}|x\rangle = \frac{1}{\sqrt{2^L}}\sum_{y=0}^{2^L-1}(-1)^{x\cdot y}|y\rangle,$$

where $|x\rangle = |x_{L-1}\rangle \otimes |x_{L-2}\rangle \otimes \cdots \otimes |x_0\rangle$ is an L-qubit computational basis state (similarly $|y\rangle$) and $x \cdot y = x_{L-1}y_{L-1} \oplus x_{L-2}y_{L-2} \oplus \cdots \oplus x_0 y_0$ is the bit-by-bit dot product of the two integers. *Hint: You may assume the two-qubit result Eq. (4.24); the method of induction might be a good thing to try.*

4. Which, if any, of the qubits appearing in the algorithms we have discussed in this chapter are entangled?

5 Period Finding and Shor's Algorithm

This chapter introduces what is the best-known quantum algorithm to date, namely Shor's algorithm for finding the factors of a number, invented by Peter Shor in 1994. If I give you a number, say 15, grade-school arithmetic tells us pretty quickly that $15 = 3 \times 5$, and thus the factors are 3 and 5. It gets a lot more difficult as the number to be factored starts to get bigger: try factoring 51, 533, or 8201 in your head (the answers are: $51 = 3 \times 17$, $533 = 13 \times 41$, and $8201 = 59 \times 139$). Thus, to solve the problem, we rapidly turn to a computer. But it turns out to be hard for computers to solve the problem as well; factoring numbers with hundreds of bits takes months on distributed multi-computer networks, and the effort required (measured by the total number of operations) grows exponentially with the size of the number.

Shor's algorithm approaches the problem by first relating the factoring problem to the periodicity of a certain mathematical function (the *modular exponential*), and then exploiting quantum parallelism to efficiently find that period. Thus, this chapter begins with a review of Fourier theory, a very good way to find the period of a function.

While a quantum factor-finder is of great interest to number theorists, it would hardly engage the excitement of the wider public were it not for the rather important fact that the difficulty of factoring large composite numbers is at the heart of the presumed security of the *RSA public key encryption scheme* (as we expand on in Section 5.3), which is widely used to ensure the security of internet communications. Or, to put it bluntly, Shor's algorithm allows you to break codes and read everyone's email.

Learning Objectives

- Learn about Fourier transforms and discrete Fourier transforms and the role of the latter in period finding.
- Learn about the quantum Fourier transform, its implementation, and its role in period finding.
- Learn concepts from number theory that are essential in understanding RSA encryption, Shor's quantum factoring algorithm, and how the latter threatens the security of the former encryption method.

5.1 Fourier Transforms

Fourier analysis is a mathematical technique for characterizing functions by writing them as linear combinations of the trigonometrical functions sin and cos. Thus, just about any function $f(t)$ can be written as a combination of complex exponential functions $\exp(-i2\pi vt) = \cos(2\pi vt) - i\sin(2\pi vt)$ as follows:

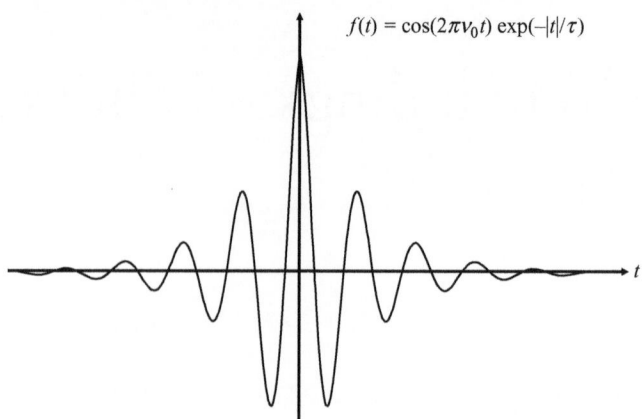

Figure 5.1   The damped cosine function $f(t) = \cos(2\pi v_0 t) \exp(-|t|/\tau)$, where v_0 is the frequency and τ is the damping time. In this particular curve $2\pi v_0 \tau = 1$.

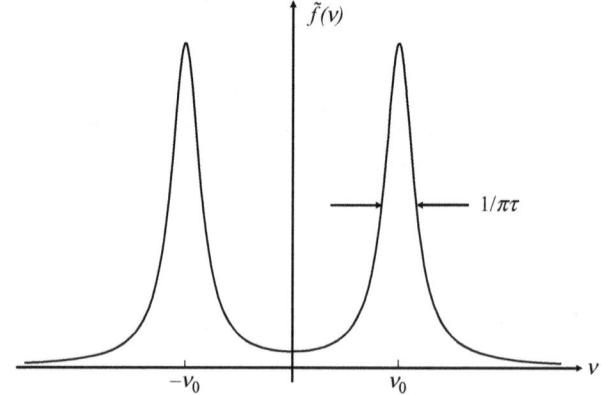

Figure 5.2   The Fourier transform of $f(t) = \cos(2\pi v_0 t) \exp(-|t|/\tau)$. For this function, the integral can be worked out exactly: $\tilde{f}(v) = \tau/[1 + \{2\pi\tau(v - v_0)\}^2] + \tau/[1 + \{2\pi\tau(v + v_0)\}^2]$. The location of the peaks in $\tilde{f}(v)$ yields the value of v_0, and hence the period of the oscillations of $f(t)$, $T = 1/v_0$. The width of the peaks (i.e., the frequency difference between the points where the Fourier transform has half the value of the peak) gives the value of τ.

$$f(t) = \int_{-\infty}^{\infty} \tilde{f}(v) \exp(-2\pi i v t) dv, \tag{5.1}$$

where $\tilde{f}(v)$ is a weighting factor showing how much of each frequency contributes to the function $f(t)$. You will find variations of this definition in the scientific literature, with the factors of 2π cropping up in different places, but the basic properties of **Fourier transforms** are standard. Note that the combination is an integral rather than a sum because *all* frequencies are possible, not just a few discrete frequencies. If $\tilde{f}(v)$ is strongly peaked about the value $v = v_0$, then it is not unreasonable to suppose that $f(t)$ will display some oscillation with a period of $1/v_0$.

It turns out that the weighting factor (called the *Fourier transform*) is a function given by an almost identical integral involving the original function $f(t)$:

$$\tilde{f}(v) = \int_{-\infty}^{\infty} f(t) \exp(2\pi i v t) dt \qquad (5.2)$$

(note the change of sign in the exponent, and the different variable of integration). The Fourier transform of a function, also called the Fourier spectrum, yields almost instantly important features of the function, such as the periodicity of any oscillations and their rate of decay. An example is shown in Figs. 5.1 and 5.2.

Because a system's response to some external stimulus often depends on the frequency with which the stimulus is applied, Fourier analysis (also called spectral analysis) is very important in mathematics, physics, and engineering. In this chapter we are going to be primarily concerned with ways to find the period of a repeating function, for which, for continuous functions at any rate, the Fourier transform is ideal.

 Interesting Facts 5.1

The Origin of Fourier Analysis. The systematic use of harmonic functions (also known as sine and cosine, which repeat in a fixed pattern over and over) to analyze problems was popularized by Jean-Baptiste Joseph Fourier (1768–1830). Besides being a renowned mathematician, Fourier was also an enthusiastic supporter of Napoleon, and accompanied him on his expedition to conquer Egypt in 1798. Legend has it that Fourier originated his study of harmonic analysis in the hope of maximizing the performance of Napoleon's cavalry horses in battle: the horse's heart beat is the periodic stimulus, and flow of blood through the horse's veins is the effect that needed to be optimized.

5.1.1 Discrete Fourier Transforms

In computing one does not have the infinitesimal separation between adjacent values of t that one has in the calculus: one must sample a function at discrete separated values of the argument t. Yes, one can make the separation of those points as small as one's computer capabilities will allow; and yes, when plotted, such discretely evaluated samples of a function can appear smooth and indistinguishable from a continuously evaluated function; nevertheless, they are a discrete series of values. Thus, in computation we must replace $f(t)$ by an array of N discrete, equally spaced values $\{f_0, f_1, f_2, \ldots, f_{N-1}\}$, where $f_k = f(t_0 + k\Delta t)$.

Suppose we have such an array and want to know the period of the discretely sampled function, like the one shown in Fig. 5.3. The plot allows us to make a good guess of the period by looking at it. But as functions get more and more complicated, and their periods longer and longer, such bootstrap methods become more and more complicated to implement. It would be nice if there were a mathematical way to figure it out directly without guess work.

The answer is provided by the **discrete Fourier transform**. Suppose we know the array of values of our function, $\{f_0, f_1, f_2, \ldots, f_{N-1}\}$. The discrete Fourier transform of this sequence of numbers is defined by

$$\tilde{f}_\ell = \frac{1}{\sqrt{N}} \sum_{k=0}^{N-1} f_k \exp\left(-2\pi i k \ell / N\right), \qquad (5.3)$$

where ℓ takes integer values $0, 1, \ldots, N-1$.

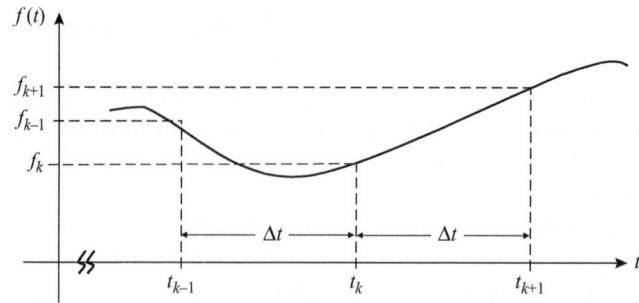

Figure 5.3   Discretely sampled functions. The continuous function $f(t)$ is sampled at a sequence of discrete points, usually equally spaced. The sampled arguments are $t_k = t_0 + k\Delta t$ (where t_0 is some suitable starting value, for example $t_0 = 0$), and the discrete function values are $f_k = f(t_k)$.

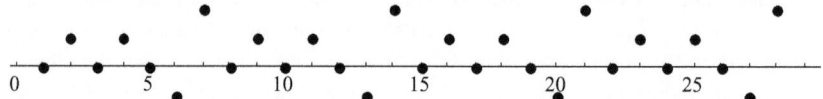

Figure 5.4   A periodic function. Here, we have the plot of a discretely sampled periodic function $f(t)$. Notice that in order to find its period, classically, what needs to be done is to evaluate the function at multiple points, then look for a pattern. In this case, one sees that the period = 7. Since the function repeats itself four times, $q = N/r = 4$.

5.1.2 Period Finding

Let us consider the specific situation where $f(t)$ is periodic, so that $f(t + T) = f(t)$, where T is the period. In terms of our discrete sample version of the function, $f_{k+r} = f_k$, where $r = T/\Delta t$ is the *non-dimensional* period (also called the **order**). If we sample the function at N points in total, then the number of complete periods showing up in our sample is $q = \lfloor N/r \rfloor$.

An example with $r = 7$, $N = 28$, and $q = 4$ is shown in Fig. 5.4.

The sum appearing in Eq. (5.3) can be simplified because of the periodicity of the f_k sequence. To illustrate how this works, let us consider a simple case: let f_k have period 2, and we have sampled $N = 7$ points, so the full sequence is $\{f_0, f_1, f_0, f_1, f_0, f_1, f_0\}$. Writing $\eta = \exp(2\pi i \ell / N)$, the sum in Eq. (5.3) is

$$
\begin{aligned}
\tilde{f}_\ell &= \frac{1}{\sqrt{7}} \left(f_0 + f_1 \eta + f_2 \eta^2 + f_3 \eta^3 + f_4 \eta^4 + f_5 \eta^5 + f_6 \eta^6 \right) \\
&= \frac{1}{\sqrt{7}} \left(f_0 + f_1 \eta + f_0 \eta^2 + f_1 \eta^3 + f_0 \eta^4 + f_1 \eta^5 + f_0 \eta^6 \right) \\
&= \frac{1}{\sqrt{7}} f_0 (1 + \eta^2 + \eta^4) + \frac{1}{\sqrt{7}} f_1 (\eta + \eta^3 + \eta^5) + \frac{1}{\sqrt{7}} f_0 \eta^6 \\
&= \frac{1}{\sqrt{7}} (f_0 + f_1 \eta)(1 + \eta^2 + \eta^4) + \frac{1}{\sqrt{7}} f_0 \eta^6 .
\end{aligned}
\tag{5.4}
$$

More generally, the sequence $\{f_0, f_1, \ldots, f_{N-1}\}$ can be broken down into q sub-sequences, each with r terms ($\{f_0, f_1, \ldots, f_{r-1}\}$, $\{f_r, f_{r+1}, \ldots, f_{2r-1}\}, \ldots, \{f_{(q-1)r}, f_{(q-1)r+1}, \ldots, f_{qr-1}\}$) plus a single sub-sequence at the end, shorter than the others and which does not contain a whole period: $\{f_{qr}, f_{qr+1}, \ldots, f_N\}$. Because $f_{k+r} = f_{k+2r} = f_{k+3r} = \cdots = f_k$, the q full-period sub-sequences can be factored out of the summation, just as we did in the $N = 7, r = 2$ case above:

$$\tilde{f}_\ell = \frac{1}{\sqrt{N}} \left(\sum_{k=0}^{r-1} f_k \eta^k \right) \left(\sum_{p=0}^{q-1} \eta^{rp} \right) + \tilde{f}_\ell^{\text{end}}, \tag{5.5}$$

where $\tilde{f}_\ell^{\text{end}} = \frac{1}{\sqrt{N}} \sum_{k=qr}^{N-1} f_k \eta^k$ is the sum over the shorter sub-sequence at the end. The first sum in Eq. (5.5) is just the discrete Fourier transform of the function, but only over one period (without the $1/\sqrt{r}$ in front). The second sum is a geometric series, which we can evaluate using the well-known formula (and writing η out in full once again):

$$\sum_{p=0}^{q-1} [\exp(-2\pi i k r/N)]^p = \frac{1 - \exp(-2\pi i k r q/N)}{1 - \exp(-2\pi i k r/N)}. \tag{5.6}$$

Let us look at the numerator of this expression. Note that $|1 - \exp(i\theta)| = 2|\sin(\theta/2)|$, hence the numerator will be a complex number whose amplitude varies between 0 and 1; it takes the value 0 whenever $\sin(\pi k r q/N) = 0$, i.e., when krq/N is an integer. However, the denominator cannot be ignored, since for some values of k it will have value 0, specifically when $\sin(\pi k r/N) = 0$, or when kr/N is an integer. Thus, both the numerator and the denominator can both be zero, so that we must calculate the ratio by l'Hôpital's rule:

$$\lim_{x \to 0} \frac{1 - \exp(-2\pi i x q)}{1 - \exp(-2\pi i x)} = q. \tag{5.7}$$

This represents a sharply defined peak: for large values of q and values of kr/N that are not integers, the right hand side of Eq. (5.6) will be small and negligible; but when kr/N is an integer, n, the right hand side of Eq. (5.6) is $q \approx N/r$.

Thus, if we calculate the discrete Fourier transform from our sample f_k, we should get

$$\tilde{f}_\ell \approx \begin{cases} 0 & \text{if } \ell \neq sN/r, \text{ where s is an integer,} \\ \\ \frac{q}{\sqrt{N}} \sum_{k=0}^{r-1} f_k \exp(2\pi i k s/r) & \text{if } \ell \approx sN/r. \end{cases} \tag{5.8}$$

An example is plotted in Fig. 5.5. The Fourier transform is, in this case, exactly zero except for some discrete, isolated values of k; the separation between these isolated values is uniform, and equal to N/r; thus from knowledge of N one can readily obtain r.

An important refinement of the discrete Fourier transform is the *fast Fourier transform*. The sum appearing in Eq. (5.3) becomes burdensome for large values of N, but if we sampled the function not for an arbitrary number of points N, but instead some power of 2, then it turns out there is a remarkably efficient numerical method involving re-arrangements of elements of the array and multiplications by phase factors that can evaluate the discrete Fourier transform without approximation. We will see elements of this philosophy reappear in the quantum Fourier transform.

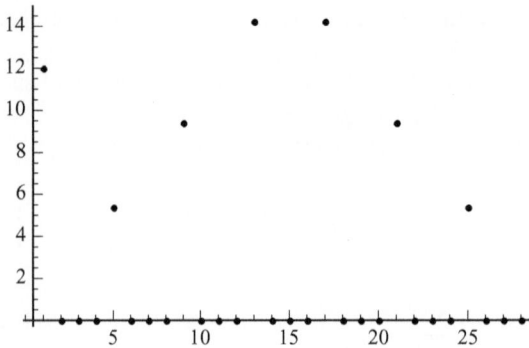

Figure 5.5   The absolute value of the Fourier transform of the periodic function in Fig. 5.4. Notice that the function is only non-zero at specific values of k, i.e., $\{0, 4, 8, 12, \ldots, 24\}$, separated by $M = 4$; since $M = N/P$, we can deduce $P = N/M = 7$.

Exercises 5.1.3

1. Suppose that a function has the Fourier transform

$$\tilde{f}(v) = \sqrt{2\pi(\Delta t)^2}e^{-(2\pi \Delta t)^2(v-v_0)^2/2}.$$

   Use Eq. (5.1) to find the function $f(t)$. *Hint: Use the integral $\int_{-\infty}^{\infty} \exp(-x^2/2)dx = \sqrt{2\pi}$ without proof.*

2. Find, simplify (in the manner of Eq. (5.4)), and sketch-plot (by hand) the discrete Fourier transform of the following sequences, determined analytically:

   (a) $\{0, 1, 0, 1\}$,
   (b) $\{0, 4, 2, 0, 4, 2, 0, 4\}$.

3. Find and plot the discrete Fourier transform of the following sequences *computationally*, and comment on the results (given the goal is to find periods):

   (a) $\{-2, 1, 2, 0, -2, 1, 2, 0, -2, 1, 2, 0\}$,
   (b) $\{2, 4, 2, 0, 5, 2, 4, 2, 0, 5, 2, 4, 2, 0, 5, 2, 4, 2\}$.

4. Demonstrate the inverse of the discrete Fourier transform Eq. (5.3) is given by

$$f_\ell = \frac{1}{\sqrt{N}} \sum_{k=0}^{N-1} \tilde{f}(k) \exp\left(2\pi i k \ell / N\right).$$

   Hint: The formula for the sum of a geometric series might come in handy: $\sum_{n=0}^{N-1} a\xi^n = a(1 - \xi^N)/(1 - \xi)$.

5. In this exercise we are going to explore the sum given in Eq. (5.6) a bit more carefully. Numerically plot the value of the following expression for the values $r = 7$ and $N = 98, 99$, and 100. *Hint: You will have to be careful in encoding the function since the denominator can be zero for certain values of k*

$$\left| \frac{1 - \exp(-2\pi i k r \lfloor N/r \rfloor /N)}{1 - \exp(-2\pi i k r/N)} \right|.$$

5.2 Quantum Period Finding

5.2.1 The Quantum Fourier Transform

The **quantum Fourier transform** is the quantum analogue of the discrete Fourier transform we encountered in Section 5.1. It is a unitary transform defined as follows:

$$\hat{U}_{\mathrm{QFT}}|\ell\rangle = \frac{1}{\sqrt{2^L}} \sum_{k=0}^{2^L-1} \exp\left(2\pi i \frac{k\ell}{2^L}\right) |k\rangle. \tag{5.9}$$

Although this looks like a daunting transform to implement, it turns out there is a quite efficient way of performing it using only one- and two-qubit gates. To see this, let us look at the simplest possible example of $L = 2$: the quantum Fourier transform applied to an element of the computational basis $|\ell\rangle$ is given by the formula

$$\hat{U}_{\mathrm{QFT}}|\ell\rangle = \frac{1}{2} \left\{ |00\rangle + \exp\left(2\pi i \frac{\ell}{4}\right) |01\rangle + \exp\left(2\pi i \frac{2\ell}{4}\right) |10\rangle + \exp\left(2\pi i \frac{3\ell}{4}\right) |11\rangle \right\}. \tag{5.10}$$

If we calculate the concurrence of this state (see Exercise 3) we find it is a separable state; hence we can factor it:

$$\hat{U}_{\mathrm{QFT}}|\ell\rangle = \frac{1}{\sqrt{2}} \left\{ |0\rangle + \exp\left(2\pi i \frac{2\ell}{4}\right) |1\rangle \right\} \otimes \left\{ |0\rangle + \exp\left(2\pi i \frac{\ell}{4}\right) |1\rangle \right\}. \tag{5.11}$$

Let us now write out ℓ in terms of its bit values $\ell = 2\ell_1 + \ell_0$; and since $\exp(2\pi i \ell_1) = 1$, the term involving ℓ_1 in the exponent in the first bracket is superfluous, and we get

$$\hat{U}_{\mathrm{QFT}}|\ell\rangle = \frac{1}{\sqrt{2}} \left\{ |0\rangle + \exp\left(2\pi i \frac{\ell_0}{2}\right) |1\rangle \right\} \otimes \left\{ |0\rangle + \exp\left(2\pi i \left[\frac{\ell_1}{2} + \frac{\ell_0}{4}\right]\right) |1\rangle \right\}. \tag{5.12}$$

Recall the Hadamard gate and what it does to a single qubit in a computational basis state:

$$\hat{H}|0\rangle = \frac{1}{\sqrt{2}} \left(|0\rangle + |1\rangle\right),$$
$$\hat{H}|1\rangle = \frac{1}{\sqrt{2}} \left(|0\rangle - |1\rangle\right). \tag{5.13}$$

These can be combined as

$$\hat{H}|\ell_k\rangle = \frac{1}{\sqrt{2}} \left\{ |0\rangle + (-1)^{\ell_k}|1\rangle \right\}$$
$$= \frac{1}{\sqrt{2}} \left(|0\rangle + \exp\left(2\pi i \frac{\ell_k}{2}\right) |1\rangle \right), \tag{5.14}$$

which looks rather suggestively similar to the terms appearing in Eq. (5.12). Let us look at QC 5.1:

QC 5.1

We find that the state just after the Hadamards is given by the expression

$$|\varphi_1\rangle = \hat{H} \otimes \hat{H}|\ell_1\ell_0\rangle$$
$$= \frac{1}{\sqrt{2}}\left(|0\rangle + \exp\left(2\pi i\frac{\ell_1}{2}\right)|1\rangle\right) \otimes \frac{1}{\sqrt{2}}\left(|0\rangle + \exp\left(2\pi i\frac{\ell_0}{2}\right)|1\rangle\right). \qquad (5.15)$$

Note, however, the term involving the $\exp\left(2\pi i\frac{\ell_1}{2}\right)$ in Eq. (5.12) is the *second* qubit rather than the first, as it is here. Thus, when the SWAP gate is added to the two qubits we get

$$|\varphi_2\rangle = \frac{1}{\sqrt{2}}\left(|0\rangle + \exp\left(2\pi i\frac{\ell_0}{2}\right)|1\rangle\right) \otimes \frac{1}{\sqrt{2}}\left(|0\rangle + \exp\left(2\pi i\frac{\ell_1}{2}\right)|1\rangle\right), \qquad (5.16)$$

which is almost, but not quite, the same as Eq. (5.12). The only difference is the phase factor $\exp(2\pi i\ell_0/4)$ multiplying the second qubit, with the added complication that the phase factor in question depends on the bit value ℓ_1 of the *other* qubit. In other words, we need a controlled phase change gate (QC 5.2):

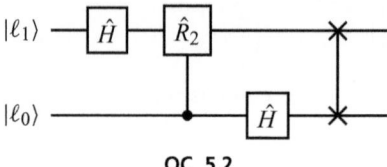

QC 5.2

Note that the controlled gate operation must take place before the Hadamard gate on qubit #0, (otherwise you would get a rather entangled mess). Here $\hat{R}_2$ is an example of a generic single qubit unitary:

$$\hat{R}_k = \begin{bmatrix} 1 & 0 \\ 0 & \exp(2\pi i/2^k) \end{bmatrix}, \qquad (5.17)$$

so that $\hat{R}_0 = \hat{I}$ and $\hat{R}_1 = \hat{Z}$. Sometimes you also see the notation $\hat{R}_2 = \hat{S}$ and $\hat{R}_3 = \hat{T}$.

L-qubit Quantum Fourier Transform
We have seen the basic idea in the two-qubit case: one can perform the quantum Fourier transform using a judicious mix of local Hadamard gates, controlled phase gates, and SWAP gates at the end. Let us see if we can figure the details out for an arbitrary number of qubits, L. We begin with the definition

$$\hat{U}_{\text{QFT}}|\ell\rangle = \frac{1}{\sqrt{2^N}} \sum_{k=0}^{2^L-1} \exp\left(2\pi i\frac{\ell k}{2^L}\right)|k\rangle. \qquad (5.18)$$

Remember, the $|k\rangle$ is really short-hand for the L qubit state $|k_{L-1}\rangle \otimes |k_{L-2}\rangle \otimes \cdots \otimes |k_0\rangle$, where k_{L-1}, $k_{L-2}, \ldots, k_0$ are the bit values for the L-bit integer $k = \sum_{m=0}^{L-1} k_m 2^m$. Thus, the term in the summation can be written as

$$\exp\left(2\pi i\frac{\ell k}{2^L}\right)|k\rangle = \exp\left(2\pi i\frac{\ell k_{L-1}2^{L-1}}{2^L}\right)|k_{L-1}\rangle \otimes \exp\left(2\pi i\frac{\ell k_{L-2}2^{L-2}}{2^L}\right)|k_{L-2}\rangle \otimes \cdots$$
$$\otimes \exp\left(2\pi i\frac{\ell k_0 2^0}{2^L}\right)|k_0\rangle$$
$$= \bigotimes_{m=0}^{L-1} \exp\left(2\pi i\ell k_m 2^{m-L}\right)|k_m\rangle_m. \qquad (5.19)$$

We have added a subscript to the ket in order to keep track of which qubit we are referring to: it might seem a bit pedantic right now, but it is going to be really important in a minute. Putting this back into Eq. (5.18), and re-writing the summation over k as a series of summations over each of the bit values, we get

$$\hat{U}_{\text{QFT}}|\ell\rangle = \sum_{k_{L-1}=0}^{1} \sum_{k_{L-2}=0}^{1} \cdots \sum_{k_0=0}^{1} \bigotimes_{m=0}^{L-1} \exp(2\pi i \ell k_m 2^{m-L})|k_m\rangle_m. \tag{5.20}$$

Now let us use the fact that multiplication is distributive over addition, i.e., for any four quantities a, b, c, and d we have $ac + ad + bc + bd = (a + b)(c + d)$. This is true of tensor multiplication as well as regular multiplication. Hence Eq. (5.20) can be re-written as

$$\hat{U}_{\text{QFT}}|\ell\rangle = \bigotimes_{m=0}^{L-1} \frac{1}{\sqrt{2}} \left\{ |0\rangle_m + \exp(i\phi_m) |1\rangle_m \right\}, \tag{5.21}$$

where $\phi_m = 2\pi \ell 2^{m-L}$. In other words, after the quantum Fourier transform unitary has been applied, each qubit is in an equal superposition state, with a "funny" phase ϕ_m, which differs from qubit to qubit and depends on the original state (specified by ℓ).

Writing ℓ in terms of its bit values, $\ell = \sum_{p=0}^{L-1} \ell_p 2^p$, we find

$$\phi_m = (2\pi) \sum_{p=0}^{L-1} \ell_p 2^{p+m-L}. \tag{5.22}$$

Well, this is still pretty hard to understand. But recall that in the $L = 2$ case we swapped the two qubits at the end; maybe that will be a useful thing to do in the general case as well? Let us try: we will swap qubit #$(L - 1)$ with qubit #0, qubit #$(L - 2)$ with qubit #1, and so on: in general, qubit #m is swapped with qubit #$(L - 1 - m)$. Since the only difference between qubit states in Eq. (5.21) is the "funny" phase, the great qubit swap is equivalent to changing the phase of qubit #m to

$$\phi'_m = 2\pi \sum_{p=0}^{L-1} \ell_p 2^{p-1-m}$$

$$= \underbrace{\frac{2\pi \ell_0}{2^{m+1}} + \cdots + \frac{2\pi \ell_{m-1}}{2^2}}_{\text{controlled } \hat{R}_k \text{ gates}} + \underbrace{\frac{2\pi \ell_m}{2}}_{\hat{H} \text{ on qubit } \#m} + \underbrace{2\pi \ell_{m+1} + \cdots + 2^{L-m-1}\pi \ell_{L-1}}_{\text{integer multiples of } 2\pi: \text{ no effect}}. \tag{5.23}$$

As can be seen in Eq. (5.23), there are three types of terms giving phase changes to qubit #m:

1. Phase terms dependent on the bit values $\ell_0, \ell_1, \ldots, \ell_{m-1}$: these can be realized by controlled-$\hat{R}_k$ operations, where k is equal to m+1-control qubit#. These controlled-$\hat{R}_k$ operations have to take place *after* qubit #m has had its Hadamard applied, but *before* the Hadamards are applied to qubits #0 to #$(m - 1)$.
2. A phase term dependent on the bit value ℓ_m, which can be realized by a Hadamard gate, just as we saw in the $L = 2$ case.
3. Terms involving $\ell_{m+1}, \ell_{m+2}, \ldots, \ell_{L-1}$, all of which give phases that are integer numbers of 2π, and can thus be discounted since $\exp(2\pi in) = 1$ where n is any integer.

Putting this all together, we get the general circuit in QC 5.3:

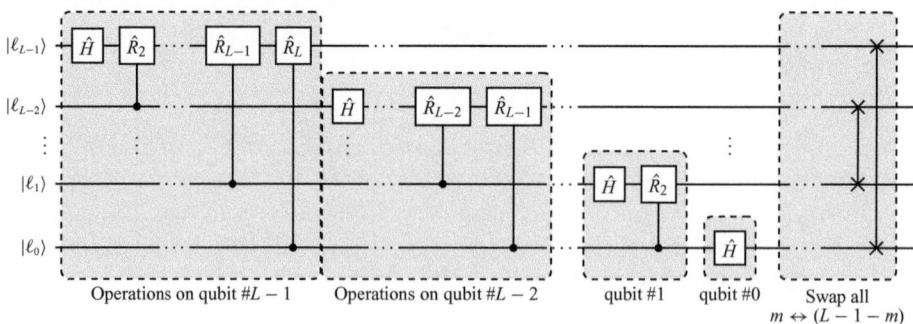

QC 5.3

5.2.2 Quantum Period Finding

The quantum algorithm for finding the period of a function is another algorithm that fits into the generic scheme we showed in Quantum Circuit 4.6. Suppose $f(x)$ is a periodic function, so that $f(x+r) = f(x)$, where r is the unknown period (also called the order in some applications). The problem is to find the period with as few evaluations of the function as possible. Classically, one needs to evaluate the function at least r times; quantumly, we can do it with only one. Quantum Circuit 5.4 shows this algorithm.

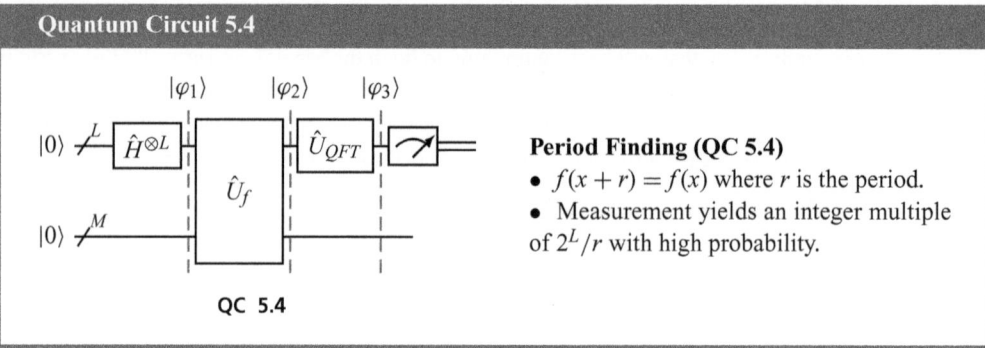

Period Finding (QC 5.4)
- $f(x + r) = f(x)$ where r is the period.
- Measurement yields an integer multiple of $2^L/r$ with high probability.

QC 5.4

In this circuit there is no preparation of the function register, which needs multiple qubits, rather than the single qubit we have used before now. The state input into the oracle $\hat{U}_f$ is

$$|\varphi_1\rangle = |\text{all}_L\rangle \otimes |0\rangle. \tag{5.24}$$

The output of the oracle is then

$$|\varphi_2\rangle = \frac{1}{\sqrt{2^L}} \sum_{x=0}^{2^L-1} |x\rangle \otimes |f(x)\rangle. \tag{5.25}$$

Now let us use the fact that $f(x)$ is periodic, with period r. As before, the number of whole periods that will "fit" inside the argument register is $q = \lfloor 2^L/r \rfloor$; there will be some additional terms at the end, just like the classical case:

$$|\varphi_2\rangle = \sqrt{\frac{q}{2^L}} \sum_{x=0}^{r-1} \left\{ \frac{1}{\sqrt{q}} \sum_{p=0}^{q-1} |x + pr\rangle \right\} \otimes |f(x)\rangle + |\varphi_2^{\text{end}}\rangle, \tag{5.26}$$

where $|\varphi_2^{\text{end}}\rangle = \sum_{x=qr}^{2^L-1} |x\rangle|f(x)\rangle$ is the part of the sum that is left over when we divided the argument register into q separate periods.

Applying the quantum Fourier transform to the term in curly brackets in Eq. (5.26):

$$\hat{U}_{\text{QFT}} \frac{1}{\sqrt{q}} \sum_{p=0}^{q-1} |x+pr\rangle = \frac{1}{\sqrt{2^L}} \sum_{k=0}^{2^L-1} \frac{1}{\sqrt{q}} \sum_{p=0}^{q-1} \exp\left[2\pi i \frac{k}{2^L}(x+pr)\right] |k\rangle$$

$$= \frac{1}{\sqrt{q2^L}} \sum_{k=0}^{2^L-1} \left\{ \sum_{p=0}^{q-1} \left[\exp\left(2\pi i \frac{kr}{2^L}\right)\right]^p \right\} \exp\left(2\pi i \frac{kx}{2^L}\right) |k\rangle. \quad (5.27)$$

The term in curly brackets is just the same geometric series we saw in Eq. (5.5); and as before, we can evaluate it using the usual formula for a geometric series, and then make the approximation for $q \gg 1$:

$$\left\{ \sum_{p=0}^{q-1} \left[\exp\left(2\pi i \frac{kr}{2^L}\right)\right]^p \right\} = \frac{1 - \exp(2\pi i krq/2^L)}{1 - \exp(2\pi i kr/2^L)}$$

$$\approx q \sum_{s=0}^{r-1} \delta_{k,s2^L/r}. \quad (5.28)$$

Thus, the state of the circuit after applying the quantum Fourier transform is

$$|\varphi_3\rangle = \hat{U}_{\text{QFT}} \otimes \hat{I}^{\otimes M} |\varphi_2\rangle$$

$$= \sqrt{\frac{qr}{2^L}} \sum_{s=0}^{r-1} |s2^L/r\rangle \otimes |\phi_s\rangle + |\varphi_3^{\text{end}}\rangle, \quad (5.29)$$

where $|\phi_s\rangle = \sum_{x=0}^{r-1} \exp(2\pi i sx/r)|f(x)\rangle$, $(s = 0, 1, \ldots, r-1)$ are the orthonormal function register states and $|\varphi_3^{\text{end}}\rangle$ is the contribution to the state from the remnant $|\varphi_2^{\text{end}}\rangle$ (and which should have a negligible effect if q is large enough). The normalization constant $\sqrt{qr/2^L}$ should be approximately equal to 1.

The final step is to make a measurement in the computational basis of the argument register. The possible outcomes are $k_{\text{meas}} = 0, 2^L/r, 2(2^L/r), \ldots, (r-1)(2^L/r)$, each with equal probability, allowing one to deduce r with a reasonable probability of success. There is a small probability of getting some other answers (due to the presence of $|\varphi_3^{\text{end}}\rangle$ in the output state), but we can be comforted by the fact that, once we have an answer for r, it is pretty easy to check.

5.2.3   Continued Fractions Give Us the Answer

As we have seen, with high probability the measurement result will be $k_{\text{meas}} = s(2^L/r)$, where s is some unknown integer, and r is the unknown period; and because r is not necessarily a factor of 2^L this will be a rational number in general. Taking the result, k_{meas}, which is an L-bit integer, we can calculate $k_{\text{meas}}/2^L$, which should give us s/r (and which will be less than 1).

If the answer was something like 0.3333, then it is pretty easy to make a guess at the value of $r = 3$; and a few other simple denominators like 2, 4, or 5 should be similarly easy to spot. But nature in general does not yield her secrets so easily: what happens if we get a number like 0.71429? Not so easy to guess r in that case. We also have to cope with the fact that $k_{\text{meas}}/2^L$ might have an error, due to the fact that, in real life, we cannot hope to get lucky and have a period that is a factor of the length of the argument register.

There is, however, a cute method of extracting a fraction from a decimal fraction (and not just by multiplying it by a power of 10). First, one separates the number into an integer part and a fractional part (trivial for numbers less than 1):

$$0.71429 = 0 + 0.71429.$$

Next, take the reciprocal of the fractional part: $0.71429 = 1/1.39999$, so we get

$$0.71429 = 0 + \frac{1}{1.39999}.$$

Now we repeat the procedure – separating the integer and the fractional part, then taking the reciprocal of the fractional part – for the decimal fraction in the denominator:

$$0.71428 = 0 + \frac{1}{1 + \frac{1}{2.50005}}.$$

And again:

$$0.71428 = 0 + \frac{1}{1 + \frac{1}{2 + \frac{1}{2}}},$$

where we have rounded up 1.99979 to 2. Since the last denominator is an integer, the procedure comes to an end. This is called the *continued fraction* expression for 0.71428 Finally, we use basic addition of fractions to find a rational number approximating 0.71428:

$$\begin{aligned}
0.71428 &= 0 + \frac{1}{1 + \frac{1}{2 + \frac{1}{2}}} \\
&= 0 + \frac{1}{1 + \frac{1}{\frac{5}{2}}} \\
&= 0 + \frac{1}{1 + \frac{2}{5}} \\
&= 0 + \frac{5}{7} \\
&= \frac{5}{7}.
\end{aligned} \tag{5.30}$$

This final fraction derived from the continued fraction expression is called the *convergence*; terminating the continued fraction after n-steps and calculating the fraction gives the n-th convergence. There is a theorem from the theory of continued fractions which states that if $x \approx s/r$ then s/r will be one of the convergences that appears in the partial fraction expansion of x (more precisely, we require $|x - (s/r)| < 1/2r^2$ for this theorem to hold). The bottom line is that the classical post-processing of the result using continued fractions should be very helpful in extracting the period r. In the example given, we find a pretty good guess for r is 7.

This procedure may be used with irrational numbers as well. As n increases, the n-th convergence will give a rational fraction that is a better and better approximation for the original irrational number.

Exercises 5.2.4

1. Suppose we encoded the values of the discretely sampled function f_ℓ as probability amplitudes of a quantum state, i.e., $|f\rangle = \sum_{\ell=0}^{2^L-1} f_\ell |\ell\rangle$. Show that

$$\hat{U}_{\text{QFT}} |f\rangle = \sum_{k=0}^{2^L-1} \tilde{f}_k |k\rangle = |\tilde{f}\rangle,$$

   where $\tilde{f}_k$ is the discrete Fourier transform of f_ℓ.

2. Prove that the quantum Fourier transform on L qubits is unitary.

3. Calculate the concurrence of the state on the right of Eq. (5.10).

4. Below (QC 5.5) is the circuit for the quantum Fourier transform on $L = 3$ qubits.

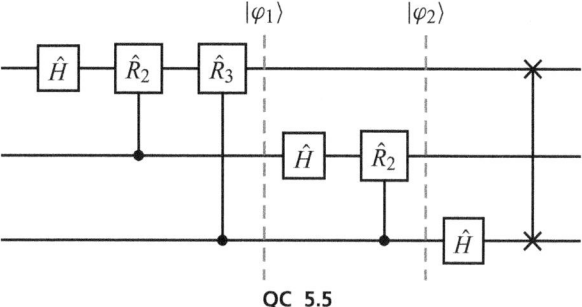

QC 5.5

   (a) Work out the states $|\varphi_1\rangle$ and $|\varphi_2\rangle$ at the slices shown, and confirm that this gives the expected final output state.

   (b) What would the output state look like if the Hadamards were all performed just before the final swap gate, i.e., QC 5.6?

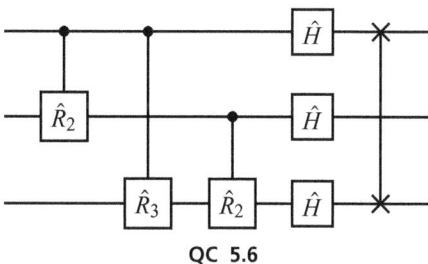

QC 5.6

   (c) Suppose that the input state of the three qubits is

$$|\varphi_{\text{in}}\rangle = \frac{1}{2} \left(|000\rangle + |010\rangle + |100\rangle + |110\rangle \right).$$

   i. Give explicit expressions for the states $|\varphi_1\rangle$, $|\varphi_2\rangle$ and $|\varphi_{\text{out}}\rangle$.

   ii. What are the possible outcomes of a measurement on the output state? Comment on their relation to the periodicity of the input state.

5. Use the continued fraction method to find fractions that approximate the following numbers: (a) 1.85714; (b) 0.73913; (c) $\sqrt{2}$.

6. Calculate the continued fraction expansion for $\pi \approx 3.14159$. Show that the first four convergences are 3, 22/7, 333/106, and 355/113.

7. If one suspects that the period of a (periodic) function is T, how many qubits are needed in order to find T?

8. Given a three-bit integer x and the function $f(x) = x/2$, let $x \in \{0, 2, 4\}$, write down all the possible ket representations for the evaluation of $f(x)$ at all the possible values of x.

5.3   RSA Encryption

🜨 **Mathematics Tips 5.2**

Useful Results from Number Theory

- *Number theory* is the study of the properties of numbers and their arithmetic relations.
- A *factor* of a number a is another number, b, for which a/b is an integer; thus, 15 has factors 3 and 5.
- A natural number is *prime* if it has no factors, other than the trivial ones of 1 and the number itself; the first prime numbers are 2, 3, 5, 7, 11, 13, 17, 19 and so on.
- Two numbers are *coprime*, also called *relatively prime*, if they have no factors in common, other than 1; thus 15 and 8 are coprime (even though neither is prime).
- The *greatest common divisor* (GCD), also called the *highest common multiple* (HCM), of two numbers a and b, denoted $\text{GCD}(a, b)$, is the largest integer that is a factor of both; thus $\text{GCD}(18, 24) = 6$. If the two numbers are relatively prime then $\text{GCD}(a, b) = 1$.
- *Modulo notation*. Let a and b be integers and m be a natural number greater than 1. If $(a - b)$ is divisible by m; i.e., $(a - b)/m$ is an integer, then we can write

$$a \equiv b \quad (\text{mod } m), \tag{5.31}$$

which is read: "*a is congruent to b modulo m*". For example, $3 \equiv 29$ (mod 2) and $11 \equiv 6$ (mod 5).

- *Euler's totient function* $\phi(m)$, where m is a positive integer, is defined to be the number of elements in the set $\{1, 2, 3, \ldots, m - 1\}$ that are relatively prime to m. For example, since 8 is relatively prime to 1, 3, 5, and 7, $\phi(8) = 4$. If p is prime, $\phi(p) = p - 1$. Further, it can be shown that if p and q are prime numbers, then the Euler's totient function of their product is $\phi(pq) = (p - 1)(q - 1)$.
- *Euler's theorem* states that if a and m (both natural numbers, with $m > 1$) are relatively prime; i.e., $\text{GCD}(a, m) = 1$, then

$$a^{\phi(m)} \equiv 1 \quad (\text{mod } m). \tag{5.32}$$

A corollary (which is a mathematician's word for "implication") of this theorem is that there exists a number b such that

$$ab \equiv 1 \quad (\text{mod } m). \tag{5.33}$$

- *Useful lemma*: if $N = pq$, where p and q are both prime and $p \neq q$, and k is an integer, then $\forall a$,

$$a(a^{k\phi(N)}) \equiv a \quad (\text{mod } N), \tag{5.34}$$

where $\forall$ stands for "*for all*".

A widely used encryption method today is the RSA (or Rivest–Shamir–Adleman) encryption system, so named after the first scientists to publish a paper on the method. Its strength is based on the near impossibility of factoring very large numbers into their prime factors. It works as follows. Suppose Salwa wants to send a message to be received by Rhys.

1. First, Rhys, the receiver, chooses two large prime numbers, p and q ($p \neq q$). No one, not even the sender, gets to know what these numbers are except Rhys.
2. Rhys then calculates the product of the two primes, $N = pq$, and the Euler function $\phi(N) = (p-1)(q-1)$.
3. Next he chooses an *encryptor* number, e, which is relatively prime to $\phi(N)$; i.e., $GCD(\phi(N), e) = 1$. He makes e public knowledge.
4. Rhys also needs to find the *decryptor* d, defined as follows: We know from (5.33) that, since $GCD(e, \phi(N)) = 1$, there exists a positive integer d, which can be found using the Euclidean algorithm for finding the GCD (or any other appropriate method), such that

$$ed \equiv 1 \pmod{\phi(N)}. \tag{5.35}$$

In other words, $(ed - 1)/k = \phi(N)$ for some positive integer k, which gives

$$ed = k\phi(N) + 1. \tag{5.36}$$

Rhys never tells (or needs to tell) anyone what d is. This makes the RSA method of encryption secure in the sense that there is no chance for anyone to intercept d. If the latter had to be sent to (or told to) someone, then there is always the chance of someone eavesdropping or copying this information without Rhys knowing. Thankfully, this is not the case here.

5. Both N and e can be sent to Salwa by any means. It does not matter if anyone else finds out the values of these numbers: provided p and q are not known (and if N is sufficiently large, they will be hard to figure out); knowledge of N and e will not help in decoding the message.
6. Salwa's message for Rhys could be a letter or some other symbol, represented by the number M. We have seen in Chapter 1 how, for example, a phrase can be re-written as a binary number. In this case we assume that $M < N$. (If Salwa wants to send Rhys a very long message, so that $M > N$, she would have to break it into smaller parts, each less than N, and send them each separately.)
7. To encode the massage, Salwa simply finds the number R (the remainder) such that

$$M^e \equiv R \pmod{N}, \tag{5.37}$$

where R is between 0 and $N - 1$. One cannot reverse this calculation without knowledge of p and q.
8. Salwa now sends R to Rhys by any means she likes: no-one will be able to decipher it except Rhys.
9. Since $GCD(M, N) = 1$, we know from (5.32) that

$$M^{\phi(N)} \equiv 1 \pmod{N}. \tag{5.38}$$

From (5.37), $M^e \equiv R \pmod{N}$ implies that

$$\begin{aligned} R^d &\equiv (M^e)^d \pmod{N} \\ &\equiv M^{ed} \pmod{N}. \end{aligned} \tag{5.39}$$

Using (5.36), the previous equation becomes

$$\begin{aligned} R^d &\equiv M^{k\phi(N)+1} \pmod{N} \\ &\equiv MM^{k\phi(N)} \pmod{N}. \end{aligned} \tag{5.40}$$

Then, using (5.34), we finally get

$$R^d \equiv M \pmod{N}. \tag{5.41}$$

Therefore, to recover the message M, the second and final step that Rhys has to take is to find the remainder that R^d leaves when divided by N. The answer to that will give him M.

We will now demonstrate how RSA encryption that we just discussed works in an example, which is chosen to be simple enough to understand how encryption works. However, keep in mind that it is too simple to be secure, since the prime numbers chosen are not big enough for the encryption to be secure; they can be guessed easily. Let $p = 2$ and $q = 5$. This means that $N = pq = 10$ and $\phi(N) = (p-1)(q-1) = 4$. Rhys needs to find e such that $(\phi(N), e) = 1$; i.e., GCD$(4, e) = 1$. He chooses $e = 3$. Recall that it is only $N = 10$ and $e = 3$ that are to be made public. He will also need d such that $ed \equiv 1 \pmod{\phi(N)}$; i.e., $3d \equiv 1 \pmod 4$. He finds $d = 3$.

Salwa wants to send Rhys the message $M = 7$. She knows that $N = 10$ and $e = 3$ and needs to compute the remainder R between 0 and $N - 1 = 9$ such that

$$\begin{aligned} M^e &\equiv R \pmod{N}, \\ 7^3 &\equiv R \pmod{10}. \end{aligned} \tag{5.42}$$

Since $7 \equiv -3 \pmod{10}$, Salwa knows that

$$7^3 \equiv (-3)^3 \pmod{10}. \tag{5.43}$$

Equation (5.43) tells Salwa that she needs to work on finding the congruent relations of multiples of -3. She finds that

$$(-3)^2 \equiv -1 \pmod{10}, \tag{5.44}$$

from which she obtains

$$7^2 \equiv (-3)^2 \equiv -1 \pmod{10}. \tag{5.45}$$

Finally, Salwa combines (5.43) and (5.45) to obtain

$$7^3 \equiv 3 \pmod{10}, \tag{5.46}$$

which gives her $R = 3$. She sends this information to Rhys.

To decrypt the message, Rhys solves the following for M:

$$\begin{aligned} R^d &\equiv M \pmod{N}, \\ 3^3 &\equiv M \pmod{10}. \end{aligned} \tag{5.47}$$

This is straightforward: Since $3^3 = 27$, he needs to subtract 7 from it to make the total divisible by 10. Therefore, setting $M = 7$ achieves that and reveals to him the message that Salwa sent.

 Interesting Facts 5.3

Who Invented RSA Encryption? The RSA encryption we discussed in this section is named after Rivest, Shamir, and Adleman, the three scientists who first published it in 1977; however, they were not the first to discover it. Clifford Cocks, a British mathematician who worked for the secret Government Communications Headquarters (GCHQ) in the UK, invented an equivalent algorithm four years earlier in 1973. However, since GCHQ clings to the notion that, when you have a secret method of communications, one of the last things you should do is tell anyone about it, he was not allowed to publish it, nor was his achievement even made public until 1997!

The Euclidean Algorithm

The Euclidean algorithm is a fast and easy way to find the *greatest common divisor* of two integers. Given two positive integers a and b, their *greatest common divisor*, $GCD(a, b)$, is the largest positive integer that divides both. For example, if $a = 9$ and $b = 3$, then $GCD(9, 3) = 3$. If $a = 16$ and $b = 12$, then $GCD(16, 12) = 4$.

One way to find $GCD(a, b)$ is to express each of a and b as a product of their prime factors, with the repeating prime numbers raised to the power of the number of times they appear. For example, $16 = 2^4$ and $12 = 2^2 \cdot 3$. What they both have in common is $2^2 = 4$. Therefore, $GCD(16, 12) = 4$. More generally, if the prime factorization of a is given by $a = p_1^{\alpha_1} p_2^{\alpha_2} \cdots p_n^{\alpha_n}$ and that of b is $b = p_1^{\beta_1} p_2^{\beta_2} \cdots p_n^{\beta_n}$, then $GCD(a, b) = p_1^{\min(\alpha_1, \beta_1)} p_2^{\min(\alpha_2, \beta_2)} \cdots p_n^{\min(\alpha_n, \beta_n)}$, where $\min(x, y)$ is the smaller number in the set $\{x, y\}$. This technique, however, can be cumbersome and impractical because, in general, finding the prime factors of numbers is a difficult problem, especially when the numbers are large. The Euclidean algorithm for finding the GCD can be more practical, so we describe it next.

Back to our positive integers a and b, where we take $a > b$. Let us write a in terms of b as follows:

$$a = bq_0 + r_0, \tag{5.48}$$

for some q_0, where $0 \leq r_0 < b$. If $r_0 = 0$, then we are done with our quest, and we know that $GCD(a, b) = b$.

However, if $r_0 > 0$, then the next step is to write b in terms of r_0 as follows:

$$b = r_0 q_1 + r_1, \tag{5.49}$$

where $0 \leq r_1 < r_0$. If $r_1 = 0$, then we know we are done with our search and that $GCD(a, b) = r_0$. However, if $r_1 > 0$, then we have to continue and, this time, write r_0 in terms of r_1:

$$r_0 = r_1 q_2 + r_2, \tag{5.50}$$

where $0 \leq r_2 < r_1$.

Again, if $r_2 = 0$, then $GCD(a, b) = r_1$. Otherwise, if $r_2 > 0$, write r_1 in terms of r_2:

$$r_1 = r_2 q_3 + r_3, \tag{5.51}$$

where $0 \leq r_3 < r_2$.

One has to keep going in this fashion until a remainder of zero; i.e., $r_i = 0$ (for some i), is obtained, where $GCD(a, b)$ is the last non-zero remainder.

Here is a simple example of $a = 30$ and $b = 8$, where the following steps helps us find their GCD:

1. $30 = 8 \times 3 + 6$,
2. $8 = 6 \times 1 + 2$,
3. $6 = 2 \times 3 + 0$.

Therefore, $GCD(30, 8) = r_1 = 2$.

Exercises 5.3.1

1. Show that:
   (a) $13 \equiv 3 \pmod{5}$,
   (b) $18 \equiv 60 \pmod{2}$,
   (c) $46 \equiv 2 \pmod{11}$.
2. Prove that: If $a \equiv b \pmod{m}$, and $c \equiv d \pmod{m}$, then $ac \equiv bd \pmod{m}$.
3. Using the result in Exercise 2 and *mathematical induction*, prove that: If $a \equiv b \pmod{m}$, then $a^n \equiv b^n \pmod{m}$ for any natural number n.
4. Prove that: If $a \equiv b \pmod{m}$, and $b \equiv c \pmod{m}$, then $a \equiv c \pmod{m}$.
5. Rhianwen chooses the two prime numbers $p = 3$ and $q = 5$, which she keeps secret. However, to the world, she announces $N = 15$ (the product of the chosen prime numbers) as well as $e = 7$. Sulaiman wants to send Rhianwen the message $M = 10$. To do so, he needs to find the remainder R such that $M^e \equiv R \pmod{N}$ and then send R to Rhianwen. Help Sulaiman find R, then help Rhianwen use this information to recover the original message M.
6. Find $\mathrm{GCD}(a, b)$ for the following:
   (a) $a = 29, b = 9$,
   (b) $a = 38, b = 12$,
   (c) $a = 60, b = 17$.
7. (a) For each case in Exercise 6, write $\mathrm{GCD}(a, b)$ in the form

$$\mathrm{GCD}(a, b) = am + bn, \tag{5.52}$$

   where m, n are integers.
   (b) For each of the expressions in part (a), using algebra, write them in the form $\mathrm{GCD}(a, b) + ar = bs$, where r and s are positive integers.
8. Sulwyn wants to send Rami the message $M = 7$. Rami had chosen his secret prime numbers to be $p = 11$ and $q = 7$. He only tells the world their product, $N = 77$, as well as his choice of $e = 17$. Help Sulwyn calculate R. Then, help Rami choose d and use it along with the value of R to decrypt Sulwyn's message.

5.4   Modular Exponentiation and Shor's Algorithm for Factor Finding

⊕ Mathematics Tips 5.4

Possible Values for the Remainder. If we divide an integer by some natural number n, the remainder will be a number in the set $\{0, 1, 2, \ldots, n - 1\}$; i.e., there are only n possible values for the remainder. Think about all the values of integers that are possible. Even though they are infinite in number, the possible remainders upon division by a given n are not infinite in number; you have a finite number of them, and they will repeat as you keep evaluating them for one integer after another.

In the last section, we introduced modular arithmetic, where the remainder between two numbers is central. For example, when we divide 11 by 2 the remainder is 1, and this can be expressed as

$$11 \equiv 1 \pmod{2}, \tag{5.53}$$

and read as *11 is congruent to 1 modulo 2*. The number 11 is not the only integer that gives a remainder of 1 upon division by the number 2; in fact, all odd numbers do; they are all *congruent to 1 modulo 2*. Suppose we are given the relationship

$$x \equiv R \quad (\text{mod } y), \tag{5.54}$$

where x and y are known. To find R, if $x < y$, set $R = x$. However, if $x \geq y$, then two possible ways to find R are to either keep subtracting y from x until a number less than y is obtained *or* compute the following: $R = x - \lfloor \frac{x}{y} \rfloor y$, where, as we saw in Chapter 4, $\lfloor z \rfloor$ is the largest whole number smaller than or equal to z.

In this section, we will focus on a function f defined to be a remainder with respect to some natural number. When we define it that way, sooner or later the values of the remainder have to repeat themselves, so f ends up being periodic. This period is central to the algorithm used to obtain the prime factorization of the odd number $N = pq$.

The most famous quantum algorithm is the factoring algorithm devised by Peter Shor in 1994. It is in fact an application of the period finding (Quantum Circuit 5.4) we discussed earlier, as we will see in this section.

Let us look at an example to illustrate how the task of finding the prime factors of the number N can be achieved by finding the period of a function. We discuss the case in which the prime factorization consists of two numbers only, so that $N = pq$. Note that in this case, the problem is very simple if N is an even number because immediately one knows that one of p or q is equal to 2, making it very easy to find the other number. That is why it is the odd N case that gives us a more challenging factoring task.

First, we define the **modular exponential function** $f_{N,C}(x)$ as follows:

$$C^x \equiv f_{N,C}(x) \quad (\text{mod } N), \tag{5.55}$$

where N is the number we want to factor, C is a number we choose such that $\text{GCD}(C, N) = 1$ (i.e., they are relatively prime: C and N are coprime), and x represents the set of *whole numbers*. This function is periodic, i.e., $f_{N,C}(x + r) = f_{N,C}(x)$, where r is the period (also called the *order*).

For example, we will take $N = 15$. This means that the options we have for C are any of the numbers in the set $\{2, 4, 7, 8, 11, 13, 14\}$. Let us take $C = 7$. This means that we will have to compute the function $7^x \equiv f_{15,7}(x)$ (mod 15) for several values of x to find its period. We will compute it for $0 \leq x \leq 8$.

For $x = 0$, we compute $f_{15,7}(0)$:

$$7^0 \equiv f_{15,7}(0) \quad (\text{mod } 15) \;\Rightarrow\; 1 \equiv 1 \quad (\text{mod } 15) \;\Rightarrow\; \boxed{f_{15,7}(0) = 1.} \tag{5.56}$$

For $x = 1$, we compute $f_{15,7}(1)$:

$$7^1 \equiv f_{15,7}(1) \quad (\text{mod } 15) \;\Rightarrow\; 7 \equiv 7 \quad (\text{mod } 15) \;\Rightarrow\; \boxed{f_{15,7}(1) = 7.} \tag{5.57}$$

For $x = 2$, we compute $f_{15,7}(2)$:

$$7^2 \equiv f_{15,7}(2) \quad (\text{mod } 15) \;\Rightarrow\; 49 \equiv 4 \quad (\text{mod } 15) \;\Rightarrow\; \boxed{f_{15,7}(2) = 4.} \tag{5.58}$$

For $x = 3$, we compute $f_{15,7}(3)$:

$$7^3 \equiv f_{15,7}(3) \quad (\text{mod } 15) \;\Rightarrow\; 343 \equiv 13 \quad (\text{mod } 15) \;\Rightarrow\; \boxed{f_{15,7}(3) = 13.} \tag{5.59}$$

For $x = 4$, we compute $f_{15,7}(4)$:

$$7^4 \equiv f_{15,7}(4) \pmod{15} \ \Rightarrow \ 2401 \equiv 1 \pmod{15} \ \Rightarrow \ \boxed{f_{15,7}(4) = 1.} \tag{5.60}$$

Therefore, $f_{15,7}(4) = 1$. Similarly, for $x = 5$, $7^5 \equiv f_{15,7}(5) \pmod{15}$ implies that $f_{15,7}(5) = 7$, for $x = 6$, $7^6 \equiv f_{15,7}(6) \pmod{15}$ implies that $f_{15,7}(6) = 4$, for $x = 7$, $7^7 \equiv f_{15,7}(7) \pmod{15}$ implies that $f_{15,7}(7) = 13$, and that for $x = 8$, $7^8 \equiv f_{15,7}(8) \pmod{15}$ implies that $f_{15,7}(8) = 1$.

Notice how the value of the function $f_{15,7}(x)$ for the first nine values of x is given by the list: $1, 7, 4, 13, 1, 7, 4, 13, 1$. This reveals that the period, r, of the function is $r = 4$, which means that $C^{r/2} = 7^2 = 49$. We can use this along with $\mathrm{GCD}(C^{r/2} \pm 1, N)$ to find the prime factors of 15 in this fashion:

$$\mathrm{GCD}(C^{r/2} + 1, N) = \mathrm{GCD}(50, 15) = 5,$$
$$\mathrm{GCD}(C^{r/2} - 1, N) = \mathrm{GCD}(48, 15) = 3. \tag{5.61}$$

This gives us (what we already know): $15 = 5 \times 3$. Note that, in general, $\mathrm{GCD}(C^{r/2} \pm 1, N)$ cannot be used to find the prime factors of N if r, the period, is odd. However, it can work in special cases when C is a perfect square, such as in $4 = 2^2$, for example.

◈ Mathematics Tips 5.5

Why $\mathrm{GCD}(C^{r/2} \pm 1, N)$ Gives Us the Prime Factors of N. After figuring out the period r of $f_{N,C}(x)$, we saw in the main text that we were able to find the prime factors of N, namely p and q. Here we present a proof to justify using $\mathrm{GCD}(C^{r/2} \pm 1, N)$ in our quest. Without loss of generality, we will assume that $p > q$.

Given $C^x \equiv f_{N,C}(x) \pmod{N}$, one has to evaluate it for several values of x, starting with $x = 0$, obtaining the value for $f_{N,C}(x)$ for each x. This is done enough times until one starts to see a pattern and figure out the period r.

In all cases, $f_{N,C}(0) = 1$. This means that one has to keep increasing x and evaluating $f_{N,C}(x)$ until again the same 1 is obtained, and $f_{N,C}(x_f) = 1$ for $x_f > 0$. The period r in this case will be $r = x_f$.

For both cases ($x = 0$ and $x = x_f$), we know that

$$C^x \equiv 1 \pmod{N}. \tag{5.62}$$

When $x = x_f$, (5.62) gives us

$$C^{x_f} - 1 = N \tag{5.63}$$

or

$$C^r - 1 = N. \tag{5.64}$$

Applying the algebraic relation $a^2 - b^2 = (a + b)(a - b)$ to (5.64) gives

$$(C^{r/2} + 1)(C^{r/2} - 1) = N. \tag{5.65}$$

The *fundamental theorem of arithmetic* tells us that the prime factorization of every integer greater than 1 is unique (except perhaps for the order of the factors); i.e., you cannot have more than one prime factorization for the same integer. Combining this, as well as our assumption $p > q$, what (5.65) tells us is that

$$\text{GCD}(C^{r/2} + 1, N) = (C^{r/2} + 1) = p,$$
$$\text{GCD}(C^{r/2} - 1, N) = (C^{r/2} - 1) = q. \tag{5.66}$$

∎

Thus, finding the period of the modular exponential function $f_{N,C}(x)$ is a systematic way to find the factors of N. First, choose a number C in the range 0 to $N - 2$ ($N - 1$ does not yield useful factors); then check that C and N are coprimes using Euclid's algorithm (if they are not coprimes, you got lucky because Euclid's algorithm immediately yields a factor of N); next you have to evaluate the modular function $f_{N,C}(x)$ multiple times in order to figure out the period r; and finally, using r you can generate the factors from the formulae in Eq. (5.61).

The period r depends on the choice of the cofactor C. Sometimes you might get lucky and happen on a cofactor for which r is a small number like 2 or 4 (it has to be an even number, otherwise the factoring step will involve $\sqrt{C}$ which will not be an integer unless C is a perfect square); sometimes you might end up with a really large period requiring a large number of evaluations; for an L-bit number N, this entails a number of evaluations of $f_{N,C}(x)$ which scales exponentially with L. This is not deemed the best known classical computational method of finding the factors of N: what is thought to be the best is an algorithm called the **number field sieve**, which also requires a number of computational steps that scales with an exponential of L.

How can a quantum computer help? If we can construct a quantum oracle $\hat{U}_f$ to evaluate the function $f_{N,C}(x)$ then the quantum factor-finding method described in Section 5.2 will allow the period r to be determined *with a single call to the oracle*. In other words, an exponential number of function evaluations are replaced by one. Furthermore, it can be shown that such oracles can be constructed with a number of CNOT gates which increases as a multiple of L^3. In addition, the quantum Fourier transform needs $(2L + 1)L/2$ CNOT gates; in other words, the circuit will be an exponential speed-up over the classical algorithms, and thus places the security of RSA encryption in question.

Exercises 5.4.1

1. Prove that the function $f_{N,C}(x)$ defined by Eq. (5.55) is periodic. *Hint: If $f_{N,C}(r) = 1$, show that $f_{N,C}(r + x) = f_{N,C}(x)$.*
2. Repeat the factoring algorithm, using the period-finding method discussed in this section, to factor $N = 15$ using:
   (a) $C = 4$,
   (b) $C = 11$,
   (c) $C = 14$.
3. For any natural number N, is $N - 1$ always coprime?
4. If you choose the coprime $C = N - 1$, what is the period r of the modular exponential function $f_{N,C}(x)$? Do you always get the trivial factors of N and 1?
5. Explain why, when $x = x_f$, Eq. (5.62) implies the equality relation $C^{x_f} - 1 = N$; i.e., why is the relation not $C^{x_f} - 1 = kN$ for some integer k such that $|k| > 0$ instead?
6. Devise a simple quantum circuit to execute the modular exponential oracle for the case $N = 15$ and $C = 7$.

5.5 CONCEPT CHECK AND EXERCISES

Concept Check

- The Fourier transform allows one to find the period of a function, but requires evaluation of the function for multiple arguments.
- The quantum Fourier transform is a unitary operation that allows one to deduce the period of a function with a single call on the quantum oracle for the function.
- The strength of the RSA encryption method relies on the difficulty in factoring very large numbers.
- Factoring integers can be reduced to evaluating the period of a modular exponential function, which is the basis for Shor's algorithm.

Chapter Exercises

1. Given the prime numbers p and q, show that the Euler function of their product is given by

$$\phi(pq) = (p-1)(q-1). \tag{5.67}$$

2. In the decimal number system, the ASCII code for the letters in the word *Quantum* are given as follows: $Q = 81$, $u = 117$, $a = 97$, $n = 110$, $t = 116$, and $m = 109$. You have agreed beforehand with your friend Bilal, who lives in a far away land, that if you are ever in dire need, instead of sending him a message with the word *Help*, you will send him one with the word *Quantum* (just in case someone had succeeded in building a powerful quantum computer that can break the RSA encryption scheme and understand what you are trying to say). Put together an RSA encryption and decryption algorithm for you and Bilal to pursue in case of an emergency. What are the details of the information he needs to give you, what are you going to send him, and how will he decode your message?

3. In this question we are going to go through all the mathematical steps needed to find the factors of the number $N = 221$. You should carry out the numerical steps on a computer using your preferred coding language.
   (a) How may bits are needed to store N? Express N as a binary number.
   (b) Let us choose $C = 11$: is this coprime to N?
   (c) Create an array of values of the modular exponential function $C_k \equiv f^k \pmod{N}$ for $k = 0, 1, \ldots, 2^{10} - 1$; plot this for the range $k = 0, 1, \ldots, 127$. Is the function periodic?
   (d) Calculate the discrete Fourier transform of f_k, $\tilde{f}_\ell$. Plot $|\tilde{f}_\ell|$ for $\ell = 0, 1, \ldots, 127$. What are the values of ℓ for which $\tilde{f}_\ell$ has maxima?
   (e) Let the smallest value of $\ell > 0$ for which $|\tilde{f}_\ell|$ is a maximum be ℓ_0. We know that $\ell_0 \approx 2^{10}/r$. Use the method of convergence of continued fractions to find r.
   (f) Calculate $\text{GCD}(C^{r/2} - 1, N)$ and $\text{GCD}(C^{r/2} + 1, N)$. Are they the factors of N?

4. Suppose we wanted to find the factors of the number $N = 247$. Using $C = 9$, find the order of the modular exponential function, and use the order to find the factors. When does Shor's algorithm work for r being an odd number?

5. Consider QC 5.7. It shows the complete factoring circuit for the case $N = 15$ and $C = 4$. The first box is the $\hat{H}^{\otimes 3}$ operation on the three-qubit argument register, while the final box is the three-qubit quantum Fourier transform. The shaded section shows the modular exponential oracle for this simple case.

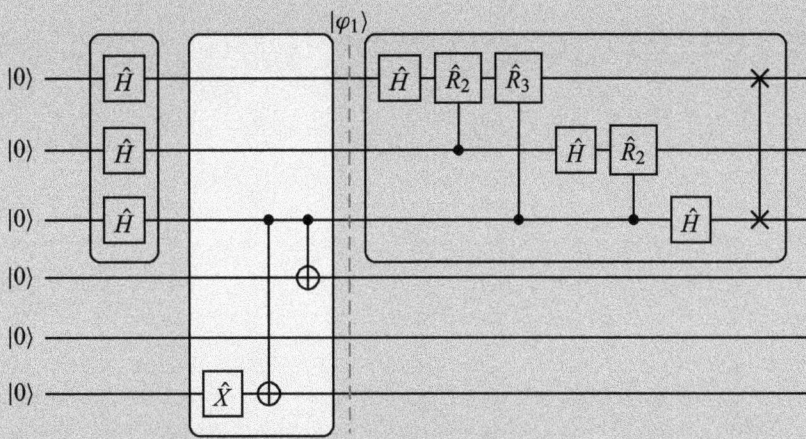

QC 5.7

(a) Calculate the modular exponential $C^k \equiv f_{N,C}(k) \pmod{N}$ for $k = 0, 1, \ldots, 7$.

(b) Find the state $|\varphi_1\rangle$ based on the above circuit, and compare it with what you would expect given the result of part (a).

(c) Find the six-qubit output state. Is the state entangled?

(d) Determine the possible outcomes and the associated probabilities of a measurement of the argument register at the end of the algorithm.

(e) Deduce the period r of f_k, and deduce the factors.

6 Realizations of Qubits: General Discussion

Conventional computers have, generally speaking, used the same semiconductor-based transistors as their building blocks for over 65 years. For quantum computers, the situation is much more fluid. In this chapter, we discuss the generic requirements for devices that can be used as qubits in quantum computers; in the following chapter we will see how some of the key candidate technologies fulfill these requirements.

Learning Objectives

- Outline the DiVincenzo criteria for building a practical quantum computer.
- Understand that the kind of quantum operations that need to be implemented in order to keep the qubit state constant depend on the natural evolution of the qubit, and be able to design them accordingly when necessary.
- Learn about the general unitary operation that changes the state of a qubit, how to select the right parameters to achieve the desired implementation, and understand the implications on the purity of the quantum state due to the technical difficulties encountered upon implementation.

 Interesting Facts 6.1

The first commercial transistor-based computer was the IBM 608 of 1957: it could perform 4500 additions a second (i.e., a processor speed of 4.5 kHz); data was read in and out using punch cards, and it sold for $83,210 (equivalent to $934,573 in 2025); today, a laptop operating with a processor speed of 2.4 GHz (more than a million times faster), but using the same basic technology of semiconductor-based electronics, sells for as little as $200.

6.1 Basic Requirements: The DiVincenzo Criteria

Like all good recipe books, we start with a list of ingredients. This list is called the DiVincenzo criteria, named for the scientist who first wrote them down. To build a practical quantum computer you will need:

- **A lot of qubits.** As we have seen in our discussion of algorithms, a quantum computer truly comes into its own when tackling large, complex mathematical problems: the more qubits, the more powerful will be the device. Because the implementation of quantum algorithms requires you to perform unitary operations on any individual qubit, each qubit has either its own distinct physical location far enough from the other qubits that one can interact with it in isolation, or its own distinctive resonant frequency; i.e., the frequency at which it responds most strongly, so that one can chose to interact with that qubit leaving the others untouched. This precludes quantum systems such as individual atoms in a gaseous or liquid state, photons in a diffuse light field, or electron spins in the conduction band of a semiconductor.

- **Initialization.** Quantum algorithms require you to start in some state such as $|0\rangle \otimes |0\rangle \otimes \cdots \otimes |0\rangle$. Physically, this means your qubits have to be very cold – approaching absolute zero temperature – or, in the case of photons, you have to be able to create pulses of light that contain one, and only one photon. Both of these capabilities are at the very sharp edge of contemporary experimental physics.

- **Reliability.** Once you create your qubits in some complicated state required by an algorithm, you want to be able to rely on it remaining in that state while you proceed with the other tasks in the algorithm. This is not as easy at it sounds: magnetic fields can drift, atomic levels undergo **spontaneous decay**, all manner of random influences in the qubits' environment can lead to the qubit states becoming mixed. Generically this is called *decoherence*, and it is so pervasive that quantum phenomena like entanglement are not seen in everyday life. Fortunately, there are ways to cope with a certain degree of decoherence and still perform quantum computation, but they do require that the decoherence not be too strong.

- **Quantum logic gates.** To perform quantum algorithms you must be able to execute quantum logic gates – unitary operations acting on the individual qubits, plus an entangling two-qubit logic gate equivalent to the controlled-Z gate we discussed in Chapter 3. It is the latter which is usually the most difficult physical operation to perform in a reliable fashion, so much so that we will discuss how to do it in a separate chapter.

- **Readout.** There is not much point in doing the very challenging quantum experiments implied by the first four items unless you can, at the end of the algorithm, read out the result. When measuring any quantum system, the key figure of merit is the *quantum efficiency* of the detector, which is simply the probability that you get the right answer. Quantum systems are notoriously shy: a lot of the time when you try to find out what state they are in, you just don't get any answer; then again, sometimes you get the wrong answer.

None of the items are immutable: serious challenges to one or more of them have been proposed over the years, and have met with varying degrees of success, as we shall see in Chapter 10. These criteria do, however, form a good place to start the discussion.

Qubits are physical systems. The two distinct configurations of the qubit – the "Heads" and "Tails" of our quantum coin back in Chapter 1, or more abstractly the $|0\rangle$ and $|1\rangle$ we used in our discussion of quantum algorithms in Chapters 2 to 5 – represent specific configurations of that physical system. The general requirements of quantum computing, listed above, place various constraints on the type of physical system we can use.

Temperature

If our qubits are made of matter (as opposed to light, as we will see), the two quantum states we select for the logical $|0\rangle$ and $|1\rangle$ will usually be configurations of the system that have different energies, E_0 and E_1, respectively. For example, a spin will have different energies depending on its alignment in a magnetic field; or an electron a different energy depending on the configuration of its bound state, and so on. We define energy difference as follows:

$$\Delta E = E_1 - E_0. \tag{6.1}$$

This quantity is important for the initialization of our qubit state. When we turn on our quantum computer, it is reasonable to expect that each qubit is initially in thermal equilibrium with its surroundings. One of the features of thermal equilibrium states is that the ratio of probabilities of being in state 1 and state 0, p_1/p_0, is given by $\exp[-\Delta E/K_B T]$, where T is the absolute temperature, measured in kelvins (so that 0 kelvin corresponds to -273.15 degrees Celsius) and K_B is Boltzmann's constant (whose value is approximately 1.38×10^{-23} joules per kelvin). To initialize the qubit in state $|0\rangle$, this ratio must be very small, so that the temperature $T \ll \Delta E/K_B$. In other words, qubits have to be *cold* or the energy difference ΔE has to be large. How we attain very low temperatures depends on the technology in question (e.g., dilution refrigerators for solid-state devices, laser cooling for atoms, etc.), but the larger the energy difference ΔE, the easier the qubit is to cool. Nor do photons, for which $\Delta E = 0$, allow for higher temperature operation: the highly sensitive detectors required to measure the photons must be cooled to very low temperatures as well.

Physical Size

We must be able to interact with the qubits in order to perform quantum gates and read out the result. This interaction almost always involves either electric or magnetic fields – none of the other forces of nature (gravitation, weak nuclear force, strong nuclear force) have been seriously proposed as a means to control qubits. An electromagnetic field resonant to the qubit transition will have a frequency $\nu = \Delta E/h$, where h is Planck's constant, and wavelength $\lambda = c/\nu = hc/\Delta E$, where c is the speed of light. As a general rule, it is very difficult to confine an electromagnetic field to a volume whose dimensions are smaller than a wavelength: for example, the area of the smallest attainable focal spot of a camera or other imaging device is roughly λ^2; waveguides and optical fibres whose physical dimensions are smaller than the wavelength will not support propagating waves, and so on.

If we want to control one qubit, leaving the others untouched, we must find some way to single out the qubit in question: it can be spatially separated from other qubits by several wavelengths (which can lead to impractically large devices) or some means of selectively changing the **resonant frequency** of a particular qubit, so that it alone is resonant with an applied field, and all the other qubits will be off-resonance and thus unaffected. The latter approach must greatly increase the problems of device design and fabrication since one must find enough space for, and then build with very high tolerances, the waveguides/electrodes needed for individual qubit control.

Natural and Fabricated Qubits

Nature provides us with a variety of two-level quantum systems. Photons have two linearly independent polarization directions; electrons and protons have an inherent **angular momentum** called spin, which can have two linearly independent orientations in a magnetic field; the same is true of a number of other types of atomic nuclei, such as carbon-13 (^{13}C) or phosphorus-31 (^{31}P). In addition to two-level systems, multi-level systems in which the levels are *not* equally spaced in energy can make very good qubits: two levels can be isolated spectroscopically by only applying control forces exactly tuned to resonance frequency $\Delta E/h$; any other transitions will be off-resonance and so the other levels will never be populated: examples of such systems are atoms and ions. All of these "natural" qubits have one distinct advantage: every calcium-40 atom in the universe is identical in *all* important physical attributes to every other calcium-40 in the universe. Natural qubits are inherently uniform, making the task of calibration of multi-qubit systems far easier.

In contrast, we can also use fabricated quantum mechanical systems to be our qubits. Remember: fundamentally, everything obeys the laws of quantum mechanics. Thus, large-scale (or *macroscopic*) phenomena, such as the electric current flowing through the wires in the kitchen to power your dishwasher can be described using the laws of quantum mechanics. Most of the time the system is at too high a temperature and there is so much dissipation present that any quantum phenomena are not observable, and simpler principles like electric current and Ohm's law are sufficient to describe what is going on. However, if we choose to make the circuit very cold, and remove dissipation, then quantum phenomena can be observed in the right circumstances, and, as we will see, can be used to make qubits. Such fabricated qubits have the great advantage of versatility – we can design qubits exactly the way we want them – but they also pose the problem that every qubit is going to be slightly different because of the flaws that will inevitably arise in the fabrication process.

Do Quantum Computers Have to Be Solid State Devices?

Integrated circuits, which were invented in the late 1950s, allow all the components of an electrical circuit, including resistors, capacitors, semiconductor diodes and transistors, to be incorporated onto a single piece of silicon. Since then the term "solid state" has become synonymous with cheap, robust, and reliable electronics. It is natural, therefore, to assume the next generation of high-technology devices must also be solid state. We should be vigilant against any prejudice however: the more complex the environment, the more prone a qubit will be to some form of decoherence, and there is a big advantage to keeping our qubits in a vacuum in which nothing can interact with them except by design. Such devices can be made robust: for example, modern atomic clocks, which are certainly not solid state devices, have been made into automatic modules for mounting on laboratory racks with other electronic equipment.

Processing Speed

It would be pointless building a quantum computer if it is going to be so slow to operate that the computation problem at hand can be solved using conventional computers. Taking the Shor's algorithm example, conventional computers are not *so* bad at factoring numbers: for example, in 2020 an 829-bit number was factored using a conventional computer algorithm called the number field sieve implemented on a large cluster of conventional computers (a total of 2700 core-years was required, each core with a processing speed of 2.1 GHz). If we make optimistic assumptions about error mitigation, Shor's algorithm, including all modular exponentiation steps, requires about $32L^3$ operations to find the factors of an L-bit number (although this estimate does not take into account operations that can be performed in parallel). For $L = 829$, this gives 1.8×10^{10} operations. To match the performance of conventional computers by factoring this number in a few months would require each operation to be performed in less than 1 microsecond (10^{-3} seconds); if a quantum computer were to be truly a disruptive technological leap forward, so they can do the task in, say, 5 minutes, we would need each quantum gate to be performed in about 10 nanoseconds (10^{-8} seconds). Processing speeds in quantum computers are limited by the inherent strength of the interaction of the qubit with its control field; and by other possible resonances of our **physical qubits** which might become excited if strong control fields and short pulses are used.

Exercises 6.1.1

1. Suppose we have a quantum computer made of identical qubits (the exact technology is not important) each with an energy difference ΔE. They will be controlled by oscillating electric fields at frequencies resonant with the qubit transition. For ease of access for the external control electronics and for readout, as well as to help with cooling, a planar chip design is adopted. A total of 20 million qubits (20 megaqubits) are to be fabricated on the chip, separated from each other by 10 wavelengths (closer than this might lead to excessive cross-talk between the qubits). What are the overall dimensions of the chip if (a) $\Delta E = 3$ electronvolts (eV) (typical of atomic and optical qubits); (b) $\Delta E = 3 \times 10^{-5}$ eV (typical of electron spin resonance and superconducting qubits); and (c) $\Delta E = 3 \times 10^{-8}$ eV (typical of nuclear-spin based qubits)? Note: An *electronvolt* is a unit of *energy* equal to 1.602×10^{-19} joules; it is commonly used in solid-state and atomic physics since it avoids writing lots of powers of 10.

2. The second DiVincenzo criterion states we need qubits prepared in a pure state (which we can assume is $|0\rangle$). Suppose we can use error correction to cope with qubits that have a chance $p_{\text{crit}} \ll 1$ of being initially excited in state $|1\rangle$ (neglect the possibility of states other than $|0\rangle$ or $|1\rangle$ of the qubit system being excited). Show that the temperature required is $T \ll T_{\text{crit}}$ where

$$T_{\text{crit}} = -\frac{\Delta E}{K_B \ln p_{\text{crit}}}.$$

(The minus sign is there because $p_{\text{crit}} \leq 1$ and so $\ln p_{\text{crit}}$ will be negative.) Supposing $p_{\text{crit}} = 0.01$, find the value of T_{crit} for the three values of ΔE given in Exercise 1.

3. Estimate the maximum number of qubits of the type in Exercise 2 that one can have and still have a 50% probability that they are all in state $|0\rangle$.

6.2   Single Qubits and Time Evolution

Suppose we have a qubit whose two quantum logic levels $|0\rangle$ and $|1\rangle$ are energy eigenstates, with energies E_0 and E_1, respectively, as we discussed above. Let's consider the dynamics of a single qubit at rest: it is just sitting there, fixed in some location in our quantum computer memory. There are no forces or external influences at play. How is it changing in time? Recall from Section 2.3, the state of the system obeys Schrödinger's equation Eq. (2.85),

$$i\hbar \frac{d}{dt}|\psi(t)\rangle = \hat{\mathcal{H}}|\psi(t)\rangle, \tag{6.2}$$

where the Hamiltonian $\hat{\mathcal{H}}$ is the operator for the total energy, i.e., for the single qubit with two states:

$$\hat{\mathcal{H}} = \begin{bmatrix} E_0 & 0 \\ 0 & E_1 \end{bmatrix}. \tag{6.3}$$

Let us write the state $|\psi(t)\rangle$ in terms of the basis states as $|\psi(t)\rangle = A(t)|0\rangle + B(t)|1\rangle$, where as before A and B are the probability amplitudes. Thus, shifting to the vector representation, Schrödinger's equation becomes

$$i\hbar \frac{d}{dt} \begin{bmatrix} A \\ B \end{bmatrix} = \begin{bmatrix} E_0 & 0 \\ 0 & E_1 \end{bmatrix} \begin{bmatrix} A \\ B \end{bmatrix} = \begin{bmatrix} E_0 A \\ E_1 B \end{bmatrix}. \tag{6.4}$$

Hence,

$$i\hbar \frac{dA}{dt} = E_0 A \quad \Rightarrow \quad A(t) = A(0)\exp(-iE_0 t/\hbar), \tag{6.5}$$

$$i\hbar \frac{dB}{dt} = E_1 B \quad \Rightarrow \quad B(t) = B(0)\exp(-iE_1 t/\hbar) \tag{6.6}$$

and thus the quantum state of the isolated, non-interacting qubit is

$$\begin{aligned} |\varphi(t)\rangle &= A(0)\exp(-iE_0 t/\hbar)|0\rangle + B(0)\exp(-iE_1 t/\hbar)|1\rangle \\ &= \exp(-i(E_0 + E_1)t/2\hbar)\left\{A(0)\exp(i\omega t/2)|0\rangle + B(0)\exp(-i\omega t/2)|1\rangle\right\}, \end{aligned} \tag{6.7}$$

where we have written $\Delta E = E_1 - E_0 = \hbar\omega$. The term $\exp(-i(E_0 + E_1)t/2\hbar)$ represents a global phase change, which is not observable, and which we will neglect: in effect, one can assume that $E_0 + E_1 = 0$.

Writing the complex exponentials in terms of sines and cosines:

$$|\varphi(t)\rangle = A(0)\{\cos(\omega t/2) + i\sin(\omega t/2)\}|0\rangle + B(0)\{\cos(\omega t/2) - i\sin(\omega t/2)\}|1\rangle$$
$$= \{\cos(\omega t/2)\hat{I} + i\sin(\omega t/2)\hat{Z}\}\{A(0)|0\rangle + B(0)|1\rangle\}. \tag{6.8}$$

Let us denote

$$\hat{U}_0(t) = \cos(\omega t/2)\hat{I} + i\sin(\omega t/2)\hat{Z}, \tag{6.9}$$

so we get

$$\boxed{|\varphi(t)\rangle = \hat{U}_0(t)|\varphi(0)\rangle.} \tag{6.10}$$

Thus, instead of a simple, static qubit which just sits there keeping its probability amplitudes nice and safe, the need for the qubit levels to have different energies means that we have a continuously varying time-dynamic qubit. This is a pretty generic problem with just about all qubits. Is it going to change anything vital?

First, the probabilities in the computational basis $\{|0\rangle, |1\rangle\}$ are not affected: from Eqs. (6.5) and (6.6), $|A(t)|^2 = |A(0)|^2$ and $|B(t)|^2 = |B(0)|^2$. Thus, if we don't actually want to do anything to our qubits besides create them in some state then measure them, the time variation due to the energy difference does not have any effect. But this would be an excessively dull algorithm: we need to perform quantum gates on our qubits in order to do anything useful. Let's suppose we have a generic unitary operation $\hat{U}$ acting on a state $|\varphi\rangle$ and we want to know the probabilities of the two outcomes when the transformed state is measured in the computational basis. Using the formula for a generic unitary gate, Eq. (2.60), we find that the state will be

$$|\varphi'\rangle = \hat{U}\hat{U}_0(t)|\varphi\rangle$$
$$= \{aA(0)\exp(i\omega t/2) + bB(0)\exp(-i\omega t/2)\}|0\rangle$$
$$+ \{-b^*A(0)\exp(i\omega t/2) + a^*B(0)\exp(-i\omega t/2)\}|1\rangle \tag{6.11}$$

and thus the probabilities for measurement in the computational basis have become explicitly time-dependent:

$$P_0' = |a|^2|A(0)|^2 + |b|^2|B(0)|^2 + 2|abA(0)B(0)|\cos(\omega t + \Phi), \tag{6.12}$$
$$P_1' = |b|^2|A(0)|^2 + |a|^2|B(0)|^2 - 2|abA(0)B(0)|\cos(\omega t + \Phi), \tag{6.13}$$

where Φ is a phase dependent on the phases of a, b, $A(0)$, and $B(0)$. This does seem more serious: how can single qubit gates be implemented physically if the probabilities of measurement outcomes are continuously changing?

The answer is that the gates themselves have to change to compensate. Instead of performing the operation $\hat{O}$, one needs to perform the operation $\hat{O}(t) = \hat{U}_0\hat{O}\hat{U}_0^\dagger$. In this case, the probability of measuring the outcome $|\xi\rangle$ will be

$$P_\xi(t) = |\langle\xi(0)|\hat{U}_0^\dagger\hat{U}_0\hat{O}\hat{U}_0^\dagger\hat{U}_0|\varphi(0)\rangle|^2 = P_\xi(0), \tag{6.14}$$

where we have taken advantage of the fact that $\hat{U}_0^\dagger\hat{U}_0 = \hat{I}$.

For example, one of the gates occurring a lot in the algorithms we have been studying is $\hat{X}$. How does this look in this new, time-dependent regime?

$$\hat{X}(t) = \hat{U}_0(t)\hat{X}\hat{U}_0^\dagger(t)$$
$$= \{\cos(\omega t/2)\hat{I} + i\sin(\omega t/2)\hat{Z}\}\hat{X}\{\cos(\omega t/2)\hat{I} - i\sin(\omega t/2)\hat{Z}\}$$

$$
\begin{aligned}
&= \cos^2(\omega t/2)\hat{X} - i\cos(\omega t/2)\sin(\omega t/2)\hat{X}\hat{Z} \\
&\quad + i\cos(\omega t/2)\sin(\omega t/2)\hat{Z}\hat{X} + \sin^2(\omega t/2)\hat{Z}\hat{X}\hat{Z} \\
&= \{\cos^2(\omega t/2) - \sin^2(\omega t/2)\}\hat{X} - 2\cos(\omega t/2)\sin(\omega t/2)\hat{Y} \\
&= \cos(\omega t)\hat{X} - \sin(\omega t)\hat{Y},
\end{aligned}
\tag{6.15}
$$

where we have used $\hat{Z}\hat{X}\hat{Z} = -\hat{X}$ and $\hat{Z}\hat{X} = -\hat{X}\hat{Z} = i\hat{Y}$ and also the trigonometry formulas for $\cos 2\theta$ and $\sin 2\theta$. In matrix form this is

$$
\hat{X}(t) = \begin{bmatrix} 0 & \cos(\omega t) + i\sin(\omega t) \\ \cos(\omega t) - i\sin(\omega t) & 0 \end{bmatrix} = \begin{bmatrix} 0 & e^{i\omega t} \\ e^{-i\omega t} & 0 \end{bmatrix}.
\tag{6.16}
$$

Similarly, the other two Pauli matrices in the time-dependent form are

$$
\hat{Y}(t) = \begin{bmatrix} 0 & -ie^{i\omega t} \\ ie^{-i\omega t} & 0 \end{bmatrix} \quad \text{and} \quad \hat{Z}(t) = \begin{bmatrix} 1 & 0 \\ 0 & -1 \end{bmatrix}.
\tag{6.17}
$$

In other words, our time-dependent qubit states are not so much of a problem provided we can compensate for their time variation by finding a way to do time-dependent quantum gates.

This does lead to a serious issue in that, to perform a time-varying operation, one must have an oscillating force acting on the qubit. Ensuring that the frequency of this oscillating force is exactly equal to the natural qubit frequency, $\omega = \Delta E/\hbar$, is one of the most difficult things that has to be done to make a practical quantum computer. Any drift between the two frequencies leads to **dephasing**, in which the relative phases of the probability amplitudes of level $|0\rangle$ and $|1\rangle$ become randomized. This is one of the most prevalent errors that can occur in quantum computing. However, as we will see in Chapter 11, there are very elegant methods of mitigating dephasing, and other types of error that can occur.

⚓ **Physics Tips 6.2**

Interaction Picture. Suppose we want to solve for the dynamics of some quantum system – a qubit, say. To do so, we must solve the Schrödinger equation:

$$
i\hbar \frac{d}{dt}|\varphi(t)\rangle = \hat{\mathcal{H}}(t)|\varphi(t)\rangle.
\tag{6.18}
$$

The Hamiltonian first will include the natural energy levels of the two qubit states, E_0 and E_1, plus any additional terms due to any other interactions – these might be time-dependent in general:

$$
\hat{\mathcal{H}} = \begin{bmatrix} E_0 & 0 \\ 0 & E_1 \end{bmatrix} + \begin{bmatrix} V_{00}(t) & V_{01}(t) \\ V_{10}(t) & V_{11}(t) \end{bmatrix}.
\tag{6.19}
$$

This does *not* look like something particularly easy to solve!

There is one useful hack which might simplify things a bit: instead of writing the state vector in the usual way, i.e., $|\varphi(t)\rangle = \alpha(t)|0\rangle + \beta(t)|1\rangle$, we can incorporate the oscillations due to the qubit energies directly:

$$
|\varphi(t)\rangle = \begin{bmatrix} \alpha_I(t)\,e^{-iE_0 t/\hbar} \\ \beta_I(t)\,e^{-iE_1 t/\hbar} \end{bmatrix} = \begin{bmatrix} e^{-iE_0 t/\hbar} & 0 \\ 0 & e^{-iE_1 t/\hbar} \end{bmatrix} \begin{bmatrix} \alpha_I(t) \\ \beta_I(t) \end{bmatrix}.
\tag{6.20}
$$

The subscript I on the probability amplitudes stands for "interaction"; often the subscripts will not be included when there is no need to distinguish between the two approaches.

Taking the time derivative we find

$$\frac{d}{dt}|\varphi(t)\rangle = \begin{bmatrix} e^{-iE_0 t/\hbar} & 0 \\ 0 & e^{-iE_1 t/\hbar} \end{bmatrix} \left(\frac{d}{dt}\begin{bmatrix} \alpha_I(t) \\ \beta_I(t) \end{bmatrix} - \frac{i}{\hbar}\begin{bmatrix} E_0 & 0 \\ 0 & E_1 \end{bmatrix}\begin{bmatrix} \alpha_I(t) \\ \beta_I(t) \end{bmatrix} \right). \tag{6.21}$$

The matrix involving E_0 and E_1 is rather signalling what is going to happen next. When we substitute from Eq. (6.21) into Eq. (6.18) and do a little re-arranging, we find

$$i\hbar\frac{d}{dt}\begin{bmatrix} \alpha_I(t) \\ \beta_I(t) \end{bmatrix} = \begin{bmatrix} V_{00}(t) & V_{01}(t)e^{-i\omega t} \\ V_{10}(t)e^{i\omega t} & V_{11}(t) \end{bmatrix}\begin{bmatrix} \alpha_I(t) \\ \beta_I(t) \end{bmatrix}. \tag{6.22}$$

Here we have written $E_1 - E_0 = \hbar\omega$. This is identical to Schrödinger's equation except we have a different – perhaps simpler – Hamiltonian to deal with. This approach to solving the dynamics is called the *interaction picture*.

Exercises 6.2.1

1. Find the time-evolved version of the Hadamard gate:

$$\hat{H}(t) = \hat{U}_0(t)\hat{H}\hat{U}_0^\dagger(t). \tag{6.23}$$

2. Consider the simple circuit in QC 6.1:

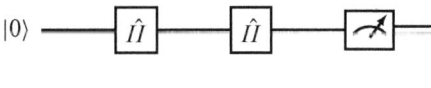

QC 6.1

   (a) If there is no time dependence of the qubit state (i.e., $\Delta E = 0$), what is the probability of obtaining "0" as the measurement outcome?

   (b) Now suppose there is a time dependence; the circuit starts at time $t = 0$, the two Hadamards are applied at times t_1 and t_2 and the final measurement at time t. If the Hadamard gates evolve in time according to Eq. (6.23), what is the probability of obtaining "0" as the measurement outcome?

   (c) Now suppose there is a time dependence for the qubit, but there is no time dependence of the Hadamard gates. Assuming the gates are applied at the same time as those in part (b), what is the probability of obtaining "0" as the measurement outcome in this case?

3. Consider the simple two-qubit circuit in QC 6.2:

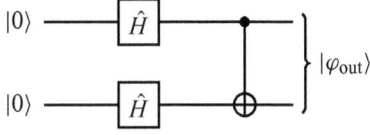

QC 6.2

(a) If there is no time dependence of the qubit state (i.e., $\Delta E = 0$), what is the concurrence of the output state $|\varphi_{\text{out}}\rangle$?

(b) Now suppose there is a time dependence; the two Hadamards are applied at time $t = 0$, the CNOT is applied at time t_1 (and we assume that the CNOT is *not* time varying); what is the concurrence of the output state $|\varphi_{\text{out}}\rangle$ at time t?

6.3 Forces and Single-Qubit Gates

6.3.1 Forces and Gates

Now let us think about the internal configuration of the qubit that gives rise to the different states $|0\rangle$ and $|1\rangle$. Potentially there are lots of different attributes of the system that could be changed – generically these are called "degrees of freedom", which makes them candidates for $|0\rangle$ and $|1\rangle$. To keep matters simple, let's just focus on one possible degree of freedom, namely the position of a particle within the qubit. Think, for example, of an electron bound to an atomic nucleus, the displacement $\hat{\mathbf{r}}$ of the electron from the nucleus being the degree of freedom. This sounds like a very specific example, but the dynamics that result are in fact generic and can be applied to other qubit architectures with little modification.

Since it is a quantum observable, the position is represented by an operator, but remember, it is an operator that lives in the two-dimensional qubit space, and we can handle 2×2 matrices. In an energy eigenstate, such as $|0\rangle$ or $|1\rangle$, the average value of the position is not changing. Writing $\langle m|\hat{\mathbf{r}}|m\rangle = \langle \mathbf{r}\rangle$ (where m stands for either 0 or 1), we find

$$\frac{d}{dt}\langle \mathbf{r}\rangle = \left(\frac{d}{dt}\langle m|\right)\hat{\mathbf{r}}|m\rangle + \langle m|\hat{\mathbf{r}}\left(\frac{d}{dt}|m\rangle\right)$$

$$= -\left(\frac{E_m}{i\hbar}\right)\langle \mathbf{r}\rangle + \left(\frac{E_m}{i\hbar}\right)\langle \mathbf{r}\rangle = 0. \tag{6.24}$$

This does not mean that the position is not changing, only that the *average* position is not changing. In any dynamics problem we are free to choose the origin where we like: in this case it will be natural to choose it so that $\langle 0|\hat{\mathbf{r}}|0\rangle = \langle 1|\hat{\mathbf{r}}|1\rangle = 0$. (The argument that the average position for *both* $|0\rangle$ and $|1\rangle$ be 0 is a bit more subtle, and relies on parity symmetry: if the system has parity symmetry, i.e., it looks the same as its mirror image, then the energy eigenstates will be eigenstates of the parity operator, which in turn implies these matrix elements will be 0.) The average position when the qubit is in a superposition state $|\psi(t)\rangle = A(t)|0\rangle + B(t)|1\rangle$ is

$$\langle \psi(t)|\hat{\mathbf{r}}|\psi(t)\rangle = |A(t)|^2\langle 0|\hat{\mathbf{r}}|0\rangle + |B(t)|^2\langle 1|\hat{\mathbf{r}}|1\rangle + A(t)^*B(t)\langle 0|\hat{\mathbf{r}}|1\rangle + A(t)B(t)^*\langle 1|\hat{\mathbf{r}}|0\rangle$$

$$= \mathbf{r}_0\cos(\omega t) + \mathbf{v}_0\sin(\omega t)/\omega, \tag{6.25}$$

where $\mathbf{r}_0 - i\mathbf{v}_0/\omega = A(0)^*B(0)\langle 0|\hat{\mathbf{r}}|1\rangle$. In words, the average position of the particle in a superposition state is moving in an elliptical orbit, whose dimensions are proportional to the matrix element $|\langle 0|\hat{\mathbf{r}}|1\rangle|$.

What happens if an external time-varying force is applied to this physical qubit? Force and energy are closely related in physics: when an object is displaced by a distance $\mathbf{r}$ due to the action of a force $\mathbf{F}(t)$ then the energy is changed by $-\mathbf{r}\cdot\mathbf{F}(t)$. The effect of an external force on our qubit is simply described by changing the Hamiltonian from $\hat{\mathcal{H}}$ to $\hat{\mathcal{H}} - \hat{\mathbf{r}}\cdot\mathbf{F}(t)$. In other words, the interaction of the qubit with an external influence changes its energy.

The presence of a force will thus alter the equations appearing in Eqs. (6.5) and (6.6) to the following:

$$i\hbar\frac{dA}{dt} = E_0 A - \langle 0|\mathbf{r}\cdot\mathbf{F}(t)|0\rangle A - \langle 0|\mathbf{r}\cdot\mathbf{F}(t)|1\rangle B, \tag{6.26}$$

$$i\hbar\frac{dB}{dt} = E_1 B - \langle 1|\mathbf{r}\cdot\mathbf{F}(t)|0\rangle A - \langle 1|\mathbf{r}\cdot\mathbf{F}(t)|1\rangle B. \tag{6.27}$$

In order to simplify this, let us take a closer look at these matrix elements.

First, the force is an external influence which is independent of the qubit. By analogy, the weather outside is independent of someone walking down the street: the pedestrian gets wet if it rains, or gets their hair messed up if the wind blows, but the rain will fall and the wind will blow regardless of whether or not the pedestrian chose to stay at home or not. Here, the force is being applied, and it is not influenced by the configuration of the qubit. Thus, the force $\mathbf{F}(t)$ in the matrix element $\langle m|\mathbf{r}\cdot\mathbf{F}(t)|n\rangle$ will factor out: $\langle m|\mathbf{r}|n\rangle\cdot\mathbf{F}(t)$, and we are left with the matrix elements of the position operator we discussed above. As we saw, $\langle 0|\hat{\mathbf{r}}|0\rangle = \langle 1|\hat{\mathbf{r}}|1\rangle = 0$, which immediately takes out one term in Eq. (6.26) and another in Eq. (6.27).

As we have seen, the values of matrix elements $\langle 0|\hat{\mathbf{r}}|1\rangle$ and $\langle 1|\hat{\mathbf{r}}|0\rangle$ (which are related by being complex conjugates of each other) determine the average displacement of the particle inside the qubit from its equilibrium location; thus, this will determine how effective the force is in re-configuring the internal arrangement of the qubit, mixing $|0\rangle$ with $|1\rangle$: in other words, just what we need to do to make a quantum gate.

To proceed, we assume that the time-varying force is a simple oscillation at the frequency $\omega = (E_0 - E_1)/\hbar$, so that $\mathbf{F}(t) = \mathbf{F}_0\cos(\omega t + \phi)$ where ϕ is a constant phase. The times at which the force reaches its maximum are given by $t_{\max} = \phi/\omega, (\phi \pm 2\pi)/\omega, (\phi \pm 4\pi)/\omega$, an so on. Since both $\mathbf{F}_0$ and $\langle 0|\hat{\mathbf{r}}|1\rangle$ are constants, we combine them as

$$-\langle 0|\hat{\mathbf{r}}|1\rangle\cdot\mathbf{F}_0 = \hbar\Omega e^{i\chi}, \tag{6.28}$$

where Ω is a real-valued parameter called the **Rabi frequency**, which combines $\mathbf{F}_0$, the strength of the applied force, and $\langle 0|\hat{\mathbf{r}}|1\rangle$, which characterizes how susceptible the qubit is to being pushed. We will also make the assumption $E_0 + E_1 = 0$ (which is equivalent to neglecting the global phase of the state), and define the new probability amplitudes:

$$A(t) = \alpha(t)\exp(i\omega t/2), \tag{6.29}$$

$$B(t) = \beta(t)\exp(-i\omega t/2). \tag{6.30}$$

Upon substituting into Eqs. (6.26) and (6.27), we find

$$i\frac{d}{dt}\alpha(t) = \Omega\cos(\omega t + \phi)\exp(-i\omega t)\beta(t), \tag{6.31}$$

$$i\frac{d}{dt}\beta(t) = \Omega\cos(\omega t + \phi)\exp(i\omega t)\alpha(t). \tag{6.32}$$

If we write the cosine term as a sum of complex exponentials, i.e.,

$$\cos(\omega t + \phi) = \frac{1}{2}\left[\exp(i\omega t + i\phi) + \exp(-i\omega t - i\phi)\right], \tag{6.33}$$

we get

$$i\frac{d}{dt}\alpha(t) = \frac{\Omega}{2}\beta(t)e^{i\phi} + \frac{\Omega}{2}\beta(t)e^{-i(2\omega t + \phi)}, \tag{6.34}$$

$$i\frac{d}{dt}\beta(t) = \frac{\Omega}{2}\alpha(t)e^{-i\phi} + \frac{\Omega}{2}\alpha(t)e^{i(2\omega t+\phi)}. \tag{6.35}$$

Look at the two terms involving $\exp[\pm i(2\omega t + \phi)]$: these are rapidly oscillating compared with the other terms present in the equation, and can, to a good approximation, be neglected. This is called the **rotating wave approximation**. Thus we have the equations:

$$\frac{d}{dt}\alpha(t) = -i\frac{\Omega}{2}\beta(t)e^{i\phi}, \tag{6.36}$$

$$\frac{d}{dt}\beta(t) = -i\frac{\Omega}{2}\alpha(t)e^{-i\phi}. \tag{6.37}$$

These can be solved as follows: First, differentiate Eq. (6.36) with respect to time:

$$\frac{d^2}{dt^2}\alpha(t) = -i\frac{\Omega}{2}\frac{d}{dt}\beta(t)e^{i\phi}. \tag{6.38}$$

Now substitute from Eq. (6.37):

$$\frac{d^2}{dt^2}\alpha(t) = -i\frac{\Omega}{2}\left(-i\frac{\Omega}{2}\alpha(t)e^{-i\phi}\right)e^{i\phi} = -\frac{\Omega^2}{4}\alpha(t), \tag{6.39}$$

which is just the equation for a simple **harmonic oscillator**. Its solution is

$$\alpha(t) = \cos\left(\frac{\Omega t}{2}\right)\alpha(0) + \sin\left(\frac{\Omega t}{2}\right)\dot{\alpha}(0)/(\Omega/2), \tag{6.40}$$

where $\dot{\alpha}$ denotes the time derivative. Using Eq. (6.36), $\dot{\alpha}(0) = -i(\Omega/2)\beta(0)\exp(i\phi)$, we find

$$\alpha(t) = \cos\left(\frac{\Omega t}{2}\right)\alpha(0) - i\sin\left(\frac{\Omega t}{2}\right)e^{i\phi}\beta(0). \tag{6.41}$$

A similar procedure, only starting with the time derivative of Eq. (6.37), yields

$$\beta(t) = \cos\left(\frac{\Omega t}{2}\right)\beta(0) - i\sin\left(\frac{\Omega t}{2}\right)e^{-i\phi}\alpha(0). \tag{6.42}$$

These two can be combined as a matrix equation:

$$\begin{bmatrix}\alpha(t)\\ \beta(t)\end{bmatrix} = \begin{bmatrix}\cos(\Omega t/2) & -i\sin(\Omega t/2)e^{i\phi}\\ -i\sin(\Omega t/2)e^{-i\phi} & \cos(\Omega t/2)\end{bmatrix}\begin{bmatrix}\alpha(0)\\ \beta(0)\end{bmatrix}, \tag{6.43}$$

or, in terms of operators $|\varphi(t)\rangle = \hat{U}_1(\Omega t, \phi)|\varphi(0)\rangle$, where

$$\hat{U}_1(\theta, \phi) = \left\{\cos(\theta/2)\hat{I} - i\sin(\theta/2)\left[\cos\phi\hat{X} + \sin\phi\hat{Y}\right]\right\}. \tag{6.44}$$

Thus, we have two parameters (θ, ϕ) which can be controlled and used to create operations acting on single qubits. For example, suppose we apply the force in pulses: the force (whose strength is proportional to Ω) is turned on, then a certain time later, turned off again. If we choose the pulse duration t so that $\Omega t = \pi$ (this is called a "π-pulse"), then

$$\hat{U}_1(\pi, \phi) = -i\left[\cos\phi\hat{X} + \sin\phi\hat{Y}\right], \tag{6.45}$$

and by controlling the phase ϕ we can choose to perform either an $\hat{X}$ gate or a $\hat{Y}$ gate. A $\hat{Z}$ gate can be performed by applying a $\hat{Y}$ followed by an $\hat{X}$, since $\hat{X}\hat{Y} = i\hat{Z}$.

A $\pi/2$ ("π-over-2") pulse is such that $\Omega t = \pi/2$; let's also choose $\phi = 3\pi/2$ so we get

$$\hat{U}_1(\pi/2, 3\pi/2) = \left(\hat{I} + i\hat{Y}\right)/\sqrt{2}, \tag{6.46}$$

which is a close relative of the Hadamard gate $\hat{H} = \left(\hat{Z} + \hat{X}\right)/\sqrt{2}$. In fact

$$\hat{Z}\hat{U}_1(\pi/2, 3\pi/2) = \hat{H}. \tag{6.47}$$

In practice, it is often easier to perform the $\pi/2$ pulse directly and re-write the algorithm than it is to do two pulses to create a Hadamard gate.

Another sort of pulse we will have occasion to use is the 2π pulse, for which $\Omega t = 2\pi$, so that

$$\hat{U}_1(2\pi, \phi) = -\hat{I}. \tag{6.48}$$

The control of qubits by external forces does place some constraints on the architecture: the "force" that is conducting the control operation needs to access the qubits. This force is always either electric or magnetic, for example time-varying voltages in waveguides connected with the qubit, or lasers directed onto the qubit by optical systems, or time-varying magnetic fields applied via coils.

6.3.2   Far Off-Resonance Interactions

We have been a bit blazé about the rotating wave approximation. Usually when we apply a rapidly oscillating force to a system it will only have an appreciable effect if the force is *on-resonance*, i.e., the oscillation frequency is close to some natural frequency of oscillation of the system, for example $\Delta E/\hbar$, the transition frequency for qubits. But what happens if you are very far off-resonance? Does the force really have no effect?

Consider the generic version of Schrödinger's equation for this situation:

$$i\hbar\frac{d}{dt}|\varphi(t)\rangle = \left(\hat{V}e^{-i\omega_0 t} + \hat{V}^\dagger e^{i\omega_0 t}\right)|\varphi(t)\rangle, \tag{6.49}$$

where ω_0 is the frequency of the force oscillations, and the operator $\hat{V}$ represents the effect of the force on the system (we have assumed that Planck's constant has been absorbed into $\hat{V}$, just as we did in Eq. (6.28)). Note that there is no time-independent term in the large brackets on the right-hand side: such a time-independent term could represent the different energy levels, and transition energies between those levels could be on-resonant with ω_0. In the language of quantum mechanics, the equation is written in the interaction picture. We will assume that the frequency ω_0 is large (that is, large compared with the eigenvalues of $\hat{V}/\hbar$), so that we are *far* off-resonance.

Let's integrate Eq. (6.49) to obtain the following *integral* equation:

$$|\varphi(t)\rangle = |\varphi(0)\rangle - \frac{1}{\hbar}i\int_0^t \left(\hat{V}e^{-i\omega_0 t'} + \hat{V}^\dagger e^{i\omega_0 t'}\right)|\varphi(t')\rangle dt'. \tag{6.50}$$

At first sight, we have only made the problem a lot more complicated. However, let's think about the integral: the integrand is oscillating very rapidly (the factor $e^{\pm i\omega_0 t'}$), modulated by a slowly varying function $|\varphi(t')\rangle$. Figure 6.1 illustrates this sort of integral: contributions from adjacent peaks and valleys of the oscillation almost exactly cancel, so that the integral in Eq. (6.50) will depend almost

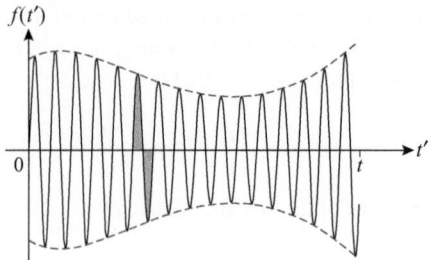

Figure 6.1   A rapidly oscillating signal modulated by a slowly varying envelope function (dashed lines). The contributions to the integral of the two shaded areas almost exactly cancel each other out; similar cancellations occur for all the other periods, so the value of the integral largely depends only on the partial periods near the end points.

entirely on the contributions of the partially completed oscillations at the beginning and the end of the integration period. Thus, integrating by parts, we find

$$
\begin{aligned}
|\varphi(t)\rangle &= |\varphi(0)\rangle + \frac{1}{\hbar\omega_0} \left[\left(\hat{V} e^{-i\omega_0 t'} - \hat{V}^\dagger e^{i\omega_0 t'} \right) |\varphi(t')\rangle \right]_0^t \\
&\quad - \frac{1}{\hbar\omega_0} \int_0^t \left(\hat{V} e^{-i\omega_0 t'} - \hat{V}^\dagger e^{i\omega_0 t'} \right) \frac{d}{dt'} |\varphi(t')\rangle dt', \\
&\approx |\varphi_0\rangle + \frac{1}{\hbar\omega_0} \left(\hat{V} e^{-i\omega_0 t} - \hat{V}^\dagger e^{i\omega_0 t} \right) |\varphi(t)\rangle,
\end{aligned}
\tag{6.51}
$$

where $|\varphi_0\rangle = [\hat{I} - (\hat{V} - \hat{V}^\dagger)/\hbar\omega_0]|\varphi(0)\rangle$. In the second expression we have neglected the remaining integral (which will be smaller than the terms we retained by a factor $1/\hbar\omega_0$). Again, the usefulness of this expression might not be immediately obvious, but what happens if we substitute it into the right hand side of Eq. (6.49):

$$
\begin{aligned}
i\hbar \frac{d}{dt} |\varphi(t)\rangle &= \left(\hat{V} e^{-i\omega_0 t} + \hat{V}^\dagger e^{i\omega_0 t} \right) |\varphi_0\rangle + \frac{1}{\hbar\omega_0} \left(\hat{V} e^{-i\omega_0 t} + \hat{V}^\dagger e^{i\omega_0 t} \right) \left(\hat{V} e^{-i\omega_0 t} - \hat{V}^\dagger e^{i\omega_0 t} \right) |\varphi(t)\rangle \\
&= \frac{1}{\hbar\omega_0} \left[\hat{V}^\dagger, \hat{V} \right] |\varphi(t)\rangle + \text{terms proportional to } e^{\pm i\omega_0 t} \text{ or } e^{\pm i2\omega_0 t}.
\end{aligned}
\tag{6.52}
$$

Thus, neglecting the oscillatory terms, we see that the effect of a highly detuned interaction is given by an *effective Hamiltonian*:

$$
\hat{\mathcal{H}}_{\text{eff}} = \frac{1}{\hbar\omega_0} \left[\hat{V}^\dagger, \hat{V} \right].
\tag{6.53}
$$

This result can be generalized to the case in which there are multiple time-harmonic terms in the Hamiltonian:

$$
\hat{\mathcal{H}}(t) = \sum_n \hat{V}_n e^{-i\omega_n t} + \text{h.a.},
\tag{6.54}
$$

where "h.a." stands for the Hermitian adjoint of the preceding term. Then, provided $\min |\omega_n - \omega_m|$ is much larger than the maximum eigenvalue of $\hat{V}_n/\hbar$, the effective Hamiltonian is

$$
\boxed{\hat{\mathcal{H}}_{\text{eff}} = \sum_n \frac{1}{\hbar\omega_n} \left[\hat{V}_n^\dagger, \hat{V}_n \right].}
\tag{6.55}
$$

This result will be quite useful, and will crop up again when dealing with interactions in the far off-resonance (also known as the dispersive regime).

Exercises 6.3.3

1. As we have seen, realistic physical qubits are continuously evolving in time, so that the state of a single qubit is $|\Phi(t)\rangle = A(t)|0\rangle + B(t)|1\rangle$. Here we are going to consider the case of single qubit operations in which the drive frequency is slightly "detuned" from the qubit resonant frequency. Let $\omega = (E_1 - E_0)/\hbar$ be the natural frequency of the qubit; when an external force, oscillating at frequency $\omega_1 = \omega + \Delta$ (where the **detuning** Δ is the amount by which the driving frequency differs from the qubit resonance frequency), is applied, the amplitudes $A(t)$ and $B(t)$ obey the following coupled differential equations:

$$
\begin{aligned}
i\dot{A} &= \quad\;\; -(\omega/2)A \;\; - \;\; \Omega_0 \cos(\omega_1 t)\, B \\
i\dot{B} &= -\Omega_0 \cos(\omega_1 t)\, A \;\; + \;\;\quad (\omega/2)B,
\end{aligned}
$$

where $\dot{f}(t)$ is the time derivative of $f(t)$, and Ω_0 is a constant proportional to the strength of the applied force.

   (a) Introducing the new variables $\alpha(t) = \exp(-i\omega_1 t/2)A(t)$ and $\beta(t) = \exp(i\omega_1 t/2)B(t)$, and using the rotating wave approximation, show that $\alpha(t)$ and $\beta(t)$ obey the following time-independent equations:

$$
\begin{aligned}
i\dot{\alpha} &= \quad -(\Delta/2)\alpha \;\; - \;\; (\Omega_0/2)\beta, \\
i\dot{\beta} &= -(\Omega_0/2)\alpha \;\; + \;\; (\Delta/2)\beta.
\end{aligned}
$$

   (b) By differentiating the first of these equations and substituting from the second, one can obtain a simple second-order ordinary differential equation for α. Solve this equation, and the similar equation for β, using the method described in Appendix A.

   (c) Using the results you obtained in part (b), show that

$$
\begin{bmatrix} \alpha(t) \\ \beta(t) \end{bmatrix} = \hat{U}(t) \begin{bmatrix} \alpha(0) \\ \beta(0) \end{bmatrix},
$$

   where $\hat{U}(t) = \cos(\Omega t/2)\hat{I} + i\sin(\Omega t/2)\{\Delta\hat{Z} + \Omega_0\hat{X}\}/\Omega$, and $\Omega = \sqrt{\Omega_0^2 + \Delta^2}$.

   (d) Suppose one started at time $t = 0$ in the state $|0\rangle$. What is the maximum subsequent probability of being in state $|1\rangle$?

6.4 CONCEPT CHECK AND EXERCISES

Concept Check

- Quantum computers need lots of physical qubits which can be reliably prepared in the state $|0\rangle$, are controllable (so logic gates can be performed), long-lived, and can be read out.
- Qubits generally need an energy gap between the qubit levels, which alters the type of gates that need to be performed.
- External forces oscillating on-resonance with the energy gap are used to perform the gates.

Chapter Exercises

1. Let's suppose we want to build a quantum computer that operates at liquid nitrogen temperatures ($T \approx 77$ kelvin) and that we can tolerate a 0.01% probability of error in initial state preparation. What is the required energy gap ΔE? What is the wavelength of radiation that corresponds to that energy? If the distance between qubits is to be at least 10 wavelengths, and the qubits are arranged on a *cubic* lattice, what would be the dimensions of a chip containing 20 million qubits (20 megaqubits)?

2. In this question we are going to examine the effect of the off-resonance terms we ignored when making the rotating wave approximation.

   (a) Show that Eqs. (6.34) and (6.35) are equivalent to a Schrödinger equation of the form

   $$i\frac{d}{dt}|\varphi(t)\rangle = \hat{\mathcal{H}}_0|\varphi(t)\rangle + \left[\hat{V}e^{-i2\omega t} + \hat{V}^\dagger e^{i2\omega t}\right]|\varphi(t)\rangle,$$

   where $|\varphi(t)\rangle = \alpha(t)|0\rangle + \beta(t)|1\rangle$ and $\hat{\mathcal{H}}_0 = \hbar\Omega(|0\rangle\langle1|e^{i\phi} + |1\rangle\langle0|e^{-i\phi})/2$. Find the expression for $\hat{V}$.

   (b) Using Eq. (6.55) find an effective Hamiltonian that describes the effect of the far off-resonance terms (we assume that $2\omega \gg \Omega_0$, and so the presence of the $\hat{\mathcal{H}}_0$ term can be neglected in the derivation of the effective Hamiltonian). What does this change correspond to physically? *Hint: The name generally given to this effect is the* Bloch–Siegert shift.

7 Realizations of Qubits: Examples

In this chapter we look at some of the actual devices that are being used to make a quantum computer. Identifying the best physical implementation of a qubit remains the subject of intense and competitive research. Here we concentrate on four basic types of qubits, all of which have demonstrated considerable promise: photons, electron or nuclear spins, trapped ions, and superconducting circuits. All have legitimate claims to primacy in some (but not all) aspects, and have been validated in proof-of-principal experiments.

Learning Objectives

- Learn the necessary physics and mathematical description of photons, electrons, atoms, and electrical circuits that make them good candidates to be used as qubits.

7.1 Photons

As we saw in Chapter 2, quantum mechanics grew out of Planck's notion that light is made up of indivisible quanta, each with an energy $h\nu = \hbar\omega$. Thus, a light beam whose wavelength is $\lambda = 6 \times 10^{-7}$ metres (which is typical of the light we see in everyday life), and which is therefore vibrating at frequency $\nu = c/\lambda = 5 \times 10^{14}$ hertz (c being the speed of light), can be thought of as a stream of photons, each of energy 3.31×10^{-19} joules.

In everyday life, the sort of light we encounter is made up of a large random number of these photons. For example, a 60 watt LED light bulb (with 90% efficiency) radiates 54 joules of light energy per second, or an *average* of about 1.6×10^{20} photons per second. However the *actual* number of photons radiated per second is random, with probability given by the Poisson distribution; the standard deviation of the number of photons radiated each second – which tells us how big the random variations about the average tend to be – is about 1.3×10^{10} photons, which is too small a variation to perceive.

In the last 50 years or so, optical physics has advanced so that experiments involving *single* photons are now routinely performed (although "routinely" should not be thought of as a synonym of "easily"). These individual quantum systems require special sources and special highly sensitive detectors which took years to develop, and they play a central role in quantum technologies.

A light beam of wavelength λ, propagating along the z-axis in some finite region of volume V, and whose electric field strength is pointing along the x-axis (Fig. 7.1) is described by the following formula:

$$\mathbf{E}(z, t) = \mathbf{e}_x E_x \sin[kz - \omega t], \qquad (7.1)$$

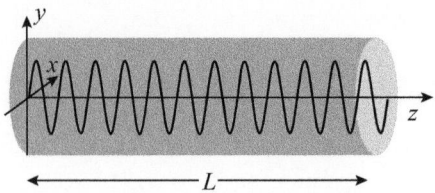

Figure 7.1   Simple model of a photon mode. A cylindrical region length L, volume $\mathscr{V}$ travelling along the z-direction at speed c; the electric field is oscillating with a polarization vector in the x–y plane.

where the wavenumber $k = 2\pi/\lambda$; the **boldface** quantities indicate that they are vectors in three-dimensional space, with $\mathbf{e}_x$, $\mathbf{e}_y$, and $\mathbf{e}_z$ being basis vectors in the x-, y-, and z-directions. The electric field strength $\mathbf{E}$ is defined so that if it acts on a charge q, it exerts a force $\mathbf{F} = q\mathbf{E}$ on the charge; thus the field $\mathbf{E}$ has units of newtons per coulomb. We assume this formula is valid in the *mode volume* $\mathscr{V}$; outside of this region (which is moving along with the wave field at speed c in the z-direction), we assume the field is effectively zero. This simplified model does ignore the effect of diffraction (which means a beam of finite cross-sectional area will grow wider as it travels along the z-direction) and that a pulse of light of finite duration (and hence a finite length L in the z-direction) cannot be oscillating at a single frequency: a more sophisticated quantum theory of light is needed to address these important issues, especially with extremely short pulses which are tightly focused, but this simple model will be sufficient to get the hang of how photons can be used in quantum technologies.

How can this classical wave, which is oscillating gently away at a fixed frequency, be compatible with Planck's idea, which describes the wave as a quantum system which has a series of equally spaced energy levels, i.e., $|0\rangle$ corresponding to 0 photons, $|1\rangle$ for 1 photon, $|2\rangle$ for 2 photons, and so on? (To avoid confusion, we are using double-struck digits ($0, 1, 2, \ldots$) to denote the photon states; when you see $|0\rangle$ and $|1\rangle$, they will always refer to levels of a qubit.)

To answer this question, consider that any interaction of the light will involve transferring a photon of light energy to something else. Analyzing such dynamics needs an operator that reduces by 1 the number of photons in the system, called the *lowering operator* (or, more dramatically, the **annihilation operator**):

$$\hat{a} = |0\rangle\langle 1| + \sqrt{2}|1\rangle\langle 2| + \sqrt{3}|2\rangle\langle 3| + \cdots . \tag{7.2}$$

As we will see, the square root factors are important. Thus, if you have a state $|n\rangle$ in which n photons are present, the state $\hat{a}|n\rangle = \sqrt{n}|n-1\rangle$ will have $n-1$ photons.

The annihilation operator is not Hermitian, so its Hermitian adjoint is a new operator called the **creation operator**:

$$\hat{a}^{\dagger} = |1\rangle\langle 0| + \sqrt{2}|2\rangle\langle 1| + \sqrt{3}|3\rangle\langle 2| + \cdots , \tag{7.3}$$

which, as you would expect, has the opposite effect of increasing the number of photons by 1. (A useful aid to memory is that the † looks a bit like a + symbol, so you can remember $\hat{a}^{\dagger}$ adds one photon.)

The number of photons, represented by operator $\hat{n}$, should be an observable quantity: we can set up our photon detector, and count them. Thus $\hat{n}$ should be a Hermitian operator; fortunately, it is quite easy to express it in terms of $\hat{a}^{\dagger}$ and $\hat{a}$:

$$\hat{a}^{\dagger}\hat{a} = \left(|1\rangle\langle 0| + \sqrt{2}|2\rangle\langle 1| + \sqrt{3}|3\rangle\langle 2| + \cdots \right) \left(|0\rangle\langle 1| + \sqrt{2}|1\rangle\langle 2| + \sqrt{3}|2\rangle\langle 3| + \cdots \right)$$
$$= |1\rangle\langle 1| + 2|2\rangle\langle 2| + 3|3\rangle\langle 3| + \cdots$$
$$= \hat{n}. \tag{7.4}$$

We have used the fact that the kets describing different numbers of photons are orthogonal, just like the "Heads" and "Tails" kets in Chapter 1: $\langle 0|1\rangle = 0$ and so on. If you reverse the order you get:

$$\hat{a}\hat{a}^{\dagger} = \left(|0\rangle\langle 1| + \sqrt{2}|1\rangle\langle 2| + \sqrt{3}|2\rangle\langle 3| + \cdots \right) \left(|1\rangle\langle 0| + \sqrt{2}|2\rangle\langle 1| + \sqrt{3}|3\rangle\langle 2| + \cdots \right)$$
$$= |0\rangle\langle 0| + 2|1\rangle\langle 1| + 3|2\rangle\langle 2| + \cdots$$
$$= (|1\rangle\langle 1| + 2|2\rangle\langle 2| + 3|3\rangle\langle 3| + \cdots) + (|0\rangle\langle 0| + |1\rangle\langle 1| + |2\rangle\langle 2| + \cdots)$$
$$= \hat{n} + 1. \tag{7.5}$$

Let's now use Planck's idea: each time we add a photon to the light beam, its energy should be increased by a quantum of $\hbar\omega$. Since we do not know what the energy of the state with 0 photons might be, we will call it E_0:

$$\hat{\mathcal{H}} = \hbar\omega\hat{a}^{\dagger}\hat{a} + E_0\hat{I}. \tag{7.6}$$

Any state of our system of photons is given by $|\varphi(t)\rangle = \sum_{n=0}^{\infty} c_n(t)|n\rangle$. Substituting into Schrödinger's equation, we find

$$i\hbar\frac{d}{dt}|\varphi(t)\rangle = \hbar\omega\hat{n}|\varphi(t)\rangle,$$
$$i\hbar\sum_{n=0}^{\infty} \dot{c}_n(t)|n\rangle = \sum_{n=0}^{\infty} (\hbar\omega n + E_0)c_n(t)|n\rangle. \tag{7.7}$$

Since the states of different photon number are orthogonal we get

$$\dot{c}_n(t) = -i(\omega n + E_0/\hbar)c_n(t), \tag{7.8}$$

which has the solution $c_n(t) = c_n(0)\exp(i\omega nt + iE_0t/\hbar)$. Thus, the **expectation value** of the operator $\hat{a}$ is

$$a(t) = \langle\varphi(t)|\hat{a}|\varphi(t)\rangle$$
$$= \sum_{n,m=0}^{\infty} c_n^*(0)c_m(0)\langle n|\hat{a}|m\rangle \exp[i\omega(n-m)t]$$
$$= \sum_{n,m=0}^{\infty} c_n^*(0)c_m(0)\sqrt{m}\langle n|m-1\rangle \exp[i\omega(n-m)t]$$
$$= \left(\sum_{n=0}^{\infty} c_n^*(0)c_{n+1}(0)\sqrt{n+1} \right) \exp(-i\omega t)$$
$$= a(0)\exp(-i\omega t). \tag{7.9}$$

Comparing Eqs. (7.1) and (7.9), we can see that, if we define an electric field operator

$$\hat{\mathbf{E}}(z) = -i\mathcal{E}_0\mathbf{e}_x \left(\hat{a}e^{ikz} - \hat{a}^{\dagger}e^{-ikz} \right), \tag{7.10}$$

where $\mathcal{E}_0$ is a constant, then

$$\mathbf{E}(z,t) = \langle \varphi(t) | \hat{\mathbf{E}}(z) | \varphi(t) \rangle, \tag{7.11}$$

where the field strength E_x, defined in Eq. (7.1), is $2\mathcal{E}_0 a(0)$ (we have assumed that $a(0)$ is real for simplicity). In other words, the optical field is a quantum system just like an atom or an electron: it has a configuration (the quantum state $|\varphi(t)\rangle$) and observable quantities represented by Hermitian operators.

The energy density for the wave field is $u = \epsilon_0 \mathbf{E} \cdot \mathbf{E} = \epsilon_0 E_x^2 (1 - \cos[2(kz - \omega t)])/2$ (where we have included the contribution of the magnetic field that accompanies any time-varying electric field). The constant ϵ_0 is called the *permittivity of free space* and it characterizes the ability of electrical charges to create electric fields; it has the value 8.85×10^{-12} coulomb^2 per newton per metre^2. Thus, the Hamiltonian for the quantum field will be

$$\begin{aligned}
\hat{\mathcal{H}} &= \epsilon_0 \int_{\mathcal{V}} \hat{\mathbf{E}}(z) \cdot \hat{\mathbf{E}}(z) d^3 r \\
&= -\epsilon_0 \mathcal{E}_0^2 \mathcal{A} \int_{-\mathcal{L}/2}^{-\mathcal{L}/2} \left(\hat{a} e^{ikz} - \hat{a}^\dagger e^{-ikz} \right) \left(\hat{a} e^{ikz} - \hat{a}^\dagger e^{-ikz} \right) dz \\
&= \epsilon_0 \mathcal{E}_0^2 \mathcal{V} \left(\hat{a}^\dagger \hat{a} + \hat{a}\hat{a}^\dagger - [\hat{a}^2 + \hat{a}^{\dagger 2}] \sin(k\mathcal{L})/k\mathcal{L} \right). \tag{7.12}
\end{aligned}$$

The cross-section area of the photon mode volume $\mathcal{V}$ is $\mathcal{A}$ and its length is $\mathcal{L}$, with the origin at the centre. The wavenumber $k = 2\pi/\lambda$, so the terms involving $\sin(k\mathcal{L})/k\mathcal{L}$ will never have a value greater than 1 and will become vanishingly small if we assume $\mathcal{L} \gg \lambda$. We have already implicitly made that assumption with our simple model for the field mode, Eq. (7.1). Thus, we are left with the following expression for the Hamiltonian:

$$\begin{aligned}
\hat{\mathcal{H}} &= \epsilon_0 \mathcal{V} \mathcal{E}_0^2 \left(\hat{a}^\dagger \hat{a} + \hat{a}\hat{a}^\dagger \right) \\
&= 2\epsilon_0 \mathcal{V} \mathcal{E}_0^2 \left(\hat{n} + \frac{1}{2} \hat{I} \right). \tag{7.13}
\end{aligned}$$

Comparing this with Eq. (7.6), we find the two are in agreement if $\mathcal{E}_0 = \sqrt{\hbar\omega/2\mathcal{V}\epsilon_0}$ and $E_0 = \hbar\omega/2$. Thus the electric field operator is

$$\boxed{\hat{\mathbf{E}}(z) = -i\sqrt{\frac{\hbar\omega}{2\epsilon_0 \mathcal{V}}} \mathbf{e}_x \left(\hat{a} e^{ikz} - \hat{a}^\dagger e^{-ikz} \right),} \tag{7.14}$$

and the Hamiltonian is

$$\boxed{\hat{\mathcal{H}} = \hbar\omega \left(\hat{n} + 1/2 \right).} \tag{7.15}$$

How can we use photons as qubits? The optical field has a well-defined set of quantum levels in the different number states such as $|0\rangle$ and $|1\rangle$, so these might be used as the two levels of the qubit. And the lowering and raising operators, which govern how photons interact, look very promising as a means to control the qubits: indeed, the combinations

$$\begin{aligned}
\hat{a}^\dagger + \hat{a} &= |1\rangle\langle 0| + |0\rangle\langle 1| + \text{some other terms}, \\
i(\hat{a}^\dagger - \hat{a}) &= i|1\rangle\langle 0| - i|0\rangle\langle 1| + \text{some other terms}
\end{aligned} \tag{7.16}$$

look like the $\hat{X}$ and $\hat{Y}$ single-qubit gates, respectively. But, unfortunately, the "some other terms" are a very large fly in this ointment. Consider the action of these operators on the state $|\varphi\rangle = \alpha|0\rangle + \beta|1\rangle$:

$$(\hat{a}^\dagger + \hat{a})|\varphi\rangle = \alpha|1\rangle + \beta|0\rangle + \sqrt{2}\beta|2\rangle,$$
$$i(\hat{a}^\dagger - \hat{a})|\varphi\rangle = i\alpha|1\rangle - i\beta|0\rangle + i\sqrt{2}\beta|2\rangle. \tag{7.17}$$

This illustrates the generic problem with trying to make qubits using a quantum system that has a multitude of *equally spaced* states: any physical interaction that you would like to use to perform quantum logic gates is just as likely to excite an unwanted state, $|2\rangle$ in this example, which takes you out of the two-level qubit space.

One way to get around this problem is to alter the qubit system by introducing some non-linear element so that the energy gap between $|0\rangle$ and $|1\rangle$ is no longer the same as the gap between $|1\rangle$ and $|2\rangle$; as we will see, this is the approach taken in the superconducting qubit using electronic circuits at very low temperatures. For light, unfortunately the weakness of non-linear interactions precludes this approach.

Fortunately, photons have another arrow in their quiver: we have not yet discussed *polarization*. The photons associated with the field in Eq. (7.1) have an electric field in the x-direction, as indicated by the $\mathbf{e}_x$. We can have photons whose field points in the y-direction as well – the total field must be perpendicular to the z-axis, which is the direction the light is propagating. The oscillations in this direction have a whole separate ladder of photon number states of their own. To avoid confusion with the use of the letters x and y in other contexts, we will refer to light polarization in the x-direction as H for "horizontal", and in the y-direction as V for "vertical". Thus, the quantum state when both horizontal and vertical photons are present in arbitrary numbers is

$$|\varphi(t)\rangle = \sum_{m,n=0}^{\infty} c_{m,n}(t)|m\rangle_H \otimes |n\rangle_V. \tag{7.18}$$

More usually, we will be interested in a state in which there is either one photon in the H mode or one in the V mode:

$$|\mathbb{H}\rangle = |1\rangle_H \otimes |0\rangle_V, \tag{7.19}$$
$$|\mathbb{V}\rangle = |0\rangle_H \otimes |1\rangle_V, \tag{7.20}$$

so that the state of a single photon is

$$|\varphi\rangle = \alpha|\mathbb{H}\rangle + \beta|\mathbb{V}\rangle. \tag{7.21}$$

In other words, we have a qubit, with $|\mathbb{H}\rangle$ being the qubit's state $|0\rangle$ and $|\mathbb{V}\rangle$ its state $|1\rangle$.

One of the big advantages of photon-polarization based qubits is that single-qubit gates are ready-made, using an optical element called the half-**wave plate**. When light enters a transparent material such as a crystal or a block of glass, its speed decreases. Since the frequency does not change – you still have the same number of wave-crests per second, regardless of speed – a reduction in velocity of a photon has the effect of shortening its wavelength from λ_0 in vacuum to $\lambda' = \lambda_0 n$ in the material, where the factor n is called the *refractive index*. It is a property of the material through which the photon is passing; typically n is between 1.3 and 1.5 (it has no units, since it is a ratio). Thus, a photon of wavelength λ_0 in vacuum which propagates through a crystal of length ℓ will acquire a phase factor $\exp[i2\pi(n-1)\ell/\lambda_0]$ relative to a photon that did not pass through the crystal.

Certain crystals have the property of *birefringence*, which means the refractive index is dependent on the polarization direction; a photon polarized in the horizontal direction $\mathbf{e}_x$ will experience a refractive index n_x, while a photon polarized in vertical direction $\mathbf{e}_y$ will experience a refractive index n_y; the difference in refractive indices is called Δn. If the length ℓ of a birefringent crystal along the direction of propagation is carefully manufactured so that $\Delta n \ell = (m + 1/2)\lambda_0$, where m is an integer, it will have the effect of changing the phase of one polarization by a factor π with respect to the other polarization: i.e.,

$$|\varphi\rangle \to |\varphi'\rangle = \hat{U}|\varphi\rangle, \tag{7.22}$$

where

$$\hat{U} = \begin{bmatrix} 1 & 0 \\ 0 & e^{i\pi} \end{bmatrix} = \hat{Z}, \tag{7.23}$$

and we have used the fact that $e^{i\pi} = -1$. Such a crystal is called a *half-wave plate*. Similarly, a *quarter-wave plate* is cut so that $\Delta n \ell = (m + 1/4)\lambda_0$, and results in a relative phase shift of $\exp(i\pi/2) = i$. These are common optical devices that have been used in optics laboratories for generations.

Furthermore, if we rotate the wave plate by an angle θ about the z-axis, along which the photons are propagating, leaving the photons themselves untouched, it will change the orientation of the two axes of the crystal relative to the photons' H and V directions. The result is a different unitary operation:

$$\hat{U}(\theta) = \begin{bmatrix} \cos\theta & -\sin\theta \\ \sin\theta & \cos\theta \end{bmatrix} \begin{bmatrix} 1 & 0 \\ 0 & -1 \end{bmatrix} \begin{bmatrix} \cos\theta & \sin\theta \\ -\sin\theta & \cos\theta \end{bmatrix} = \begin{bmatrix} \cos 2\theta & \sin 2\theta \\ \sin 2\theta & -\cos 2\theta \end{bmatrix}. \tag{7.24}$$

Choosing $\theta = \pi/4$, we find $\hat{U}(\pi/4) = \hat{X}$. The other local unitaries we require for quantum information can be implemented similarly using rotated half- or quarter-wave plates.

7.1.1 Assessment

One can envision how photons could form the qubits of a quantum computer. One would need multiple beams of light, controlled, for example, by optical fibres or other waveguides. The photons would travel down these, like trains in a complicated railway network; they would pass wave plates, like the signal gantries for the trains, which could be used to perform single-qubit gates; if it can be contrived, two photons might be brought together to interact, affecting each other's configurations, allowing two-qubit gates to be performed; and finally, the photon arrives at the end of the line and gets detected, like the train arriving at a terminus.

There are three major problems with implementing this concept.

- Creating single-photon states is one of the key difficulties with realizing quantum computers with photons. As mentioned, the light sources we use in everyday life are random: on average they emit a certain number of photons per second, but the actual number emitted in any second is a random number governed by **Poisson statistics** (see Exercise 4). This uncertainty simply isn't going to cut it for a quantum computer, since each qubit must be represented by a single object. Nor can reducing the rate of emission help: reducing the probability of emitting two or more photons to manageable levels means that the probability of producing zero photons goes way up. One way to produce highly **anti-bunched light**, e.g., light in which one, and only one, photon is emitted in a given interval, is to use an atom. An excited atom will decay to its ground state usually within a few nanoseconds, emitting a single photon. However, such a photon can go in just about any direction, rather than along the direction we desire. This can be remedied by having the atom in an **optical cavity** formed by two highly reflecting mirrors, which will have the effect of controlling both the wavelength and the direction of the photon the atom emits. In practice, it is easier to use systems that are similar to atoms but which can be implanted directly in an optical waveguide, for example excitons (which is an electron bound to a hole, or the absence of an electron, in a semiconductor) or **nitrogen-vacancy colour centres** in a diamond (these are impurity/defect locations in a diamond lattice, which act like atoms by trapping electrons).

- Detection of single photons has always been a challenge: each photon is a rather puny thing, and only by deploying them in large numbers can they usually hope to have a noticeable effect. However, the **avalanche photo-diode** is a device that has been used for many years to detect individual photons with some degree of success. An incident photon excites a single electron, which is then accelerated by a static electric field; this in turn excites more electrons, which, like a snow-ball provoking an avalanche on a snowy mountain, results in a large pulse of electric current, which can be easily detected. More recently, *superconducting edge detectors* have become a standard method of detecting single photons with very high efficiency: these work by having a superconductor kept at a temperature very close to its critical temperature: if the conductor gets even a very small amount hotter, it transitions to a normal metal, and acquires an electrical resistance, resulting in a detectable voltage change. The only disadvantage is that they require cooling to low temperatures (though not to the ultra-cold micro-kelvin temperatures requiring a dilution refrigerator).

- Two-qubit gates are the biggest hurdle for photons to surmount if they want to be the qubits of a quantum computer. The issues will be discussed more in Sections 8.1 and 10.2, but in brief: the electric field strength of a single photon is too puny for two photons to have a strong enough interaction on their own, although cleverly designed waveguides (which make the effective mode volume $\mathcal{V}$ as small as possible) may be able to remedy this. Instead of direct photon–photon interaction, quantum interference has been used to perform non-deterministic photon–photon quantum logic gates, which have a one-in-nine chance of working: these are described in Chapter 10. Other approaches under active consideration use states with large numbers of photons to represent qubits, though this approach has other difficulties to overcome.

The great advantage of photons as qubits is the potential for room-temperature operation, which no other proposed implementation can offer. And thus efforts to overcome these difficulties continue.

⚓ **Physics Tips 7.1**

Superconductors. Electrons in a metal in many ways behave like a gas. How can this be, when they are all negatively charged particles and repel each other strongly? First, all those negative charges are balanced by the positive charges of the atoms that make up the lattice, which forms the scaffolding for our crystal. But, repulsive charges can cancel each other out as well: think of a crowd of impatient and bad-mannered commuters at a subway station. They are constantly pushing each other out of the way, like the electrons repelling one another; but this does not decrease the density. If Al pushes Bert into Chuck, Chuck immediately pushes Bert back toward Al, and so on: all the repulsive behaviour has no net effect. In electrodynamics, the analogous effect is called *screening*.

Now, in addition to all this screened-off mutual repulsion, each electron is interacting with the atomic lattice. The atoms are a bunch heavier than the electrons, but nevertheless they do vibrate about their average positions – and even at zero temperature, these vibrations don't go away: remember, the energy of the $|0\rangle$ state of the photons is $\hbar\omega/2$. In a crystal the analogous waves are called **phonons**, which can be thought of as particles that constantly bombard the electrons from all sides. Remember we are constantly under bombardment from air molecules, but we don't notice it, because for every collision that would push you one way, there will be another to push you back again. So it is with the zero-point phonons: they have no net effect on isolated electrons (see Fig. 7.2(a)).

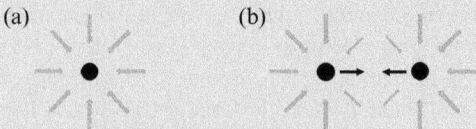

Figure 7.2 Zero-point phonons.

Now think about two electrons reasonably close together: electron A will absorb a portion of the phonons headed for B and vice versa (see Fig. 7.2(b)). This causes a net *attraction*. Thus, remarkable as it may seem, it is possible for two electrons to form a weakly bound state: these are called *Cooper pairs*.

Cooper pairs are bosons – two spin-half electrons bound together counter-align and make a spin-zero composite particle. Unlike fermions, more than one boson can occupy the same quantum state; at cold temperatures the Cooper pairs can accumulate in the lowest possible energy state, a phenomenon akin to Bose–Einstein condensation. This condensate of Cooper pairs acts like a large amorphous blob of charge, which can be accelerated by voltages and which will be untouched by the puny collisions with impurities or thermal background phonons which can impede the progress of independent electrons and thus give rise to the usual resistance to electric current. The only way for the blob to be impeded in its progress is for a collision to be energetic enough to start dissociating the Cooper pairs. This is the superconducting state.

This effect requires very low temperatures: the Cooper pairs are only weakly bound and will separate if they become too hot. The critical temperature, which is an important characteristic of superconductors, is the temperature above which the superconducting state disappears. The earliest superconductors had critical temperatures below 10 kelvin, but in the last 30 years, superconducting compounds with critical temperatures of 100 kelvin or higher have been discovered.

Exercises 7.1.2

1. What are the energies of the photons in the following?
   (a) A beam of X-band radiation of wavelength $\lambda = 3$ centimetres. (This wavelength is typical of microwave ovens.)
   (b) The infra-red TV remote control of wavelength $\lambda = 1$ micrometre.
   (c) A dental X-ray of wavelength $\lambda = 80$ picometres.
2. How many microwave photons need to be absorbed to heat a cup of water (volume 0.25 litres) from $20\,°C$ to $100\,°C$? (It requires 4.2 joules of energy to heat 1 millilitre of water by $1\,°C$).
3. How can you use a wave plate to implement the following?
   (a) A $\hat{Y}$ gate.
   (b) A $\hat{S} = \hat{R}_2$ gate (see Eq. (5.17)).
   (c) A Hadamard gate.
4. Suppose a source of photons emits on average λ photons per second. The probability that it produces n photons per second is given by the *Poisson distribution law*:

$$p_n = \lambda^n \exp(-\lambda)/n!.$$

(a) For what value of λ is the probability of emitting one photon the greatest?

(b) For what value of λ is there an equal chance of emitting both zero and one photons? What is the probability of emitting two or more photons in this case?

(c) If $\lambda \ll 1$ show that the chance of emitting two or more photons is $\lambda^2/2$.

(d) Estimate the rate of photon emission λ for which the chance of emitting two or more photons per second is 0.01. What is the probability of emitting zero photons with that value of λ?

7.2 Spin-1/2 Particles

Besides photons, nature provides us with another simple system that seems to be an absolutely perfect qubit: spin-1/2 particles such as the electron or the proton. Electrons are fundamental particles, which means they are not known to be constituted of other particles, in the way atoms are made of electrons, protons, and neutrons. When they were discovered by Joseph Thomson in 1897, it was a remarkable revelation, since before that time all materials were thought to be made of atoms (the word "atom" comes from the Greek "a" meaning "not" and "tomos" meaning "cut"; thus "atom" means "indivisible"). Electrons were the first sub-atomic particles to be discovered, and their discovery marked the start of three decades of remarkable advances in physics, which transformed the comprehension of the world and enabled the modern technological age.

Every electron has the same mass (approximately 9.11×10^{-31} kilograms), and the same charge $-e$ (where e is the fundamental unit of charge $\approx 1.60 \times 10^{-19}$ coulombs). One fascinating property of each electron is that it has an *angular momentum* called the *spin*. The electron's spin angular momentum is a quantum observable $\hat{\mathbf{S}}$, and its component along any particular direction (conventionally chosen as the z-axis), $\hat{S}_z$, can take only two possible values: $-\hbar/2$ or $+\hbar/2$: we will denote the kets associated with these two possibilities as $|\downarrow\rangle$ and $|\uparrow\rangle$, respectively. Thus electrons are natural two-level systems: how might we exploit this to implement a qubit?

First, let us suppose we can trap an electron so it is at a well-defined location: this is possible in vacuum using charged-particle traps (which employ ingenious webs of electric and/or magnetic fields), in an atom or a molecule (where electrons are naturally confined to well-characterized quantum states called **orbitals**), or in specially constructed semiconductor devices with electrodes designed to confine the electron at specific locations.

The key to making a qubit from the electron spin is its magnetic properties. On a large scale, magnetism arises from electric currents moving in circular loops. It follows that on the microscopic scale, a rotating charged particle should also generate a magnetic field as well. The strength of this contribution is quantified by the magnetic dipole moment $\hat{\boldsymbol{\mu}}$, which, for an electron spin is given by

$$\hat{\boldsymbol{\mu}} = -g\mu_B \hat{\mathbf{S}}/\hbar, \tag{7.25}$$

where $\mu_B = e\hbar/2m_e$ is the **Bohr magneton** ($\approx 9.3 \times 10^{-24}$ joules per tesla) and the g-factor ≈ 2. In a constant magnetic field $\mathbf{B}_0 = B_z \mathbf{e}_z$, the Hamiltonian of the electron is given by the formula

$$\begin{aligned}
\hat{\mathcal{H}} &= -\hat{\boldsymbol{\mu}} \cdot \mathbf{B}_0 \\
&= \frac{g\mu_B B_z}{\hbar} \frac{\hbar}{2} (|\uparrow\rangle\langle\uparrow| - |\downarrow\rangle\langle\downarrow|) \\
&= \left(\frac{g}{2}\right) \mu_B B_0 \hat{Z}.
\end{aligned} \tag{7.26}$$

In other words, $|\uparrow\rangle$ is the state $|1\rangle$ for the qubit, $|\downarrow\rangle$ is the state $|0\rangle$, and the energy difference is $\Delta E \approx \mu_B B_0$. Thus, in a magnetic field, an electron is a qubit with an energy gap dependent on the strength of the magnetic field: for a magnetic field of about 10 tesla (which is a very strong field, though one which can be attained routinely with the correct apparatus), $\Delta E = \mu_B B_0 \approx 10^{-22}$ joules. This means that, in order to initiate the qubit in state $|0\rangle$, the temperature of the electron must be much lower than 7 kelvin (or $-266\,°\text{C}$); such temperatures are attainable in a helium cryostat.

Instead of electrons, one might consider other spin-1/2 particles such as a proton, which is a positively charged particle, about 1836 times heavier than an electron, and forms the nucleus of a hydrogen atom; together with neutral particles of similar mass called neutrons, protons are the building blocks of all atomic nuclei. For nuclei, μ_B in Eq. (7.25) is replaced by the nuclear magneton, $\mu_N = ge\hbar/2m_p$, where $m_p \approx 1836 m_e$; there is also considerable variation of the g-factor: for a proton $g \approx 5.6$, for a carbon ^{13}C nucleus $g \approx 2.4$, and for a phosphorus (^{31}P) nucleus $g \approx 3.9$. Thus, the cooling requirements for nuclei are more daunting, with temperatures far below 4×10^{-3} kelvin being required.

7.2.1 Addressability

So, everything looks pretty good with using electrons as qubits so far: they are two-level systems, with a nice energy difference ΔE, making the cooling requirements not too daunting. However, when we want to perform single-qubit gates, we must apply an oscillating force. In the case of spin-1/2 charged particles, a convenient oscillating force is a magnetic field along the x-axis, perpendicular to the z-axis, which is the direction of the strong magnetic field $\mathbf{B}(t) = \mathbf{e}_x B_{\text{rf}} \cos(\omega t)$. The subscript "rf" stands for "radio-frequency": the oscillating field frequencies typically used for **electron spin resonance (ESR)** or **nuclear magnetic resonance (NMR)** experiments of this kind are in the range of radio-frequencies (3×10^3 to 3×10^{11} hertz). Instead of $-\hat{\mathbf{r}} \cdot \mathbf{F}$, which we saw in Section 6.3, the interaction Hamiltonian is $\hat{\boldsymbol{\mu}} \cdot \mathbf{B}(t) = -\mu_B B_{\text{rf}} \hat{S}_x/\hbar$: after that, the dynamics is very similar, with the Rabi frequency (recall Eq. (6.28)) is now given by the expression $\Omega = \mu_B B_{rf}/\hbar$.

What happens when you want to have more than one such qubit in your quantum computer? To operate a quantum computer, you must be able to address each qubit in isolation, or in other words, apply a unitary operation to one qubit, while leaving the other qubits untouched. The problem with radio-frequency fields is that they have wavelengths of the order of centimetres, and focusing any wave field into an area smaller than a square wavelength is difficult. No matter how you might try, the wavelength is a very strong limit to the spatial resolution of any wave field used for sensing, measurement, or control. Lenses or antennas always have a finite aperture, and the effect of diffraction means that the focus spot size in any camera or microscope will be at least about a wavelength across; using a very small sub-wavelength aperture or source to localize the field only results in most of the incident power being reflected or absorbed, while the small amount of radiation that may be transmitted will spread out very rapidly. To control an electron qubit in isolation, one needs all the other qubits to be several centimetres away – which is a very long distance in micro-electronics. Not only will any qubit–qubit interaction, which we will need to exploit to perform entangling quantum gates, be weak at that separation, there is also the problem of ensuring every qubit has the same static magnetic field strength, B_0, and any variation leads to dephasing.

If we are going to use spin-1/2 particles as our qubits, we must control the qubits' immediate environment, by implanting them either in some natural host (like an atom or molecule, or a vacancy defect in a crystal) or in some artificial host (like a semiconductor **quantum dot**, i.e., a solid state device designed to trap and control single electrons). These approaches overcome the fundamental issue of addressability by radically changing the dynamics of the electron, but they substitute a whole host of other experimental problems such as initialization/cooling and fabrication, which must be overcome.

As we write, in the spring of 2025, a variety of such approaches are under intensive development; to date, the most successful has been using electrons confined in trapped atomic ions, which we will discuss in detail next.

⚓ **Physics Tips 7.2**

Angular Momentum. A rigid body is, by definition, very difficult to deform. When no external force acts, its motion can be divided into its linear motion, which can be characterized by the position and velocity of the body's centre of mass, and its rotation, in which all of the parts of the body describe circles whose centres lie along a common axis of rotation (which passes through the centre of mass). It is convenient to quantify the angular motion for the body by a vector called the angular velocity $\boldsymbol{\omega} = \mathbf{n}\omega$, whose direction is along the axis of rotation (along the unit vector $\mathbf{n}$), and whose magnitude is the rate of rotation ω, in radians per second. Analogous to the linear momentum is the angular momentum which, at least for bodies with simple shapes, is given by $I\boldsymbol{\omega}$. The constant of proportionality I is called the moment of inertia, which quantifies how easy it is to increase the angular velocity. For example, a slingshot whirled above the head moves in a circle and has angular momentum that depends on the length of the sling, the mass of the stone in the sling, and the speed of rotation. The moment of inertia is equal to the mass times the square of the length of the sling.

In quantum physics, angular momentum, $\hat{\mathbf{J}}$, is quantized: in any direction, its cartesian components can only have certain discrete values, and the smallest difference between any two of these allowed values is Planck's constant $\hbar$. Thus, the component of $\hat{\mathbf{J}}$ in any direction – for the sake of definiteness, this is the z-direction – can only take the values $\hbar j$, $\hbar(j-1)$, $\hbar(j-2)$, ..., $\hbar(j-N)$, where $\hbar j$ is the maximum possible value and $N+1$ is the total number of possible values (which must be a finite positive integer). Since the choice of direction is arbitrary, we can equally consider the component along the exact opposite direction and should get consistent results: thus, the largest value must be the negative of the smallest value: $(j-N) = -j$, which implies $j = N/2$. Thus, j can only take the values $\{0, 1/2, 1, 3/2, 2, ...\}$, and the number of possible values of $\hat{J}_z$ will be $N+1 = 2j+1$.

Quantization of the components of angular momentum also has implications for the length of the vector: suppose a vector $\mathbf{V}$ has a fixed length but random orientation, then $\mathbf{V} \cdot \mathbf{V} = 3\langle V_z^2 \rangle$, where $\langle V_z^2 \rangle$ is the average of the square of the z-component averaged over all possible orientations. For the quantum angular momentum, with all $2j+1$ possible values of $\hat{J}_z$ equally likely (which is equivalent to assuming all directions are equally likely) we find

$$\mathbf{J}^2 = 3\langle \hat{J}_z^2 \rangle = \frac{3\hbar^2}{2j+1} \sum_{m=-j}^{j} m^2 = \hbar^2 j(j+1), \tag{7.27}$$

where we have used the formula $\sum_{m=-j}^{j} m^2 = j(j+1)(2j+1)/3$, which is valid if j is either an integer or a half-integer.

Fundamental particles have an intrinsic angular momentum called the *spin*, which is denoted by $\hat{\mathbf{S}}$ and s. While the orientation of $\hat{\mathbf{S}}$ can change, s is a fixed property, like the charge and the rest mass; there is experimental evidence for fundamental particles with spin $s = 0$, $s = 1/2$, and $s = 1$, although other values are certainly theoretical possibilities; composite particles such as atomic nuclei can have higher values. In addition, a particle can have an angular momentum due to its motion (or orbit), generally denoted by $\hat{\mathbf{L}}$ and ℓ. Orbital angular momentum is not constant: the particle can transition from one trajectory to another with a different value of ℓ; however, ℓ can only take integer values.

Exercises 7.2.2

1. Find the energy splitting, ΔE, for a spin-based qubit in a 1 tesla magnetic field for (a) an electron; (b) a proton; and (c) a ^{13}C nucleus.
2. What is the resonance frequency and corresponding wavelength for a spin-based qubit in a 1 tesla magnetic field for (a) an electron; (b) a proton; and (c) a ^{13}C nucleus?
3. A vector $\mathbf{V}$ has a fixed length but a completely random orientation in the three dimensions, with all directions having equal probability. Suppose you can measure the component of $\mathbf{V}$ along the z-axis. Show that the average value of V_z^2 is equal to $\mathbf{V} \cdot \mathbf{V}/3$.
4. Prove that, for any integer or half-integer value of j,

$$\sum_{m=-j}^{j} m^2 = j(j+1)(2j+1)/3.$$

Hint: Try the method of mathematical induction.

7.3 Atomic Ions

Following Thomson's discovery of the electron, scientists immediately pursued the problem of how electrons fit together to make different types of atoms. Thomson himself favoured a "plum pudding" model, in which the atom was a squishy, positively charged sponge with electrons implanted like plums in a cake. His former student, Ernest Rutherford, then at Manchester, famously had another idea: based on scattering experiments conducted by two of his associates, Hans Geiger and Ernest Marsden, in 1911 he proposed that electrons rotated around a compact, positively charged, atomic nucleus, just like planets orbit the Sun. While this explained the Geiger–Marsden results, it did not really explain the observed optical properties of atoms, which are spherically symmetric and which radiate light at a few discrete wavelengths rather than at all wavelengths (which is what you would expect from two charged particles orbiting each other). In 1913, Niels Bohr had another idea.

7.3.1 Atomic Structure

To form a stable circular orbit of radius r, the electron must have an *orbital angular momentum* $L = m_e r^2 \omega$ so that the centripetal force $m_e r \omega^2 = L^2/m_e r^3$ will balance the Coulomb attraction of the nucleus:

$$\frac{L^2}{m_e r^3} = \frac{e^2}{4\pi \epsilon_0 r^2}. \tag{7.28}$$

Thus, we require $r = L^2 4\pi \epsilon_0/m_e e^2$. Bohr's idea was simple: suppose L is also required to be some integer (the *principal quantum number*, n) times Planck's constant: $L = n\hbar$. This implies that the orbital radius will be given by $r = n^2 a_0$, where $a_0 = \hbar^2 4\pi \epsilon_0/m_e e^2$ is a constant (the Bohr radius) approximately equal to 5.3×10^{-11} metres. The electron's total energy (kinetic and electrostatic) must be

$$E(n) = -\frac{E_I}{n^2}, \tag{7.29}$$

where $E_I = e^2/8\pi\epsilon_0 a_0^2$ is a constant (the ionization energy of hydrogen), whose value is approximately 2.2×10^{-18} joules (you often see it quoted as 13.6 electronvolts, where the electronvolt is a unit of energy given by 1.6×10^{-19} joules; using this unit in atomic physics means we get numbers without too may big exponents). Negative energy implies that the electron is in a bound state, circling the atomic nucleus, rather than an unbound state, getting deflected while whizzing by the atomic nucleus. Thus, Bohr's idea implied that only certain discrete orbits are allowed, with well-defined radii and energies. When an electron jumps from one of these orbits (of quantum number n) to another with lower **principal quantum number** (n'), it emits a photon of energy:

$$\hbar\omega = E(n) - E(n'). \tag{7.30}$$

Until that time, Planck's ideas about quanta were thought only to apply to light: Bohr took a leap into the unknown by supposing quantization applied to matter as well. His idea was vindicated because his formula for the photon energy, Eq. (7.30), agreed exactly with the formula that had been worked out from experimental data 25 years previously by Johannes Rydberg for the frequencies of light emitted by excited hydrogen gas.

Arnold Sommerfeld subsequently refined Bohr's ideas, generalizing the result to elliptical orbits as well as circular ones, showing that the principal quantum number is a sum $n = \ell + k$, where ℓ is the orbital angular momentum and k is another integer related to the eccentricity of the orbit. One subtlety requires careful thought: even if the electron has no orbital angular momentum, it can still form a stable configuration around the nucleus. One would expect the positively charged nucleus to just suck in the negatively charged electron until they merge: but, like the photon, the electron refuses to be confined in a small volume. The wave theory of matter, developed by Erwin Schrödinger some 10 years after Bohr, fully explained this in a quantitative manner.

These semi-classical models give good agreement with experimental data, but they should not be taken too seriously. First of all, atoms are spherically symmetric, while the Solar System, in which all the planets orbit in the same direction in a single plane called the ecliptic, certainly is not spherically symmetric. Bohr's orbits can be in any plane, and in either direction (clockwise or anti-clockwise): each is a configuration in the sense of the word we used in Chapter 1. The overall quantum state is then a superposition of these different states, which gives an overall spherically symmetric quantum state. Further, the electrons do not follow a well-defined elliptical path, rather the classical elliptical orbit is the average path, and the full wave theory of matter is required to figure out the actual probability amplitude describing the particle's path. The basic idea, however, remains sound: the atomic electrons exist in one of a number of configurations, each with a well-defined energy, and thus they have the potential to be qubits for a quantum computer.

Multi-Electron Atoms, Closed Shells and Quantum Defects

Hydrogen is not an atomic species used to make qubits (the wavelengths are not particularly convenient for laser systems, and neutral atoms are more tricky to trap than ions). Multi-electron atoms have much more complicated atomic structure, since one has to deal with not only electrons interacting with the nucleus but also their interactions with each other. Fortunately, electrons like to get cosy with each other and form **closed shells**, which can be thought of as spherically symmetric spheres of electrons packed closely together around the nucleus. For certain atoms, all of their electrons are in such closed shells (at least when the electrons are not excited out of their ground state); these atoms turn out to be very unresponsive to chemical reactions: the electrons really do not want to be moved from their closed shells. Such atoms are called *inert*; examples are helium (2 electrons), neon (10 electrons), argon (18 electrons), krypton (36 electrons), xenon (54 electrons), and radon (86 electrons). An atom that has all of its electrons in a closed shell *except for one* is effectively a single-electron atom; examples are

Table 7.1 Quantum defect and spin–orbit splitting data for low-lying quantum states of singly-ionized calcium

n	ℓ	$\delta_n(\ell)$	$\zeta(n, \ell)$ (eV)
4	0	1.85949	0
5	0	1.82734	0
4	1	1.50386	0.018423
5	1	1.46796	0.006468
3	2	0.687874	0.003009
4	2	0.641771	0.000952

lithium (3 electrons), sodium (11 electrons), potassium (19 electrons), rubidium (37 electrons), and cesium (55 electrons). These all have a particularly simple atomic structure closely related to that of hydrogen, and are often the neutral atoms used in modern atomic physics experiments in which atoms are cooled and trapped. For quantum computing applications, singly ionized atoms have proven even more useful. These are atoms that have had one electron stripped off, so that the atom has a net positive charge. This charge is very useful for trapping and cooling ions and also for performing controlled quantum interactions between two or more ions: in other words, two of the essential requirements of a quantum computer. Examples of single-electron ions are beryllium (3 electrons), magnesium (11 electrons), calcium (19 electrons), strontium (37 electrons), and barium (55 electrons), all of which have been used as qubits in quantum information experiments.

The energy levels of a one-electron atom obey the following rule, similar to Bohr's result:

$$E(n) = -\frac{Z^2 E_I}{(n - \delta_n(\ell))^2}, \tag{7.31}$$

where $+Ze$ is the net charge on the nucleus and closed shells ($Z = 1$ for neutral atoms; $Z = 2$ for singly ionized atoms). The quantity $\delta_n(\ell)$ is the **quantum defect**, which quantifies how much the electron orbit penetrates the inner core electrons. For orbits with low angular momentum, the electrons will be pretty close to the nucleus, and the effective electro-static attraction will be quite strong; for orbits with larger angular momentum, the effect is much weaker. Quantum defects can be determined from experiments by fitting spectroscopic data, or they can be calculated using solutions of the Schrödinger wave equation. As an example, quantum defect data for singly ionized calcium is shown in Table 7.1; for calcium, the orbits with $n = 1, \ell = 0$, $n = 2, \ell = 0$, $n = 2, \ell = 1$, $n = 3, \ell = 0$, and $n = 3, \ell = 1$ are all filled by the 18 inner core electrons, so the lowest energy state has $n = 4, \ell = 0$, and has energy $-4E_I/(4 - 1.85949)^2 = -11.9$ eV, while the $n = 4, \ell = 1$ level has energy $-4E_I/(4 - 1.50386)^2 = -8.73$ eV, and so on.

Fine Structure

So far we have ignored the spin of atomic electrons. As far as we are concerned, looking at the atom, the electrons are whizzing around the atomic nuclei. But the electrons will have a different perspective on things: they will see a charged atomic nucleus doing the whizzing around. Charges moving in closed loops generate magnetic fields, and the magnetic field generated by the nucleus's motion will interact with the electron's magnetic moment. As before, an electron in a magnetic field can align itself in only two ways: either parallel or anti-parallel to the magnetic field. These two configurations have a *total angular momentum* $\hat{\mathbf{J}} = \hat{\mathbf{L}} + \hat{\mathbf{S}}$, and so $j = \ell + 1/2$ or $j = \ell - 1/2$. When the spin and orbital angular momenta are parallel, the energy of the configuration is shifted up by a small amount $\zeta(n,\ell)\ell/2$; conversely, when they are anti-parallel, the energy is shifted by an amount $-\zeta(n,\ell)(\ell+1)/2$. The value of $\zeta(n,\ell)$ can be determined using the wave theory with good accuracy, though we can treat it as an empirical constant as well. Thus, referring to the table again, we see that the $n=4, \ell=1$ level (energy -8.73 eV) will be split into two levels: one with $j=3/2$ and energy $(-8.73+0.018423/2)=-8.72$ eV and another with $j=1/2$ and energy $(-8.73-0.018423)=-8.75$ eV.

These energy levels are still degenerate: the component of the total angular momentum $\hat{J}_z$ can take any value from $\hbar j$ to $-\hbar j$ in steps of $\hbar$, which is a total of $2j+1$ values.

Spectroscopic Notation
It is standard in atomic spectroscopy to use a special notation for these levels, which encodes n, ℓ, and j in a special symbol, as follows:

$$n\ ^{2s+1}\mathcal{L}_j. \tag{7.32}$$

Here n is the principal quantum number as before. In the superscript, which is called the *multiplicity* (it is *not* indicating n is raised to the power $2s+1$), s is the total spin angular momentum ($s=1/2$ for single-electron atoms, so if you see any value other than 2 for this subscript, you are dealing with two or more electrons); when being read out, $2s+1=1$ is called the "singlet", $2s+1=2$ is called the "doublet", $2s+1=3$ is called the "triplet" and so on. The capital letter $\mathcal{L}$ stands for a letter denoting the orbital angular momentum: if $\ell=0, \mathcal{L}=$ S (for "sharp"); $\ell=1, \mathcal{L}=$ P (for "principal"); $\ell=2, \mathcal{L}=$ D (for "diffuse"); $\ell=3, \mathcal{L}=$ F (for "fundamental"); for higher values of ℓ, the letters G, H etc. are used without any special names attached. These names originate from the observed properties of atomic spectral lines, pre-dating the discovery of quantum mechanics. The final subscript is just the value of j; for a single-electron atom, j is always either $\ell+1/2$ or $\ell-1/2$. Thus, for example, the $n=4, \ell=1, j=3/2$ level (energy -8.72 eV) is denoted $4\ ^2P_{3/2}$, and is pronounced "four, doublet-P, three-halves" (and often the principal quantum number is omitted).

Zeeman Effect: Isolating Two Levels to Make a Qubit
We are getting close now: we have the different energy levels of the electron bound to our ion. But these are still degenerate: in each level there are $2j+1$ states, all with the same energy. Qubit levels should not be degenerate at all. What can we do? This degeneracy is a result of the spherical symmetry of the atom; if we break that symmetry by, for example, applying an external static magnetic field $\mathbf{B}_0=B_0\mathbf{e}_z$, just like we did with the spin-based qubits in the previous section, what will happen? The component of the total angular momentum $\hat{\mathbf{J}}$ along the direction of $\mathbf{B}_0$ can take $2j+1$ values $j, j-1, j-2, \ldots, -j+1, -j$; each value will correspond to a single *non-degenerate* state; we will denote the value of this component by M_J. The energy shifts of these states can be calculated in just the same manner as we did for the free electrons, combining the orbital and the spin components of the angular momentum to give a net magnetic moment for the electron in its orbit: $\Delta E = g_J \mu_B B_0 M_J$. The coefficient g_J is called the Landé g-factor; its value depends on ℓ and j:

$$g_J = \begin{cases} 1+\dfrac{1}{2\ell+1} & \text{if } j=\ell+1/2, \\[2em] 1-\dfrac{1}{2\ell+1} & \text{if } j=\ell-1/2. \end{cases} \tag{7.33}$$

So finally we have what we seek: a set of non-degenerate atomic levels, with distinct energies from which we can choose two levels to form the two states of our qubit.

Atoms as Qubits

While individual atoms make pretty good qubits, we still have the problem of controlling their location and managing their interactions to make two-qubit quantum logic gates. There are various approaches to this problem which are under active consideration:

- **Neutral Atoms**: These can be trapped using lasers and magnetic fields, and many fascinating physics experiments using the ultra-cold gases of atoms have been carried out over the past quarter-century. However, isolating *single* atoms to make qubits is more challenging: lasers in standing waves can form an **optical lattice** of potential wells, rather like the dimples in an egg-carton. However, for this to be a quantum computer, one must address the problem of ensuring that one, and only one, atom occupies each dimple.
- **Quantum Dots**: We can also build artificial atoms inside crystals by ingeniously varying the crystal's constituent atoms in some particular region to create a quantum well of lower electric potential. If quantum wells confine the electron in all three directions they are called quantum dots. These have the virtue that their location is well-characterized, but at the expense of considerable difficulty of actually making these devices in a precise and repeatable manner.
- **Nitrogen-Vacancy Colour Centres**: Naturally occurring features such as a point defect in a crystal have many atom-like characteristics. An important example is the nitrogen-vacancy defect in diamond: two adjacent carbon atoms are absent from the lattice, one being replaced by a single nitrogen atom, the other site being left vacant; the resulting potential can trap a single electron in an atom-like state. Controllable multi-qubit gates involving large numbers of qubits remain a challenge.
- **Trapped Ions**: This the most successful way to use atoms to create qubits to date, with dozens of qubits being successfully entangled with high precision. Although ultimately it might prove limited in the numbers of qubits, and in the speed of processing, we will describe its working in some detail.

7.3.2 Ion Trapping

Earnshaw's Theorem

Suppose we want to trap and hold a charged particle in a precise position. Atoms are too small to be picked up and manipulated with something mechanical like a pair of tweezers. The obvious next thing to do is to try to devise an array of electrodes to set up a static (i.e., not changing in time) electric field which can do the job. This simple approach is, unfortunately, impossible due to the nature of the electric field. The static electric force on a charge q is given by the gradient of the electrostatic potential $\Phi(\mathbf{r})$: $\mathbf{F} = -q\nabla\Phi(\mathbf{r})$. A nice way to think about this is to imagine yourself on a snowy mountain, with $\Phi(\mathbf{r})$ being the height of the piece of land where you are standing: the gradient will dictate both the direction you will start to slide down and the amount of acceleration you will experience. At an equilibrium point, there is zero force on the particle, and so the derivatives in all three directions will be zero: $\partial\Phi/\partial x = \partial\Phi/\partial y = \partial\Phi/\partial z = 0$. In our mountain analogy, this is like being on a narrow ledge which is flat, so you do not immediately start sliding in any direction; but you must be careful, since the ledge might be very narrow, and if lose your balance and fall a little bit to one side, then you might end up sliding down a steep slope. If the equilibrium point is stable, it must represent a minimum in all directions, so the second derivatives must all be positive; this is like being at the bottom of some bowl-shaped feature on the snowy mountain: if you tripped and fell in any direction, you would always slide back to the bottom of the bowl.

Unfortunately, this is where the mountain analogy fails: unlike the height of a mountain slope, the electrostatic potential $\Phi(\mathbf{r})$ obeys Laplace's equation, $\partial^2\Phi/\partial x^2 + \partial^2\Phi/\partial y^2 + \partial^2\Phi/\partial z^2 = 0$, which (although it does not appear immediately obvious) is just another way of writing down Coulomb's law. Suppose $\partial^2\Phi/\partial x^2$ and $\partial^2\Phi/\partial y^2$ are both positive, so that the potential is a minimum with respect to any displacement in those two directions; in the third direction, z, it follows from Laplace's equation that $\partial^2\Phi/\partial z^2$ must be negative. Any equilibrium point of the electrostatic field will at best be a saddle point: stable in two directions but unstable in the third, meaning that even the smallest disturbance will result in the charged particle falling out of the saddle – i.e., it will start accelerating away from the equilibrium point along the direction in which the second derivative of Φ is negative.

Paul Traps

This problem can be overcome by oscillating the saddle point. Suppose the charged particle begins to fall out of the saddle. Before it gets too far, the potential is flipped, so the direction in which the potential was a maximum now becomes the direction in which it is a minimum. The potential then pushes the displaced particle back to the equilibrium point. Meanwhile, the particle might start falling out of the saddle in the new unstable direction; to stop that happening, the potential is flipped back to its original configuration. Repeating this procedure over and over, you are in effect varying the electric field potential at some frequency: mathematically, one can determine which frequencies (usually MHz range) will result in an effective static trapping potential.

Using a combination of radio-frequency Paul traps and conventional electrostatic trapping (which can work in two out of three directions), one can design and build a linear ion trap, in which the ions are strongly confined in two out of the three directions, and weakly confined in the third. Since the ions are all positively charged, they repel each other and will naturally line up along the weak trapping axis, forming a linear crystal. Typically, the ions are about 10 micrometres apart, so that it is feasible to focus a laser on individual ions without illuminating their neighbours. Further, an oscillation of an ion about its equilibrium position will be strongly coupled to the other ions in the chain. This coupling can best be described by a set of *normal modes*, in which the ions are all oscillating at a common frequency, albeit with differing amplitudes and directions. These normal modes are exploited to perform two-qubit quantum gates, as we will see in the next chapter.

 Interesting Facts 7.3

The Invention of Radio-Frequency Ion Traps. The impossibility of confining a charged particle with a *static* electric field was discovered by the Reverend Samuel Earnshaw (1805–1888); despite being the top physics and mathematics student of his year at Cambridge, he pursued an ecclesiastical career, becoming a vicar in Sheffield. Methods to overcome the challenge of Earnshaw's theorem perplexed many scientists over the years, and it was solved by Wolfgang Paul (1913–1993). One morning he was taking his wife breakfast in bed. The breakfast was hard-boiled eggs. He realized, climbing the stairs with the eggs balanced on a plate, that it was much easier to carry them if he continuously rocked the plate back and forth. Thus comes the idea of an oscillating potential producing an *effective* trapping potential, which can confine charged particles without violating Earnshaw's theorem.

Laser Cooling, Control and Read-Out

We have the essentials of a quantum computer: individual atomic qubits with well-characterized locations. For the sake of argument, let us choose two states of the calcium ion to be our qubit: $4^2S_{1/2}$, $M_J = -1/2$ will be the $|0\rangle$ of our qubit and $3^2D_{5/2}$, $M_J = -3/2$ will be the $|1\rangle$. A spectrally narrow laser beam, precisely tuned to the transition frequency between these two levels (with all other transitions off-resonance) will ensure two-level dynamics takes place: only $|0\rangle$ and $|1\rangle$ will ever have a non-negligible probability amplitude. Further, optical pumping is a standard technique by which lasers can prepare these atomic qubits in the state $|0\rangle$, then laser pulses with precisely controlled durations can be used to perform the single-qubit gates in a manner similar to what we saw in Section 6.3. There is one subtlety: with these two specific states, the matrix element $\langle 0|\hat{\mathbf{r}}|1\rangle = 0$ and the transition is dipole forbidden; this occurs when the change in total angular momentum is greater than $\pm\hbar$; in this case $j = 1/2$ for $|0\rangle$ and $j = 5/2$ for $|1\rangle$, so $\Delta j = 2$. However, it can be shown that $\langle 0|\hat{x}_i\hat{x}_j|1\rangle \neq 0$, so that quadrupole transitions are allowed; the practical implications of this are that more laser power is required to drive the transition, and also the spontaneous decay rate of the qubit is very long. The wavelength of laser required is approximately 7.29×10^{-7} metres, which means we can focus the laser on individual ion qubits without requiring the ions to be excessively far apart.

Suppose one applied a laser precisely tuned to the transition wavelength between the $|0\rangle = 4^2S_{1/2}$, $M_J = -1/2$ and $4^2P_{1/2}$, $M_J = -1/2$ levels. The $4^2P_{1/2}$, $M_J = -1/2$ will be excited, but will almost immediately decay, emitting a photon and, at least most of the time, the electron returns to the initial $4^2S_{1/2}$, $M_J = -1/2$ state, allowing the process to be repeated (there are some subtleties: one must take steps to negate the effect of the atom decaying to one of the other atomic levels, requiring more than one laser wavelength). In other words, the ion can scatter thousands of photons if, *and only if*, the qubit was initially in state $|0\rangle$; if it was in state $|1\rangle$, nothing happens: since the scattered photons can be imaged using conventional optics, we have a highly efficient means to read out each qubit.

The only thing missing is the ability to perform two-qubit logic gates, which will be discussed in the next chapter.

7.3.3 Assessment

Ion traps were arguably the first truly practical quantum computing technology to be proposed, and it is no coincidence it was based on ultra-high precision atomic spectroscopy. Entangled quantum states are like an exotic tropical flower, so delicate that it shrivels to nothing unless it is handled in precisely the correct manner; similar meticulousness is the hallmark of optical frequency standards. Early experiments proving the principle of quantum entanglement engineering, such as deterministic teleportation, were first performed with trapped ions, and dozens of trapped ion qubits are routinely realized in this technology. Ambitious proposals for up to a thousand qubits are being aggressively pursued.

Exercises 7.3.4

1. Draw a diagram showing the energies of the low-lying $4s$, $3d$, and $4p$ configurations of the calcium ion. The ground state is the $4\,^2S_{1/2}$ (although beware, since it is possible that some states with principal quantum number $n < 4$ might be more energetic than the ground state). You should consider S, P, and D states, and include the fine structure splitting.

2. Find the laser wavelengths needed to excite transitions between ground state and the other levels you found in Exercise 1. Which of the transitions are "dipole allowed" and which are "dipole forbidden"?

3. Suppose a magnetic field of 0.01 tesla is applied to a trapped calcium ion. Calculate the change in resonance frequency of the $3^2D_{5/2}$ sub-levels (express you answer in hertz).

4. Two ions, both of charge e, are trapped in a Paul trap. Suppose that there is a very strong trapping potential in the x, and y-directions, so that the ions lie along the z-axis. Each ion's potential energy due to the trap is $U(z) = \kappa z^2/2$.

(a) What is the trapping force on an ion located at position z?

(b) The Coulomb force between two charges acts along the line joining the two charges and has strength $F = e^2/4\pi\epsilon_0 d^2$, where d is the separation distance of the two charges. Find the separation of two trapped ions when the net force on both ions is zero.

(c) Suppose one of the ions were displaced by a small amount δz. Find expressions for the forces acting on both ions.

7.4 Electrical Circuits

The qubits we have considered so far are all naturally occurring, and one has to adapt the technology to cope with nature's foibles. Admittedly, in some cases such as quantum dots, there is a lot of fabrication effort being made to create the qubit, but they still rely on the basic physical properties of a fundamental particle. But what if we could design and build a qubit entirely from scratch? Obviously we will need a system whose behaviour is governed by the principles of quantum mechanics we discussed in Chapter 2. This is not as difficult as you might think: fundamentally *all* systems are governed by quantum mechanics; the "classical" way of looking at their behaviour is nothing more than a particular limiting case of quantum mechanics, in which systems are continuously interacting with their environments and are subject to random thermal stimuli, with the result that one particular configuration has an overwhelmingly high probability, while all other possibilities become virtually impossible. Thus, to see quantum behaviour we need to cool and isolate our system; both things are more easily achieved with simple, microscopic systems; but simplicity and smallness are not pre-requisites for quantum behaviour.

 Physics Tips 7.4

Current Electricity

- **Current and Potential Difference.** An electrical circuit is a collection of electronic devices connected by conductors. The electric current I is the rate at which charge passes any point in a circuit, $I = \dot{Q}$. A potential difference $\Delta V = V_1 - V_2$ (units: volts) between points 1 and 2 in the circuit implies that the energy required to move a charge Q between 1 and 2 is $Q\Delta V$ (or, equivalently, an average electric force of $-Q\Delta V/\ell$ is acting, where ℓ is the distance between 1 and 2). Any two points connected by a good conductor will be at the same potential.
- **Resistors**

$$V_1 \;\overset{R}{\wedge\!\wedge\!\wedge}\; V_2$$

Figure 7.3 Resistor.

A resistor is a device that impedes the flow of current (Fig. 7.3). The average electric force $-Q\Delta V/\ell$ will accelerate charge; however, collisions with various impurities in the conductor slow the charges down again, so a constant average drift velocity is rapidly attained. This equilibrium is expressed by Ohm's law: $\Delta V = IR$, where R is the resistance (units: ohms, or volts per amp).

- **Capacitors**

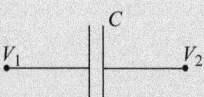

Figure 7.4 Capacitor.

Capacitors stop the flow of current and thus accumulate charge (Fig. 7.4). The charge stored depends on the potential difference: $Q = C\Delta V$, where C is the capacitance (units: farads). For a parallel plate capacitor $C = \epsilon_0 \mathscr{A}/d$, where $\mathscr{A}$ is the area of overlap of the plates and d is their separation. Energy stored $E_C = Q^2/2C$.

- **Inductors**

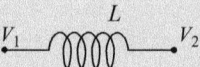

Figure 7.5 Inductor.

A current flowing through a closed loop of any shape (it doesn't have to be circular) will induce a magnetic field at points both inside and outside the loop (Fig. 7.5). The magnetic flux density **B** will vary from point to point; the *magnetic flux* Φ (units: webers, equivalent to tesla metre-squared) is the integral of the perpendicular component of **B** over any area enclosed by the loop (it doesn't matter which area, so long as it is bounded by the loop). If the loop has N turns (i.e., the wire is wound around the loop N times), then the *flux-linkage* Ψ is defined as $N\Phi$. The flux-linkage is proportional to the current I flowing through the closed loop $\Psi = LI$, where the inductance L (units: henries) depends on the size and shape of the coil. The law of magnetic induction states that a change in flux-linkage induces a voltage difference as follows: $\Delta V = -\dot{\Psi} = -L\dot{I}$. Further, since the power is $P = \Delta VI$, integrating we find the energy stored in the inductor is $E_I = LI^2/2 = \Psi^2/2L$.

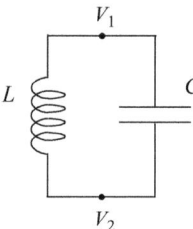

Figure 7.6 LC circuit.

Circuit electronics is used in an awful lot of modern technology: might we use it to create a qubit? Let us consider a simple circuit made of an inductor L and a capacitor C (see Fig. 7.6). If the capacitor is initially charged and is then connected to the inductor, a current I will start to flow to discharge the capacitor; but that current induces a flux-linkage $\Psi = LI$ in the inductor, which in turn induces a voltage $\Delta V = -\dot{\Psi} = L\dot{I}$. The result is that by the time the capacitor has discharged there is a strong voltage by which to re-charge it: in other words there will be an oscillation in which the charge is converted to flux-linkage and then the flux-linkage is converted back into charge.

In equations: the voltage induced by the inductor is $\Delta V(t) = -\dot{\Psi}(t)$; and the charge on the capacitor is $\Delta V(t) = Q(t)/C$, hence the charge on the capacitor is related to the net flux in the inductor by

$$Q(t) = -C\dot{\Psi}(t). \tag{7.34}$$

By definition, the current is the time derivative of the charge, so $I(t) = \dot{Q}(t)$, and the flux-linkage in the inductor is related to the current by $\Psi(t) = LI(t)$; thus, we also have the relation

$$\Psi(t) = L\dot{Q}(t). \tag{7.35}$$

Thus, eliminating $Q(t)$, we find $\Psi(t) = L\dot{Q}(t) = -LC\ddot{\Psi}(t)$; or, re-writing:

$$\ddot{\Psi}(t) + \omega^2 \Psi(t) = 0, \tag{7.36}$$

where $\omega = 1/\sqrt{LC}$.

This is the ordinary differential equation for a simple harmonic oscillator, so the solution is

$$\Psi(t) = \Psi(0)\cos(\omega t) + \dot{\Psi}(0)\sin(\omega t)/\omega. \tag{7.37}$$

The total energy stored in the circuit is the sum of the energy in the inductor and the energy in the capacitor:

$$E = \frac{1}{2L}\Psi^2 + \frac{1}{2C}Q^2. \tag{7.38}$$

Such resonant LC circuits have been in use for many years, and have many applications: for example, they are very important for the analogue radio and TV broadcasting used for much of the twentieth century: such a resonant circuit can pick up a carrier radio signal at a known frequency, while filtering out other frequencies, and the modulation of that signal was used to transmit the sound and/or pictures of your favourite radio or TV shows. But how can such a thoroughly classical concept be "quantum"?

The clue is in the quantum analysis of photons we saw in Section 7.1. Just like the photon, a quantum harmonic oscillator has a set of discrete, equally spaced energy levels, which we denote $|0\rangle$, $|1\rangle$, $|2\rangle$ and so on; these have energies $\hbar\omega(1/2)$, $\hbar\omega(1 + 1/2)$, $\hbar\omega(2 + 1/2)$, respectively. The annihilation operator $\hat{a}$ and its Hermitian conjugate, the creation operator $\hat{a}^\dagger$, can be used to express the observable quantities as Hermitian operators (specifically, the flux-linkage and the charge):

$$\hat{\Psi} = \sqrt{\frac{\hbar\omega L}{2}}(\hat{a}^\dagger + \hat{a}), \tag{7.39}$$

$$\hat{Q} = i\sqrt{\frac{\hbar\omega C}{2}}(\hat{a}^\dagger - \hat{a}). \tag{7.40}$$

The energy becomes the Hamiltonian operator:

$$\hat{H} = \frac{1}{2C}\hat{Q}^2 + \frac{1}{2L}\hat{\Psi}^2$$
$$= \hbar\omega(\hat{a}^\dagger\hat{a} + 1/2). \tag{7.41}$$

Writing the quantum state of the oscillator in terms of the number states $\{|0\rangle, |1\rangle, |2\rangle, \ldots\}$:

$$|\varphi(t)\rangle = \sum_{n=0}^{\infty} c_n \exp[-i\omega(n + 1/2)t]|n\rangle, \tag{7.42}$$

we find that the expected value of the flux-linkage is

$$\langle\varphi(t)|\hat{\Psi}|\varphi(t)\rangle = \sqrt{2\hbar\omega L}\,(\mathrm{Re}\{a\}\cos(\omega t) + \mathrm{Im}\{a\}\sin(\omega t)), \tag{7.43}$$

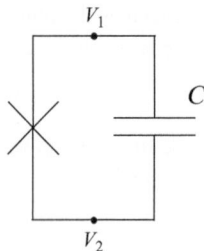

Figure 7.7   *LC* circuit with a Josephson junction (symbolized by an X) as the inductor.

where $a = \sum_m \sqrt{m+1} c_{m+1} c_m^*$. Thus, if we make the identification $\Psi(0) = \sqrt{2\hbar\omega L}\mathrm{Re}\{a\}$ and $\dot{\Psi}(0)/\omega = \sqrt{2\hbar\omega L}\mathrm{Im}\{a\}$ the classical and quantum predictions are in complete accord.

We know both the flux-linkage and charge in the circuit are quantum mechanical variables – simply because *everything* is a quantum mechanical variable. If the conditions are right, then the quantum mechanical properties will become manifest. In this case, for the circuit to be in this regime, we will need the temperature T to be much colder than $\hbar\omega/K_B$; if we put in typical values for the impedance and capacitance of, say 10^{-8} henries and 10^{-12} farads, respectively, then $\omega = 10^{10}$ radians per second, which means the temperature must be much lower than 0.076 K: in other words, if the *LC* circuit is cooled to milli-kelvin temperatures, then quantum behaviour should be seen.

Unfortunately, as we saw in Section 7.1, quantum harmonic oscillators do not make good qubits, because all their energy levels are equally spaced. However, if we modify the *LC* circuit by using a **Josephson junction** in place of the inductor (Fig. 7.7), the relation between the "flux" and the current is now

$$I(t) = I_c \sin\left(\Phi_b(t)/\Phi_0\right), \tag{7.44}$$

where I_c is the critical current of the junction and $\Phi_0 = \hbar/2e$ is the fundamental unit of flux (some authors prefer defining this as $h/2e$, with the result that extra factors of 2π start cropping up). The quantity Φ_b is called the *branch flux*: it is *not* the magnetic flux-linkage in an inductor; rather it is proportional to the quantum phase difference acquired in the tunnelling of Cooper pairs through the junction. It is still related to the voltage across the Josephson junction by an equation similar to Faraday's law, i.e., $\Delta V = \dot{\Phi}_b$. In the low-flux limit ($\Phi_b \ll \Phi_0$) the effective inductance of such a device is $L_J = \Phi_0/I_c$, which is used in the definition of the branch flux operator $\hat{\Phi}_b(t)$ analogous to Eq. (7.39).

Integrating the power formula $P = \Delta VI$ we find the energy stored in the Josephson junction to be

$$\int \Delta V(t) I(t) dt = I_c \int \dot{\Phi}_b(t) \sin(\Phi_b(t)/\Phi_0) dt$$

$$= I_c \int \sin(\Phi_b(t)/\Phi_0) d\Phi_b$$

$$= -I_c \Phi_0 \cos(\Phi_b(t)/\Phi_0) + E_0. \tag{7.45}$$

The constant of integration can be set to $E_0 = I_c\Phi_0$ so that the energy is zero when $\Phi_b = 0$; thus, the Hamiltonian for the quantum variables becomes

$$\hat{\mathcal{H}} = \frac{1}{2C}\hat{Q}^2 + I_c\Phi_0[1 - \cos(\hat{\Phi}_b/\Phi_0)]$$

$$\approx \frac{1}{2C}\hat{Q}^2 + \frac{1}{2L_J}\hat{\Phi}_b^2 - \frac{1}{24L_J\Phi_0^2}\hat{\Phi}_b^4 + \text{(terms with } \hat{\Phi}_b^6 \text{ and higher powers)}, \quad (7.46)$$

where $L_J = \Phi_0/I_c$ is the effective inductance of the Josephson junction in the low-flux limit ($\Phi \ll \Phi_0$), so that the resonance frequency of the modified circuit will be $\omega = \sqrt{I_c/C\Phi_0}$. One can find the energy eigenvalues and eigenstates exactly using the theory of *Mathieu functions* (which, curiously, also play a central part in the analysis of trapped ions); however, for our purposes, a simple perturbation theory calculation will suffice: the energy difference between the n and $n + 1$ eigenstate becomes

$$E_{n+1} - E_n \approx \hbar\omega - \frac{\hbar^2\omega^2}{8I_c\Phi_0}(n + 1). \quad (7.47)$$

Thus, we have attained our goal: a practical qubit, with non-equally spaced qubit transitions.

This simple superconducting qubit presented here is called the **Cooper pair box**, because it involves Cooper pairs – the weakly bound pairs of electrons that carry current in superconductors – moving back and forth, tunnelling through the Josephson junction or charging up the capacitor, rather like particles rattling around in a box. There are a number of variations of qubits based on resonant Josephson junction circuits, designed to optimize resilience against different sources of noise, but they all rely on the same basic building block. The adaptability of this technology, allowing us to engineer qubits to optimize control and minimize the effects of noise has made it into the current leading contender for practical large-scale quantum computers.

In order to control, read out, and couple superconducting qubits with other qubits, one must connect the qubit to an external waveguide, along which sinusoidally varying voltages can be applied to the circuit. If the frequency of the varying voltage is exactly tuned with the frequency corresponding to the difference $E_1 - E_0 \approx \hbar\omega - \hbar^2\omega^2/8I_c\Phi_0$, then the state dynamics will be completely analogous to what we saw in Section 6.3, and controlling the duration and phase of the varying voltage allows any single-qubit operation to be performed. Two (or more) qubits can be coupled to the same waveguide, enabling multi-qubit gates, Further, phase of the voltage variation in the waveguide can be effected by the state of the qubit, enabling readout.

7.4.1 Assessment

Superconducting circuits are currently one of the most promising architectures for building a large-scale quantum computer. Experiments with large-scale circuits with hundreds of qubits have been performed, and devices with thousands of qubits – perhaps up to a million – are being pursued. The versatility of the architecture, granted the fact that all its components are fabricated, leads to many advantages; however, this is a two-edged sword, in that it implies very low tolerances in manufacturing. Further, superconducting qubits must be maintained at very low temperatures, requiring milli-kelvin cryogenics on an unprecedented scale.

🗼 Looking a Little Bit Further 7.5

Josephson Effect. Figure 7.8 shows a schematic of a Josephson junction: an insulating layer, nanometres thick, between two superconducting electrodes with a voltage difference V between the two electrodes. A single Cooper pair has two possible configurations: it can be either on the left of the barrier |*left*⟩ or on the right of the barrier |*right*⟩. Thus, its quantum state will be

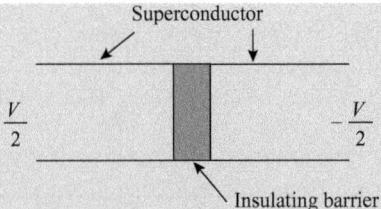

Figure 7.8 Josephson junction.

$$|\varphi_{\mathrm{cp}}\rangle = \sqrt{p_L}\,e^{-i\phi/2}|left\rangle + \sqrt{p_R}\,e^{i\phi/2}|right\rangle. \tag{7.48}$$

Though this might look a bit unfamiliar, it is just the same sort of quantum state that we had before, with the complex coefficients written out explicitly in terms of the absolute value of the probability amplitudes (i.e., the square roots of the probabilities p_R and p_L) and their phase difference ϕ. The Hamiltonian is

$$\hat{\mathcal{H}} = \begin{bmatrix} eV & E_J \\ E_J & -eV \end{bmatrix}, \tag{7.49}$$

where E_J is the energy needed to cross the insulating barrier and we have used the fact that the Cooper pair has a charge $2e$, so that the electrostatic energy difference for the Cooper pair to be on the left and for it to be on the right is $2eV$.

Substituting the state vector Eq. (7.48) and the Hamiltonian Eq. (7.49) into Schrödinger's equation we get the following pair of equations:

$$\begin{aligned} \dot{p}_L + i2p_L\dot{\phi} &= -i\frac{2eV}{\hbar}p_L - i\frac{2E_J\sqrt{p_L p_R}}{\hbar}e^{i\phi}, \\ \dot{p}_R + i2p_R\dot{\phi} &= i\frac{2eV}{\hbar}p_R - i\frac{2E_J\sqrt{p_L p_R}}{\hbar}e^{-i\phi}. \end{aligned} \tag{7.50}$$

Taking the real and imaginary parts of these equations we get

$$\begin{aligned} \dot{p}_L &= 2E_J\sqrt{p_L p_R}\sin\phi/\hbar & &\approx (E_J/\hbar)\sin\phi, \\ \dot{\phi} &= 2eV/\hbar - E_J(p_L - p_R)\cos\phi/\hbar\sqrt{p_L p_R} & &\approx 2eV/\hbar, \end{aligned}$$

where $\dot{p}_R = -\dot{p}_L$ and the approximation assumes that $p_L \approx p_R$. Writing $\Phi_0 = \hbar/2e$ (which is a unit of magnetic flux) and defining the branch flux $\Phi_b = \Phi_0\phi$ we get the relation

$$\dot{\Phi}_b = V. \tag{7.51}$$

The current across the barrier is proportional to $2e\dot{p}_R = -2e\dot{p}_L$, with the constant of proportionality depending on device-dependent quantities such as the area of the junction and the surface density of Cooper pairs. Hence, again assuming $p_L \approx p_R \approx 1/2$ we obtain the following relation between the current and the branch flux:

$$I = I_c\sin(\Phi_b/\Phi_0), \tag{7.52}$$

where I_c is a device-dependent parameter called the *critical current*.

Exercises 7.4.2

1. How would Eq. (7.36) change if there were a resistor R in series with the inductor and the capacitor? In general terms, what would be the effect of a resistor on the qubit?

2. Electric charge is naturally "quantized" in the sense that electric currents are streams of indivisible electrons, each with a charge $-e$; in superconducting circuits, current is carried by Cooper pairs, each of charge $-2e$. Do the quanta of a quantized LC circuit correspond to these concepts? Classically, what is the energy of a capacitor storing a charge $q = -\mathcal{N}e$, and how does this compare with the charge energy stored in a quantized circuit when it is in the quantum state $|\mathtt{m}\rangle$?

3. Using $\hat{a}^\dagger\hat{a} = \hat{n}$ and $\hat{a}\hat{a}^\dagger = \hat{n} + 1$ show that for any positive integer p:
   (a) $\hat{a}^p\hat{n} = (\hat{n} + p)\hat{a}^p$.
   (b) $\hat{a}^p\hat{a}^{\dagger p} = (\hat{n} + p)(\hat{n} + p - 1)\cdots(\hat{n} + 1)$.
   (c) $\hat{a}^{\dagger p}\hat{a}^p = \hat{n}(\hat{n} - 1)\cdots(\hat{n} - p + 1)$.
   Hint: Mathematical induction might be a good thing to try.

4. From Eq. (7.46), we see that the change in energy due to replacing a conventional inductor with a Josephson junction is, to the fourth order in the flux,

$$\Delta\hat{\mathcal{H}} = \frac{1}{24L_J\Phi_0^2}\hat{\Psi}_b^4.$$

   (a) Using Eq. (7.39), show that

$$\Delta\hat{\mathcal{H}} = \left(\frac{\hbar^2\omega^2}{96\Phi_0 I_c}\right)(\hat{a}^\dagger + \hat{a})^4.$$

   (b) Using the results from Exercise 3, show that

$$(\hat{a}^\dagger + \hat{a})^4 = \hat{a}^{\dagger 4} + \hat{a}^4 + \hat{a}^{\dagger 2}(4\hat{n} + 6) + (4\hat{n} + 6)\hat{a}^2 + 6\hat{n}^2 + 6\hat{n} + 3.$$

   (c) For a Cooper-pair box circuit in state $|n\rangle$ calculate the expected value of the energy shift, $\Delta E = \langle n|\Delta\hat{\mathcal{H}}|n\rangle$. Hence prove Eq. (7.47).

7.5 CONCEPT CHECK AND EXERCISES

Concept Check

Types of qubits:

- photons,
- spin-1/2 particles,
- atoms,
- superconducting circuits.

Chapter Exercises

1. Let's consider the dynamics of a simple harmonic oscillator in the quantum realm. As in the case of photons, the general recipe is: introduce operators to represent observables and ensure the wavefunction's time variation gives the correct classical time variation for the expected values of the observable. We know that, treated classically, the displacement of a harmonic oscillator of mass m and angular frequency ω is given by

$$x(t) = x_0 \cos(\omega t) + (p_0/m\omega) \sin(\omega t)$$
$$= \zeta \exp(-i\omega t) + \zeta^* \exp(i\omega t), \tag{7.53}$$

where $\zeta = (x_0 + ip_0/m\omega)/2$ and x_0 and p_0 are constants.

(a) Find the momentum $p(t)$ and the total energy $\mathcal{E} = p(t)^2/2m + m\omega^2 x(t)^2/2$ in terms of ζ.

(b) Quantum mechanically, we assume, as we did for photons of light, that the quantum mechanical state is a superposition of equally spaced energy eigenstates $\{|0\rangle, |1\rangle, |2\rangle, \ldots\}$, just like the case with photons. If $\hat{x} = \ell(\hat{a} + \hat{a}^\dagger)$, where $\ell = \sqrt{\hbar/2m\omega}$, show that $\hat{\mathcal{H}} = \hbar\omega(\hat{n} + 1/2)$.

(c) If the state of the system is $|\varphi(t)\rangle = \sum_n c_n \exp[-i(n+1/2)\omega t]|n\rangle$, show that

$$\zeta = \ell \sum_n c_n^* c_{n+1} \sqrt{n+1} = \ell \langle \varphi(0)|\hat{a}|\varphi(0)\rangle.$$

(d) Now suppose a time-dependent force $f(t) = f_0 \cos(\omega t)$ oscillating exactly at the resonance frequency is applied to the harmonic oscillator. The energy in this case becomes

$$\hat{\mathcal{H}} = \hbar\omega(\hat{n} + 1/2) - \hbar\Omega_0 \cos(\omega t)(\hat{a} + \hat{a}^\dagger).$$

Using the rotating wave equation show that the probability amplitudes c_n are now time-varying functions which obey the following equation:

$$\frac{d}{dt}c_m(t) = i\frac{\Omega_0}{2}\sqrt{m+1}c_{m+1}(t) + i\frac{\Omega_0}{2}\sqrt{m}c_{m-1}(t).$$

(e) Assuming the initial state is $|0\rangle$ show that the probability amplitudes are given by the formula

$$c_m(t) = \frac{(i\Omega_0 t/2)^m}{\sqrt{m!}} \exp\left[-\frac{(\Omega_0 t/2)^2}{2}\right].$$

What is the probability that the state $|m\rangle$ is occupied at time t? Do you recognize the probability distribution? Comment on the implications of this result for using harmonic oscillators as qubits.

8 The Physics of Two-Qubit Gates

Arguably the most challenging item on DiVincenzo's list is the requirement to perform two-qubit quantum logic gates, such as the CNOT or the CZ gates we encountered in Section 3.2. One must perform a prescribed unitary operation on one qubit – the target – conditional on the state of a second qubit – the control – *without* disturbing the quantum state of the control qubit by, for example, measuring it.

There are two basic approaches to this problem. First, and most simply, one brings the two qubits into close proximity and allows their natural interaction to do their work. As we will see, just about all possible interactions can produce a unitary evolution which is universal for quantum computing. However, the need for the qubits to be brought into close proximity makes controlling such interactions very difficult. The alternative is to mediate the interactions between the two bits via a quantum degree of freedom shared by them both – for example, a common vibrational mode or a photon with which they both interact. This **quantum data bus** approach has many advantages.

Learning Objectives

- Understand how to implement two-qubit gates based on direct qubit–qubit interaction between qubits that are close to each other.
- Understand how to implement two-qubit gates based on the idea of a quantum data bus, where qubits that are not close to each other indirectly interact through interacting with intermediate qubits.

8.1 Two-Qubit Gates from Short-Range Interactions

Two-qubit gates such as the controlled-NOT (CNOT) gate we saw in Chapter 3 are, technologically speaking, one of the most difficult steps in creating a quantum computer. Recall what is needed: supposing we have two qubits A and B, the state of B has to be altered dependent on the state of A, and this must be a reversible, unitary operation (so simply measuring A and acting on B dependent on the result will not work). Therefore, a Hamiltonian interaction of some kind is required.

If we place the two qubits, A and B, close together they will usually interact in some manner (there are exceptions, like photons, which we'll discuss separately). Since we are dealing with two 2-level systems, the Hilbert space must have dimension 4 and the Hamiltonian $\hat{\mathcal{H}}_{AB}$ will be a 4×4 Hermitian matrix. We make the assumption that the trace of $\hat{\mathcal{H}}_{AB}$ is 0 (if it is not zero, the value of the trace would only appear in a global phase term multiplying all probability amplitudes, which would have no observable effect). We can write the Hamiltonian in the following form:

$$\hat{\mathcal{H}}_{AB} = \hat{\mathcal{H}}_A \otimes \hat{I} + \hat{I} \otimes \hat{\mathcal{H}}_B + \hat{\mathcal{V}}_{AB}, \tag{8.1}$$

where $\hat{\mathcal{H}}_A$ is the Hamiltonian for qubit A , and similarly $\hat{\mathcal{H}}_B$ is the Hamiltonian for qubit B, while $\hat{\mathcal{V}}_{AB}$ is the interaction term. We can use the partial trace operation Eq. (3.6) to separate out the interactions that would apply only to one or other of the qubits; in other words:

$$\hat{\mathcal{H}}_A = \frac{1}{2}\text{Tr}_B\{\hat{\mathcal{H}}_{AB}\}, \tag{8.2}$$

$$\hat{\mathcal{H}}_B = \frac{1}{2}\text{Tr}_A\{\hat{\mathcal{H}}_{AB}\}, \tag{8.3}$$

where the factors of $1/2$ appear to compensate for $\text{Tr}\{\hat{I}\} = 2$. Thus, the interaction part of the Hamiltonian, $\hat{\mathcal{V}}_{AB}$, will have the following properties:

$$\text{Tr}_A\{\hat{\mathcal{V}}_{AB}\} = \text{Tr}_B\{\hat{\mathcal{V}}_{AB}\} = 0. \tag{8.4}$$

We choose the computational basis for the two qubits so that the states $|0\rangle_A$ and $|1\rangle_A$ are eigenstates of $\hat{\mathcal{H}}_A$, and $|0\rangle_B$ and $|1\rangle_B$ are the eigenstates of $\hat{\mathcal{H}}_B$, so that $\hat{\mathcal{H}}_A = -\hbar\omega_A\hat{Z}/2$ and $\hat{\mathcal{H}}_B = -\hbar\omega_B\hat{Z}/2$ (we have used the convention that $|0\rangle$ has lower energy than $|1\rangle$, and that $\hat{Z} = |0\rangle\langle0| - |1\rangle\langle1|$). Thus, the Hamiltonian can be written as

$$\hat{\mathcal{H}}_{AB} = -\frac{\hbar\omega_A}{2}(\hat{Z} \otimes \hat{I}) - \frac{\hbar\omega_B}{2}(\hat{I} \otimes \hat{Z}) + \hat{\mathcal{V}}_{AB}, \tag{8.5}$$

where Eq. (8.4) implies that the interaction Hamiltonian $\hat{\mathcal{V}}_{AB}$ has the following form:

$$\hat{\mathcal{V}}_{AB} = \begin{bmatrix} v_{00} & v_{01} & v_{02} & v_{03} \\ v_{01}^* & -v_{00} & v_{12} & -v_{02} \\ v_{02}^* & v_{12}^* & -v_{00} & -v_{01} \\ v_{03}^* & -v_{02}^* & -v_{01}^* & v_{00} \end{bmatrix}. \tag{8.6}$$

If we try to solve Schrödinger's equation for this Hamiltonian, i.e.,

$$i\hbar\frac{d}{dt}|\Phi(t)\rangle = \hat{\mathcal{H}}_{AB}|\Phi(t)\rangle, \tag{8.7}$$

where $|\Phi(t)\rangle$ is the state of the two-qubit system, a standard technique is to change to the interaction picture, in which the separate time variation of the individual qubits (as described by $\hat{\mathcal{H}}_0 = -\frac{\hbar\omega_A}{2}\left(\hat{Z} \otimes \hat{I}\right) - \frac{\hbar\omega_B}{2}\left(\hat{I} \otimes \hat{Z}\right)$) is removed by writing $|\Phi(t)\rangle_{AB} = \exp(-i\hat{\mathcal{H}}_0t/\hbar)|\varphi(t)\rangle_{AB}$. Substituting into Eq. (8.7), we find the equation for $|\varphi(t)\rangle_{AB}$:

$$i\hbar\frac{d}{dt}|\varphi(t)\rangle_{AB} = \exp(i\hat{\mathcal{H}}_0t/\hbar)\,\hat{\mathcal{V}}_{AB}\exp(-i\hat{\mathcal{H}}_0t/\hbar)|\varphi(t)\rangle_{AB}. \tag{8.8}$$

Thus, in the interaction picture, the new state $|\varphi(t)\rangle_{AB}$ evolves with the following time-dependent Hamiltonian:

$$\exp(i\hat{\mathcal{H}}_0t/\hbar)\,\hat{\mathcal{V}}_{AB}\exp(-i\hat{\mathcal{H}}_0t/\hbar) =$$
$$\begin{bmatrix} v_{00} & v_{01}e^{-i\omega_Bt} & v_{02}e^{-i\omega_At} & v_{03}e^{-i(\omega_A+\omega_B)t} \\ v_{01}^*e^{i\omega_Bt} & -v_{00} & v_{12}e^{-i(\omega_A-\omega_B)t} & -v_{02}e^{-i\omega_At} \\ v_{02}^*e^{i\omega_At} & v_{12}^*e^{i(\omega_A-\omega_B)t} & -v_{00} & -v_{01}e^{-i\omega_Bt} \\ v_{03}^*e^{i(\omega_A+\omega_B)t} & -v_{02}^*e^{i\omega_At} & -v_{01}^*e^{i\omega_Bt} & v_{00} \end{bmatrix}. \tag{8.9}$$

By itself, this is not a particularly useful way to try to solve the dynamics of the two-qubit system: in general, time-dependent Hamiltonians tend to be rather tricky to solve and should usually be avoided if possible. However, Eq. (8.9) is useful in that it allows us to determine which terms are rapidly oscillating, and thus have negligible contribution to resonant behaviour. In other words, we again make the rotating wave approximation: the state $|\varphi(t)\rangle_{AB}$ is already slowly varying, with the rapid variation of $|\Phi(t)\rangle_{AB}$ factored out; if we average the dynamics over a cycle (i.e., a time $2\pi/\omega_A \approx 2\pi/\omega_B$) then $|\varphi(t)\rangle_{AB}$ should not change, while the rapidly oscillating terms will average to zero. The terms involving $e^{\pm i\omega_A t}$, $e^{\pm i\omega_B t}$, and $e^{\pm i(\omega_A+\omega_B)t}$ can be neglected, provided $\hbar\omega_A$ and $\hbar\omega_B$ are very much greater than any of the amplitudes of the matrix elements v_{nm}. Thus, we are left with the following Hamiltonian:

$$\hat{\mathcal{V}}_{AB}^{(RWA)} = \begin{bmatrix} v_{00} & 0 & 0 & 0 \\ 0 & -v_{00} & v_{12}e^{-i(\omega_A-\omega_B)t} & 0 \\ 0 & v_{12}^*e^{i(\omega_A-\omega_B)t} & -v_{00} & 0 \\ 0 & 0 & 0 & v_{00} \end{bmatrix}$$

$$= v_{00}(\hat{Z} \otimes \hat{Z}) + v_{12}e^{-i(\omega_A-\omega_B)t}\left(\hat{\sigma}_+ \otimes \hat{\sigma}_-\right) + v_{12}^*e^{i(\omega_A-\omega_B)t}\left(\hat{\sigma}_- \otimes \hat{\sigma}_+\right), \quad (8.10)$$

where $\hat{\sigma}_+ = (\hat{X} + i\hat{Y})/2 = |0\rangle\langle 1|$ is the qubit "raising" operator and $\hat{\sigma}_+^\dagger = \hat{\sigma}_- = (\hat{X} - i\hat{Y})/2 = |1\rangle\langle 0|$ is the "lowering" operator (the convention that $\hat{Z}|0\rangle = |0\rangle$ and $\hat{Z}|1\rangle = -|1\rangle$ leads to this rather topsy-turvy nomenclature). Reverting to the time-independent (or Schrödinger) picture, the Hamiltonian under the rotating wave equation becomes

$$\boxed{\hat{\mathcal{H}}_{AB} = -(\hbar\omega_A/2)(\hat{Z} \otimes \hat{I}) - (\hbar\omega_B/2)(\hat{I} \otimes \hat{Z}) + \hbar J(\hat{Z} \otimes \hat{Z}) + \hbar T\left(\hat{\sigma}_+ \otimes \hat{\sigma}_- + \hat{\sigma}_- \otimes \hat{\sigma}_+\right),} \quad (8.11)$$

where we have written $v_{00} = \hbar J$ and $v_{12} = \hbar T$ (we have made the simplifying assumption that v_{12} is real-valued, though this can be justified by setting the relative phases of $|0\rangle$ and $|1\rangle$ for the two qubits when we find the eigenstates of $\hat{\mathcal{H}}_A$ and $\hat{\mathcal{H}}_B$). Note that $\hat{\sigma}_+ \otimes \hat{\sigma}_- + \hat{\sigma}_- \otimes \hat{\sigma}_+$ is equivalent to $\hat{X} \otimes \hat{X} + \hat{Y} \otimes \hat{Y}$.

Writing the Hamiltonian in terms of the Pauli operators allows us to identify the physical processes underlying the interaction: J is the frequency shift due to the qubit–qubit coupling and T is the rate at which excitations are swapped between the two qubits (i.e., the operator $\hat{\sigma}_+ \otimes \hat{\sigma}_-$ indicates qubit A is lowered while qubit B is raised, and vice versa). To re-iterate, the only approximation made in deriving this Hamiltonian is the rotating wave approximation; if that is valid, then this Hamiltonian is completely general.

Refocusing and Two-Qubit Gates

This general two-qubit interaction can in most cases be made to implement a universal quantum gate, although it requires additional single-qubit pulses called *refocusing* pulses, which have the effect of "turning off" selected terms in the Hamiltonian. To illustrate how this works, let us consider a simplified model, in which T, the rate of hopping between the two qubits, is zero, and we only have the spin–spin frequency shift.

As a simple example, let's consider a single qubit in isolation evolving in time as usual according to

$$|\varphi(t)\rangle = \exp(i\omega t\hat{Z})|\varphi(0)\rangle. \quad (8.12)$$

We know that $\hat{X}\hat{Z}\hat{X} = -\hat{Z}$, and so $\hat{X}\exp(i\omega t\hat{Z})\hat{X} = \exp(-i\omega t\hat{Z})$. In other words, by applying a pair of $\hat{X}$ gates, the time evolution of a qubit is made to act in reverse. As a consequence, if we allow the qubit to evolve in time for a duration $t/2$ then apply the first $\hat{X}$-gate, after which the qubit evolves for another $t/2$ followed by a second $\hat{X}$-gate, we find

$$|\varphi(t)\rangle = \hat{X}\exp(i\omega t\hat{Z}/2)\hat{X}\exp(i\omega t\hat{Z}/2)|\varphi(0)\rangle$$
$$= |\varphi(0)\rangle. \tag{8.13}$$

Thus, the $\hat{X}$-gates applied at time $t/2$ and t have effectively turned off the evolution.

To see how this can be used to convert the dynamics based on $\hat{\mathcal{H}}_{AB}$ given in Eq. (8.11), let's look at the simplified case in which hopping dynamics is negligible (i.e., $T = 0$), so the Hamiltonian is given by the formula

$$\hat{\mathcal{H}}_J = (\hbar\omega_A/2)(\hat{Z}\otimes\hat{I}) + (\hbar\omega_B/2)(\hat{I}\otimes\hat{Z}) + \hbar J(\hat{Z}\otimes\hat{Z}). \tag{8.14}$$

Using the following identities:

$$(\hat{X}\otimes\hat{X})(\hat{Z}\otimes\hat{I})(\hat{X}\otimes\hat{X}) = -(\hat{Z}\otimes\hat{I}),$$
$$(\hat{X}\otimes\hat{X})(\hat{I}\otimes\hat{Z})(\hat{X}\otimes\hat{X}) = -(\hat{I}\otimes\hat{Z}),$$
$$(\hat{X}\otimes\hat{X})(\hat{Z}\otimes\hat{Z})(\hat{X}\otimes\hat{X}) = (\hat{Z}\otimes\hat{Z}), \tag{8.15}$$

we find

$$(\hat{X}\otimes\hat{X})\exp(-i\hat{\mathcal{H}}_J t/2\hbar)(\hat{X}\otimes\hat{X})\exp(-i\hat{\mathcal{H}}_J t/2\hbar) = \exp(-iJt(\hat{Z}\otimes\hat{Z})). \tag{8.16}$$

In other words, the two refocusing $(\hat{X}\otimes\hat{X})$-gates have turned off the single-qubit dynamics, leaving only the spin–spin frequency shift J. If the system is well calibrated, we can select the time $t = \pi/4J$, so that the time evolution operator is

$$\exp\left[-i(\pi/4)\left(\hat{Z}\otimes\hat{Z}\right)\right] = e^{i\pi/4}\begin{bmatrix} -i & 0 & 0 & 0 \\ 0 & 1 & 0 & 0 \\ 0 & 0 & 1 & 0 \\ 0 & 0 & 0 & -i \end{bmatrix}. \tag{8.17}$$

This operation is a universal gate, and by applying the local one-qubit operation $\exp[-i(\pi/4)\hat{Z}]$ to each qubit, one obtains the controlled-Z gate (up to a global phase).

Things are not, however, as rosy as they seem. Whatever the physical source of the interaction between the qubits – dipole–dipole forces between atoms, spins or electronic current loops, the exchange interaction between identical particles, etc. – their strength drops off dramatically as the distance between the qubits increases. Thus, practically one is usually dealing with interactions between the nearest neighbours; but in quantum computers one needs to perform universal gates between any pair of qubits in the qubit array. This can be performed by multiple SWAP gates, transmitting the states of the two qubits in question onto two qubits that are adjacent; then SWAP-ing the transformed states back to their original positions after the gate interaction. This increases the number of two-qubit gates required for an algorithm rather dramatically. Alternatively, one could move the qubits physically until they are adjacent – although only a few architectures like ions or photons actually have the ability to physically move the qubits.

Another issue is that there is no way to turn off the interaction between the qubits. Ideally one would like to wave a magic wand to turn on the interaction between the qubits, allow them to evolve in time, with the help of refocusing pulses, to perform the controlled-Z gate, then turn the interaction off again. This can in fact be done in certain types of qubit architectures – notably cold trapped atoms can be excited into special Rydberg atomic states that have a strong dipole–dipole interaction. But if that is not possible, one has to time the pulses controlling the qubits and their interaction precisely. These extra unitary gates greatly increase the complexity of quantum algorithms. In the next section we will consider the ideal solution: a *quantum data bus*.

8.1.1 Example: Photons and Non-Linear Optics

Photons are a notable exception to the idea that two qubits in close proximity will interact. Light beams, when propagating in vacuum, will not generally exert a force on other light beams (except in extreme physical conditions such as in the Large Hadron Collider, where photon–photon scattering was observed in 2019). However, in certain special crystals, non-linear optical effects can occur and are routinely used in many applications in photonics. But individual photons, unfortunately, have weak electric fields, and non-linear effects at the single-photon level are too weak to be a useful quantum logic gate. As we will see in Section 10.2, there are ingenious means to overcome this difficulty at the cost of abandoning determinism and accepting a quantum gate that will *probably* work, rather than one that works all the time.

⌂ Looking a Little Bit Further 8.1

Entanglement and Indistinguishability. In physics, there are situations in which an entangled state arises naturally. For example, in the helium atom, two electrons are bound to a single nucleus of charge $+2|e|$. In the ground state, i.e., the state when the electrons have the lowest energy, both of the electrons have the same spatial wave function (called an *orbital* in atomic physics) close to the nucleus. In other words, they are indistinguishable: there are two electrons there, but it is impossible to tell which one is which. This is one of those fundamental properties of quantum mechanics, not known in classical mechanics, in which, by assumption, the trajectory of each particle is distinct and well-defined at all times.

A very basic result of physics requires that the quantum state of any two electrons must change sign if we swap the identity of two indistinguishable particles. Since the two electrons in helium are in identical spatial states (so swapping their identities will change nothing), this implies the electrons' spins must be in an anti-symmetric state, which changes sign when you swap them. There is only one such state of two spin-1/2 particles, namely the singlet state $(|\uparrow, \downarrow\rangle - |\downarrow, \uparrow\rangle)/\sqrt{2}$; assigning $|\uparrow\rangle = |1\rangle$ and $|\downarrow\rangle = |0\rangle$, this is equivalent to the maximally entangled state $|\beta_2\rangle$ (see Eq. (3.37)).

Wow! the electron spins in the ground state of helium are in a maximally entangled state. Does that mean that all we have to do to create entanglement is get a bunch of helium atoms, and . . . there you are!

Well, no. The problem with indistinguishable qubits is you cannot tell which one is which (!), thus, if you try to do anything to one of them, for example by applying a force to implement a unitary operation, you'll be doing exactly the same thing to the other one. As we have seen (Exercise 2 of Section 3.3), when you apply the same unitary to both qubits of a singlet state, the result is no change: you get the singlet state again. Thus, it is impossible to change or to measure the individual spins in such a singlet state. This sort of entanglement is not useful for performing the sort of protocols required in quantum information.

Here are the opening few sentences of Erwin Schrödinger's famous 1935 paper in which he first discussed the notion of entanglement:

When two systems, of which we know the states by their respective representatives [i.e. their states $|\varphi\rangle$], enter into temporary physical interaction due to known forces between them, and when after a time of mutual influence the systems separate again, then they can no longer be described in the same way as before, viz. by endowing each of them with a representative of its own. I would not call that *one* but rather *the* characteristic trait of quantum mechanics, the one that enforces its entire departure from classical lines of thought. By the interaction the two representatives (or ψ-functions) have become entangled.

Of particular note in this context is his requirement that "after a time of mutual influence the systems separate again": in other words, distinguishability is implicit in the definition of quantum entanglement as used in this book, implying that the ground state of helium is not entangled in any way useful to quantum information.

Exercises 8.1.2

1. The evolution of the state $|\varphi_{AB}(t)\rangle = \alpha(t)|00\rangle + \beta(t)|01\rangle + \gamma(t)|10\rangle + \delta(t)|11\rangle$ of two interacting qubits has the following form:

$$
i\hbar \frac{\partial}{\partial t}
\begin{pmatrix} \alpha \\ \beta \\ \gamma \\ \delta \end{pmatrix}
= \hbar J
\begin{pmatrix}
0 & 0 & 0 & 0 \\
0 & -1 & 1 & 0 \\
0 & 1 & -1 & 0 \\
0 & 0 & 0 & 0
\end{pmatrix}
\begin{pmatrix} \alpha \\ \beta \\ \gamma \\ \delta \end{pmatrix},
$$

   where J is a real constant.
   (a) Solve the equation of motion to find the state after a time t.
   (b) At what time t does the state correspond to a $\sqrt{\text{SWAP}}|\varphi_{AB}(0)\rangle$?
   (c) Suppose the initial state was $|\varphi_{AB}(0)\rangle = (|00\rangle + |00\rangle - |00\rangle - |00\rangle)/2$. Find an expression for the concurrence as a function of time.
2. Suppose that a single qubit is evolving in time according to $|\varphi(t)\rangle = \hat{U}(t)|\varphi(0)\rangle$ and $\hat{U}(t) = \exp(i\omega_0 t\hat{a})$ and $\hat{a} = \mathbf{a} \cdot \hat{\boldsymbol{\sigma}}$ (see Eq. (3.44)). Find an operator $\hat{b}$ such that $\hat{b}\hat{U}(t)\hat{b} = \hat{U}(-t)$.
3. Suppose a two-qubit system is evolving with the Hamiltonian $\hat{\mathcal{H}}_J$ given by Eq. (8.14). Find the sequence of refocusing pulses needed to "turn off" this evolution completely. *Hint: Try using* $(\hat{X} \otimes \hat{I})$ *and* $(\hat{I} \otimes \hat{X})$ *as well as* $(\hat{X} \otimes \hat{X})$; *you will need four refocusing pulses in all.*

8.2 Two-Qubit Gates Using a Quantum Data Bus

To overcome the shortcomings of two-qubit quantum gates based on the always-on nearest-neighbour interactions between qubits discussed in the previous section, what is really needed is a quantum "data bus" into which the quantum state of one qubit can be transferred, then transported to a second qubit so that a conditional quantum operation can be performed, and then returned back to the original qubit.

So far we have treated our qubits as if they were firmly anchored in their location in the qubit array, and that when an external force is applied, it only affects the internal configuration of the qubit. But what if the qubits could recoil when they absorb the momentum being fed into them during a quantum gate operation? And if the qubits' motion is strongly coupled – like some sort of microscopic spring is connecting them – then the recoil of one qubit will be passed on to all the others in the array. We assume that the qubits remain bound, so the motion is oscillatory and of small amplitude – they are not going to move very far from their original position, nor will they start colliding with each other. The recoil motion will be quantized, and the momentum can only be absorbed in discrete amounts. We thus have two *external* levels for the qubits: one in which they are all at rest, one in which they are moving due to absorbing a single quantum of kinetic energy; we will refer to these states as $|r\rangle$ and $|m\rangle$, respectively.

In order to exploit this motional degree of freedom into a quantum gate, we need to assume the physical qubits have at least one more accessible quantum level beyond the two, $|0\rangle$ and $|1\rangle$, that are employed to store the quantum information. We denote this extra level as $|2\rangle$. Then each qubit can be thought of as having six levels: $|0, r\rangle$, $|1, r\rangle$, and $|2, r\rangle$ when the qubit is at rest; $|0, m\rangle$, $|1, m\rangle$, and $|2, m\rangle$ when moving. When two qubits, A and B, are involved we must take into account that the motion is collective: if one qubit is in motion, both are in motion; if one is at rest, both are at rest. Thus there are 18 levels in total: $|0\rangle_A \otimes |0\rangle_B \otimes |m\rangle \equiv |00, m\rangle$, $|01, m\rangle$, $|02, m\rangle$, $|10, m\rangle$, $|11, m\rangle$, $|12, m\rangle$, $|20, m\rangle$, $|21, m\rangle$, $|22, m\rangle$, $|00, r\rangle$, $|01, r\rangle$, $|02, r\rangle$, $|10, r\rangle$, $|11, r\rangle$, $|12, r\rangle$, $|20, r\rangle$, $|21, r\rangle$, and $|22, r\rangle$.

The pulse sequence required to make a controlled-Z gate is shown in Figs. 8.1–8.3. The pulses are the standard application of an oscillating force of the type discussed in Section 6.3, but with frequencies carefully tuned to different resonances to select specific operations (we always assume that fields not on-resonance with any transition will have negligible effects).

Let us assume qubits A and B begin in a state in which neither qubit has any probability amplitude to be in level $|2\rangle$, and both qubits are at rest:

$$|\varphi_0\rangle = \alpha|00, r\rangle + \beta|01, r\rangle + \gamma|10, r\rangle + \delta|11, r\rangle. \tag{8.18}$$

Figure 8.1   Pulse 1.

Figure 8.2   Pulse 2.

Figure 8.3   Pulse 3.

The first pulse (Fig. 8.1) is a "π-pulse", Eq. (6.45), with the phase ϕ chosen to be 0; the resultant state is

$$|\varphi_1\rangle = \alpha|00, r\rangle + \beta|01, r\rangle - i\gamma|00, m\rangle - i\delta|01, m\rangle. \tag{8.19}$$

In other words, the state $|1\rangle$ of qubit A has been transposed onto the collective motional state.

The second pulse (Fig. 8.2) is applied to qubit B and is a "2π-pulse", Eq. (6.48), which takes the state $|0\rangle_B|m\rangle$ on a round trip to $|0\rangle_B|r\rangle$ and back again: in doing this, it acquires a minus sign:

$$|\varphi_2\rangle = \alpha|00, r\rangle + \beta|01, r\rangle + i\gamma|00, m\rangle - i\delta|01, m\rangle. \tag{8.20}$$

Note that the term with probability amplitude α is not affected by this, even though it involves qubit B being in $|0\rangle$. This is the only time that $|2\rangle_B$ is involved in the gate (level $|2\rangle_A$ is not needed at all).

The final pulse (Fig. 8.3) is identical to the first, and reverses its effect, albeit with another sign showing up (because two π-pulses are the equivalent of a 2π-pulse):

$$|\varphi_3\rangle = \alpha|00, r\rangle + \beta|01, r\rangle + \gamma|10, r\rangle - \delta|11, r\rangle. \tag{8.21}$$

The significance of the minus sign acquired by the term involving γ in pulse 2 is now apparent: it cancels the minus sign the term acquired from the two π-pulses. There was no such cancellation for the δ term, and hence we have a controlled-$\hat{Z}$ gate.

8.2.1 Trapped-Ion Quantum Computing

This scheme was originally proposed by Cirac and Zoller in 1995, the same year as Shor's algorithm. Historically, that year was a watershed in the history of quantum computing: a compelling application and a realistic proposal for a technology to implement it inspired the wider scientific community as well as funding agencies to begin taking this topic seriously.

Specifically, Cirac and Zoller had trapped ions in mind as the technology in which to perform this gate. Recall that ions are atoms with one or more electrons stripped off, leaving a positively charged atom with various internal quantum states just like a regular neutral atom. In Section 7.3 we saw that a challenge in ion trapping arises due to the well-established result in electrostatics called Earnshaw's theorem. The theorem, in effect, states that trapping charged particles such as ions is impossible with electrostatic fields. At a point of equilibrium for the electrostatic potential Φ, where $\nabla\Phi = 0$, there is no force on the charged particle (otherwise it would accelerate, and not be in equilibrium). However, the potential Φ obeys Laplace's equation, $\nabla^2\Phi = 0$, which implies that at least one of the second derivatives of the potential must be negative, and therefore a maximum of the potential, implying that the equilibrium will be unstable. This difficulty with the instability of the saddle point of the ion trapping potential was circumvented by Wolfgang Paul, who figured out that it can be dealt with by oscillating it. If the saddle point of potential, with its stable and unstable equilibria, is continuously reversed, so that the minima become the maxima and vice versa, one can in fact obtain a stable trapping potential for the ions. Using such a Paul trap, multiple ions can be confined and cooled to form a linear crystal, with the ions separated by a few micrometres. This apparatus was developed as an optical frequency standard. An individual ion's optical transition frequencies are very well-defined, and if the ion can be trapped and cooled, there will be no **Doppler effect** to contaminate the frequency reference, so that the transition frequency provides a highly precise standard against which to compare optical fields generated by other means. But this high precision and stability are just the sort of characteristics required for a quantum computer.

The ions' mutual repulsion forms the means by which the recoil motion is communicated to all the ions. In effect they behave like a rigid crystal, rocking back and forth in the trapping potential (there are other modes of oscillation, in which the ions move relative to each other; these have found only limited application in quantum information). A single quantum of this harmonic motion – usually referred to as a *phonon* – is the $|m\rangle$ motional state used in the gate protocol given above. This protocol is very sensitive to imperfectly cooled ions: it will not work very well if you do not start in the rest state $|r\rangle$. There are, however, protocols that are resilient against such imperfections, but at the cost of not being as fast to perform.

Today, ion traps are one of the most successful of all quantum computing technologies, with chains of dozens of ions being routinely employed in small-scale computations. A number of important early proof-of-principle experiments, such as the first quantum gate or the first demonstration of teleportation with deterministic creation of entanglement, were carried out using these devices. They are limited to about 100 qubits, simply because the dynamics of a large number of trapped ions, with strongly coupled vibrational modes, becomes increasingly difficult to manipulate with lasers as the number of ions increases. Schemes for scaling up to hundreds of qubits require multiple electromagnetic traps which must be coupled deterministically. Another longer-term drawback is the relatively slow speed (tens of microseconds) at which quantum gates can be performed, again limited by the dynamics of the ions. All of these have potential fixes which are a subject of active research.

8.2.2 Superconducting Qubits and Resonant Transmission Lines

Another very successful qubit technology that employs the quantum data bus idea is superconducting circuits. Instead of the qubits being coupled by their motion, the data bus is provided by photons of microwave electric fields confined in a resonant waveguide: thus, instead of the "rest" and "moving" states of the data bus, one can have the 0 and 1 photon states. For atoms the coupling with single photons is very weak, so much so that it has proven difficult to observe individual photons interacting with single atoms. However, in superconducting circuits one can design the coupling strengths to be much stronger; and multiple qubits can be connected to the same waveguide, which furnishes the quantum data bus.

Exercises 8.2.3

1. **The Jaynes–Cummings Model.** We have not yet examined just how the quantum data bus might be realized. In both trapped ions and superconducting qubits, the data bus is a harmonic oscillator: the collective rocking motion for ions, or photons in a waveguide for superconductors.

When the qubit transitions from $|0\rangle$ to $|1\rangle$ it must absorb a quantum (either a phonon, or a photon – the quantum mechanics is very similar, even though the physics is different), and when it goes the opposite way, it will emit a quantum. Thus, the interaction-picture Hamiltonian describing the situation is as follows:

$$\hat{\mathcal{H}}_{JC} = \hbar\omega\hat{a}^\dagger\hat{a} - \frac{\hbar\omega_0}{2}\hat{Z} + i\hbar g\left(\hat{a}^\dagger\hat{\sigma}_- - \hat{\sigma}_+\hat{a}\right),$$

where $\hat{\sigma}_+ = |1\rangle\langle0| = (\hat{X} - i\hat{Y})/2$ is the qubit raising operator, $\hat{\sigma}_- = |0\rangle\langle1| = (\hat{X} + i\hat{Y})/2$ is the qubit lowering operator, ω is the photon/phonon frequency, and ω_0 the qubit transition frequency. This model was first studied by Edwin Jaynes and Fred Cummings in 1963 to understand the quantum properties of the then newly invented laser: instead of the qubit states, they considered an atom in an optical cavity. As far as we are concerned, the coupling coefficient g is something that can be controlled externally, so this interaction can be turned on and off at will.

(a) Let the state of the qubit/bus system be

$$|\varphi(t)\rangle = \sum_{n=0}^{\infty} \alpha_n(t)e^{-i(n\omega-\omega_0/2)t}|0,\mathbb{n}\rangle + \beta_n(t)e^{-i((n-1)\omega+\omega_0/2)t}|1,\mathbb{n}-\mathbb{1}\rangle,$$

where the first symbol in the ket refers to the qubit and the second (with the double-struck argument) to the state of the harmonic oscillator data bus. Note that $\beta_0(t) \equiv 0$, since there is no harmonic oscillator state $|-\mathbb{1}\rangle$ with a negative number of quanta. Using Schrödinger's equation, find the equations of motion for the probability amplitudes $\alpha_n(t)$ and $\beta_n(t)$. *Hint: The equations should only involve probability amplitudes with the same values of the subscript n.*

(b) Solve the equations of motion and determine the rate at which the qubits change between $|0\rangle$ and $|1\rangle$. *Hint: You will need to differentiate the equations of motion found in the previous question, and then algebraically obtain two second-order differential equations, one involving $\alpha_n(t)$ and its derivatives, the other $\beta_n(t)$ and its derivatives.*

(c) Suppose we can control the quantity g so that the interaction can be turned on and off at will. For what duration should the interaction be left on in order to implement the first pulse needed in a CZ gate (Fig. 8.1).

(d) How would the performance of the interaction change if initially there were some random number of quanta in the harmonic oscillator? What would be the fidelity of the first pulse needed for the CZ gate in this circumstance?

2. **The Mølmer–Sørensen Quantum Gate**

(a) In Section 6.3 we found the equations of motion, Eqs. (6.36) and (6.37), for the probability amplitudes of a single qubit interacting with an on-resonance force. Show that these equations are equivalent to the Schrödinger equation for the wavefunction $|\varphi\rangle = \alpha|0\rangle + \beta|1\rangle$ with the Hamiltonian

$$\hat{\mathcal{H}} = \frac{\hbar\Omega}{2}|1\rangle\langle0|e^{-i\phi} + \text{h.a.},$$

where "h.a." stands for the Hermitian adjoint of the preceding term.

(b) Now suppose the qubit is free to vibrate in a trapping potential. We assume the displacement of the qubit is along the x-axis and is given by the operator $\hat{x}(t) = x_0(\hat{a}^\dagger e^{i\omega_x t} + \hat{a}e^{-i\omega_x t})$, where $x_0 = \sqrt{\hbar/2M\omega_x}$, ω_x being the trapping frequency and M the mass of the qubit. Assuming that the spatial variation of the force applied to the qubit is a plane wave, show that the Hamiltonian in this case is

$$\hat{\mathcal{H}} = \frac{\hbar\Omega}{2}|1\rangle\langle 0|e^{i\omega\cos\theta\hat{x}(t)/c} + \text{h.a.}$$

$$\approx \frac{\hbar\Omega}{2}|1\rangle\langle 0|\{1 + i\eta(\hat{a}^\dagger e^{i\omega_x t} + \hat{a}e^{-i\omega_x t})\} + \text{h.a.},$$

where θ is the angle between the direction of the plane wave and the x-axis and the Lamb–Dicke parameter is $\eta = \omega\cos\theta x_0/c = \hbar\omega\cos\theta/c\sqrt{2M\hbar\omega_x}$ (we assumed $\eta \ll 1$). Note that here M is the total mass of the ions that are vibrating: if it's just one qubit, $M = M_1$, the mass of one qubit; for a chain of N qubits all coupled together, $M = NM_1$.

(c) Let's generalize this interaction to the case were the force is applied on two qubits (whose motion is coupled together so that they both undergo the same displacement), and, as an additional refinement, the Rabi frequency Ω is modulated at frequency $\delta > \omega_x$, i.e., $\Omega \rightarrow \cos(\delta t)\Omega$. Show that

$$\hat{\mathcal{H}} = \frac{\hbar\Omega}{2}\cos(\delta t)\hat{J}^{(+)}\left\{1 + i\eta\left(\hat{a}e^{-i\omega_x t} + \hat{a}^\dagger e^{i\omega_x t}\right)\right\} + \text{h.a.}$$

$$= \frac{\hbar\Omega}{4}e^{-i\delta t}\hat{J}_x + \frac{\hbar\Omega\eta}{4}e^{-i(\delta-\omega_x)t}\hat{a}^\dagger\hat{J}_y + \frac{\hbar\Omega\eta}{4}e^{-i(\delta+\omega_x)t}\hat{J}_y\hat{a} + \text{h.a.}$$

In this equation $\hat{J}^{(+)} = |1\rangle\langle 0|\otimes\hat{I} + \hat{I}\otimes|1\rangle\langle 0| = (\hat{J}_x - i\hat{J}_y)/2$ is the collective raising operator for the two qubits, and $\hat{J}^{(-)} = \hat{J}^{(+)\dagger}$.

(d) Use Eq. (6.55) to show that the effective Hamiltonian is

$$\hat{\mathcal{H}}_{\text{eff}} = \frac{\hbar\Omega^2\eta^2}{16(\delta+\omega_x)}\left[\hat{J}_y\hat{a}, \hat{a}^\dagger\hat{J}_y\right] + \frac{\hbar\Omega^2\eta^2}{16(\delta-\omega_x)}\left[\hat{a}^\dagger\hat{J}_y, \hat{J}_y\hat{a}\right]$$

$$\approx \frac{\hbar\Omega^2\eta^2}{8(\delta-\omega_x)}(\hat{I}\otimes\hat{I} + \hat{Y}\otimes\hat{Y}).$$

As we saw in Section 8.1, this is the Hamiltonian for a universal gate for quantum computing. This is called the Mølmer–Sørensen gate, after its inventors. In effect, by applying the force with the $\cos(\delta t)$ modulation, a qubit–qubit interaction has been turned on between two qubits (not just nearest neighbours), and can be turned off again simply by removing the force. This gate is particularly useful because it is *independent* of the occupation number of the phonon modes, and so, unlike the Cirac–Zoller gate, it will not be degraded by imperfectly cooled qubits. However, this comes at the price of being slower than the Cirac–Zoller gate.

8.3 CONCEPT CHECK AND EXERCISES

Concept Check

- Two-qubit gates can usually be created from the inherent short-range interactions of two qubits.
- Refocusing gates help with the circuit engineering, in particular in removing the effects of always-on interactions.
- Using a quantum data bus avoids the difficulties of short-range, always-on interactions, although they are more tricky to engineer.

Chapter Exercises

1. Show that the unitary operator

$$\exp[i(\pi/4)(\hat{Z} - \hat{I}) \otimes (\hat{Z} - \hat{I})]$$

is equivalent to the controlled-Z gate.

2. Show that the unitary operator

$$(\hat{H} \otimes \hat{I}) \exp[i(\pi/4)(\hat{X} - \hat{I}) \otimes (\hat{X} - \hat{I})](\hat{H} \otimes \hat{I})$$

is equivalent to the controlled-NOT gate.

3. Let

$$\hat{W}(\theta) = \exp[-i(\theta/4)(\hat{X} \otimes \hat{X} + \hat{Y} \otimes \hat{Y} + \hat{Z} \otimes \hat{Z} - \hat{I} \otimes \hat{I})].$$

Show that
   (a) $\hat{W}(\pi/2) = \sqrt{\text{SWAP}}$.
   (b) $(\hat{R}_2 \otimes \hat{R}_2)(\hat{Z} \otimes \hat{I})\hat{W}(\pi/2)(\hat{I} \otimes \hat{Z})\hat{W}(\pi/2) = CZ$, where $\hat{R}_2 \equiv \hat{S} = e^{i\pi/4}(\hat{I} - i\hat{Z})$.

4. Use the effective Hamiltonian for detuned systems, Eq. (6.55), to find an effective Hamiltonian if there is a big mismatch between ω and ω_0 in the Jaynes–Cummings model. How might such an interaction be used to measure the state of the qubit?

9 Device Characterization

When building the prototype of a conventional piece of hardware, engineers require a whole variety of meters and probes to measure and characterize the components that they are putting together for the first time, and to check the reliability of manufactured devices. In circuit electronics, for example, devices such as voltage or current meters and oscilloscopes are invaluable, but they measure quantities like currents, voltages, or frequencies. What does one want to measure in order to assess how "quantum" one's qubits are?

For small-scale devices with a few qubits, the answer is a technique called *quantum state tomography* by which the quantum state, as characterized by the density matrix, can be estimated from a series of experiments. In this chapter we will introduce this topic.

Learning Objectives

- Reinforce the necessary foundations in statistics and probability theory to understand quantum tomography.
- Combine statistics and the application of different quantum gates to estimate the Stokes parameters and construct the density matrix representation of the qubit.
- Apply the concepts from the one-qubit case to the two-qubit case to construct the density matrix of the two-qubit system.
- Apply concepts from regression analysis to deal with the effects of noise that renders the constructed density matrix non-physical.
- Become familiar with the basics of how noisy intermediate-scale quantum devices are tested.

9.1 How to Estimate Probabilities from Measurements

Let's suppose you are in a back-alley down by the docks and it's 3 a.m., and someone you met suggests you play a nice game of "tossing the coin". Given the circumstances, it might not be wholly unwise to consider the possibility that your new acquaintance might have a coin that isn't exactly fair: maybe the probability for it to land on "heads", $p \neq 0.5$, as one might expect. How can you test if the coin is fair? In other words, how can you measure the true probability p?

The obvious thing to do (other than reassessing your choice of nocturnal pastimes) would be to try tossing the coin a bunch of times. Let's say you do it three times, and twice you get heads, one time you get tails: does that imply it is twice as likely to land on heads as it is to land on tails? Obviously not, since a fair coin in three tosses will never give an equal number of heads and tails. But how many times should one toss it, and what confidence might one have in the outcome of all those trials?

Let us suppose the probability of landing on heads is p and that you perform N trials. The probability that it lands on heads n times will be

$$P(n) = \binom{N}{n} p^n (1-p)^{N-n},\tag{9.1}$$

where the coefficient $\binom{N}{n}$ or "N choose n" is the binomial coefficient, which is the number of different ways you can choose n items from a collection of N things. For example, suppose you wanted to choose three distinct letters, in any order, from the first five letters of the alphabet: there are 10 possible ways to do this (i.e., ABC, ABD, ABE, ACD, ACE, ADE, BCD, BCE, BDE, and CDE; remember the order is not important, so CBA counts the same as ABC, and we are counting distinct choices of letters, so no letters are repeated). In general, $\binom{N}{n} = N!/n!(N-n)!$, where $n! = n(n-1)(n-2)\cdots 1$ is the factorial; $5!/3!2! = 120/(6 \times 2) = 10$. Another use of $\binom{N}{n}$ is in the binomial expansion:

$$(a+b)^N = \sum_{n=0}^{N} \binom{N}{n} a^n b^{N-n}.\tag{9.2}$$

Putting $a = p$ and $b = 1 - p$ this shows $\sum_{n=0}^{N} P(n) = 1$, which is as it should be: in N tosses of the coin the number of heads you get will *always* be between 0 and N.

The *expected number* of heads after N tosses will therefore be

$$E[n] = \sum_{n=0}^{N} P(n)n.\tag{9.3}$$

We can perform this summation using a cute trick: let us introduce a dummy variable x and invent the function

$$f(x) = \sum_{n=0}^{N} P(n)x^n = (xp + 1 - p)^N.\tag{9.4}$$

If we put $x = 1$, we find $f(1) = \sum_{n=0}^{N} P(n) = 1$, which doesn't tell us anything new. However, we can take the derivative:

$$\frac{df(x)}{dx} = \sum_{n=0}^{N} P(n)nx^{n-1} = Np(xp + 1 - p)^{N-1},\tag{9.5}$$

and then setting $x = 1$ we find that $E[n] = Np$, and so

$$E[n]/N = p.\tag{9.6}$$

Thus, we have the result that the probability is the expected value of the number of heads divided by the number of tosses, which is a common-sense result; and, as we would expect, the best way to measure the probability of our questionable back-alley coin is to perform N trials and divide the resultant number of heads, n, by N: $p_{est} = n/N$.

But what confidence can we have in this: a simple way to assess how accurate a measurement is is to find the *variance*, which is defined by $\text{Var}[n] = E[(n - E[n])^2]$. If we repeat the set of N trials and each time we get a nearly identical answer, then the variance is small; if we get a wildly different answer each time, it will give a large variance. We can calculate the variance using the same trick: the second derivative of $f(x)$ is

$$\frac{d^2 f(x)}{dx^2} = \sum_{n=0}^{N} P(n)n(n-1)x^{n-2} = N(N-1)p^2(xp+1-p)^{N-2}, \tag{9.7}$$

and so, putting $x = 1$ we get $E[n^2] - E[n] = N(N-1)p^2$, and thus

$$\begin{aligned}
\mathrm{Var}[n] &= E[(n - E[n])^2] \\
&= E[n^2] - 2E[n]E[n] + E[n]^2 \\
&= E[n^2] - E[n]^2 \\
&= N(N-1)p^2 + Np - (Np)^2 \\
&= Np(1-p).
\end{aligned} \tag{9.8}$$

The square root of the variance is called the *standard deviation*, σ_n and is a good way to quantify the error in a measurement of probability. As a good rule of thumb, for a large number of trials (say $N > 10$), the value of n will be within two standard deviations of $E[n]$ 95% of the time.

$$p = p_{\mathrm{est}} \pm 2\sigma_n/N = n/N \pm 2\sqrt{p(1-p)/N}. \tag{9.9}$$

Since $p(1-p) \leq 1/4$, the error is always smaller than $1/\sqrt{N}$.

As fans of politics know, opinion polls try to figure out the probability that someone will vote for a given candidate. The technique is to ask a sample of voters whether they prefer Candidate A or Candidate B: if there are only two options, it is just like finding the probability p of a coin landing on heads. The poll results are quoted as something like "A has 53% of the support, B has 47% with 3% margin of error". The margin of error depends on the number of people surveyed: for example, if 1000 people were asked their opinion (and they all gave helpful responses), the formula above gives the margin of error as $1/\sqrt{N} \approx 3.16\%$. If you wanted to get a more accurate assessment, a survey of 10,000 people gives a margin of error of 1%, but that would be 10 times as expensive, since more interviews would be conducted (also statistical errors are only one source of errors in such a survey; complicating factors such as multiple candidates and ensuring the sample of people fairly reflects the overall population make the error calculation more nuanced).

Returning to the back-alley down by the docks: when we toss the coin three times, the error can be as large as $1/\sqrt{3} \approx 58\%$, so that getting two heads in three tosses is consistent with a fair coin ($p = 50\%$), but it's equally consistent with a very unfair coin (say $p = 1\%$), so maybe a larger number of trials is called for. If we toss it 100 times, the error is down to 10% (which is getting a bit better); but to ensure the error is less than 1%, we have to perform 10,000 trials (which can take a bit of time).

Exercises 9.1.1

1. Let's think about a game using two six-sided dice, with faces numbered 1 to 6, and both of which are fair (so that there is an equal chance of landing with each face upwards). The main goal of the game is to *avoid* rolling a combined score of 7.
   (a) What is the most probable combined score?
   (b) What are the mean and standard deviation of the score?
   (c) What is the probability of *not* rolling the most probable combined score?

(d) Usually the probability of rolling a double (so both dice have the same number uppermost) is 1/6. Suppose that the dice are somehow controlled so that it becomes twice as likely to score a double; what is the most likely score now?

(e) If the dice has been tampered with so that it is twice as likely to roll a 1 or a 6 than any of the other scores, what is the mostly likely result, and what is its probability?

9.2   Quantum States and Multi-Qubit Stokes Parameters

Single qubits, as we saw in Chapter 1, are the quantum version of a simple coin; thus, it should not be terribly surprising that measuring their probability amplitudes (rather than the probabilities) follows a similar strategy: we perform N trials on a qubit, each time prepared in the same state, allowing us to estimate the probability amplitudes. For example, consider the following simple circuit in QC 9.1:

QC 9.1

We prepare the single qubit state $\hat{\rho}$ (since it could be mixed, we use the density operator instead of the ket), then measure it to see if we get the result "0". The probability of obtaining this result is $\langle 0|\hat{\rho}|0\rangle = \rho_{00}$. We repeat this N times, and count the number of times we get "0": call this number n_3 (the reason for choosing this subscript will become clear in a minute). As we have seen, we can then estimate the probability of being in state $|0\rangle$ to be $p_3 = n_3/N$.

In the case of a classical coin, a single probability is enough to forecast the statistics of any outcome; however, for a qubit we need the complete density matrix:

$$\hat{\rho} = \begin{bmatrix} \rho_{00} & \rho_{01} \\ \rho_{10} & \rho_{11} \end{bmatrix}. \tag{9.10}$$

The quantity n_3/N is an estimate of $p_3 = \rho_{00}$; and, since $\rho_{00} + \rho_{11} = 1$, that gives us an estimate of ρ_{11} as well. But to fully specify the state, we need to know the complex-valued off-diagonal term ρ_{01}. How can we get our hands on that quantity?

To proceed, it will be useful to introduce a different way of writing $\hat{\rho}$. In 1852 George Stokes introduced a set of parameters that fully characterize the polarization of classical light beams. As we have seen, photon polarization makes a pretty good qubit: Stokes, unwittingly, discovered a way to fully characterize the quantum state of these qubits which actually pre-dates the discovery of quantum mechanics by 50 years!

Let's introduce the following three quantities:

$$S_i = \text{Tr}\{\hat{\rho}\hat{\sigma}_i\} \quad (i = 1, 2, 3), \tag{9.11}$$

where $\hat{\sigma}_1$, $\hat{\sigma}_2$, and $\hat{\sigma}_3$ are just the three Pauli matrices $\hat{X}$, $\hat{Y}$, and $\hat{Z}$ using a different notation (so we can include them in summation). These are just the same as the components of the *Bloch vector* we have seen already when looking at qubit dynamics (see Section 2.7, Exercise 16). In the case that Stokes was investigating, the equivalent of the density matrix was not normalized, so he had to introduce an additional parameter $S_0 = \text{Tr}\{\hat{\rho}\hat{I}\}$; but, for a qubit, this parameter is always equal to 1.

If you multiply any Pauli matrix by itself, you always obtain the identity matrix, whose trace is 2. If you multiply two different Pauli matrices, you will always get either i or $-i$ times the third Pauli matrix; and the trace of any of the Pauli matrices is always 0. Thus, we have the relation

$$\mathrm{Tr}\{\hat{\sigma}_i\hat{\sigma}_j\} = 2\delta_{ij}, \tag{9.12}$$

where the symbol δ_{ij} is equal to 1 if $i = j$ and to 0 if $i \neq j$. Thus, using the fact that the Pauli matrices are linearly independent (i.e., it is impossible to write $\hat{Z}$ as a sum of $\hat{X}$ and $\hat{Y}$) and, with the identity matrix, they span the space of 2×2 matrices, we can write the density matrix in terms of the $\mathcal{S}_i$ as follows:

$$\hat{\rho} = \frac{1}{2}\left(\hat{I} + \sum_{i=1}^{3}\mathcal{S}_i\hat{\sigma}_i\right). \tag{9.13}$$

Thus, if we can estimate $\mathcal{S}_1$, $\mathcal{S}_2$, and $\mathcal{S}_3$, we can estimate the complete state of the qubit. Estimating the complete density matrix from such measured data is called *quantum state tomography*.

 Interesting Facts 9.1

Where Does the Word "Tomography" Come From? The Ancient Greek words *tomos* ($\tau o\mu o\varsigma$) meaning "slice" and *graphein* ($\gamma\rho\alpha\phi\eta\nu$) meaning "drawing" (in the sense of creating a visual representation) seem to have first been combined to invent the word *tomography* by the German medical researcher H. H. Chaoul in 1935, as a name for a new X-ray technique to create an image that represented a slice through a lung. The word *tomos* is also the root of the word *atom*, meaning something that is un-sliceable or indivisible (the prefix "a" meaning "not"). The use of X-rays to produce slice-like images of the inside of a body became fully realized using digital computers to reconstruct the images, and was first used in a hospital in 1971; Allan MacLeod Cormack and Godfrey Hounsfield won the Nobel Prize in Medicine in 1979 for the invention.

Although scientists have been measuring the quantum states of various systems since the early 1990s, the use of the word *tomography* in connection with measurements of quantum states originated with the work on photon modes by Michael Raymer and his collaborators in 1993. Their quantum states can be represented as a two-dimensional function called the Wigner function, and the mathematical techniques, specifically the inverse Radon transform, used to determine the Wigner function from interference experiments were remarkably similar to those developed for the X-ray scanners in medicine. The measurement of qubit states as a diagnostic for quantum computing technology came a few years later, and did not use the same mathematics at all; however, the term "quantum state tomography" seems to have stuck.

We already have the recipe for finding $\mathcal{S}_3$. Introducing the projector operator $\hat{\Pi}_3$,

$$\hat{\Pi}_3 = |0\rangle\langle 0| = \frac{1}{2}(\hat{I} + \hat{Z}), \tag{9.14}$$

we find

$$\begin{aligned}\rho_{00} &= \mathrm{Tr}\{\hat{\Pi}_3\hat{\rho}\}\\ &= \frac{1}{2}(1 + \mathcal{S}_3).\end{aligned} \tag{9.15}$$

Thus, $\mathcal{S}_3 = (2n_3/N) - 1$.

Now let's consider the slightly modified measurement circuit shown in QC 9.2:

QC 9.2

We have changed the state before performing the measurement, so the probability of obtaining a "0" outcome in this case is

$$\langle 0|\hat{H}\hat{\rho}\hat{H}|0\rangle = \text{Tr}\{\hat{\Pi}_1\hat{\rho}\}$$
$$= \frac{1}{2}(1 + \mathcal{S}_1), \qquad (9.16)$$

where we have used $\hat{H}\hat{Z}\hat{H} = \hat{X}$ and the permutation property for the traces of products of matrices: $\text{Tr}\{\hat{A}\hat{B}\} = \text{Tr}\{\hat{B}\hat{A}\}$. The new projection operator is $\hat{\Pi}_1 = \hat{H}\hat{\Pi}_3\hat{H} = \frac{1}{2}(\hat{I} + \hat{X})$.

Similarly, with the circuit in QC 9.3:

QC 9.3

where $\hat{S}$ is the phase-shift gate:

$$\hat{S} = \begin{bmatrix} 1 & 0 \\ 0 & i \end{bmatrix}, \qquad (9.17)$$

and $\hat{S}\hat{H}\hat{Z}\hat{H}\hat{S}^\dagger = \hat{Y}$. Thus, the probability of obtaining "0" is equal to $\text{Tr}\{\hat{\Pi}_2\hat{\rho}\} = (1 + \mathcal{S}_2)/2$, where $\hat{\Pi}_2 = \frac{1}{2}(\hat{I} + \hat{Y})$.

Using these three different types of measurement, we have a simple recipe for measuring the quantum state $\hat{\rho}$.

- Create the input state; measure without any unitary; repeat N times. The number of "0" outcomes is n_3.
- Create the input state, again; apply the Hadamard gate $\hat{H}$ then measure; repeat N times. The number of "0" outcomes is n_1.
- Once again, create the input state; apply the gates $\hat{S}^\dagger$ and $\hat{H}$ and then measure; repeat N times. The number of "0" outcomes is n_2.
- The Stokes parameters are $\mathcal{S}_i = (2n_i/N) - 1$. Once we have estimates of the Stokes parameters, they can be assembled into an estimate of the state itself:

$$\rho_{\text{est}} = \frac{1}{2}\left(\hat{I} + \sum_{i=1}^{3} \mathcal{S}_i\hat{\sigma}_i\right). \qquad (9.18)$$

Here's a simple worked example. Let us suppose we want to test the reliability of a single-qubit device by creating the state $|\varphi_0\rangle = (|0\rangle + \sqrt{3}|1\rangle)/2$ and performing tomography on the output. The probabilities of obtaining "0" for the three separate measurements are $\text{Tr}\{\hat{\Pi}_1\hat{\rho}\} = (2 + \sqrt{3})/4 \approx 0.9330$, $\text{Tr}\{\hat{\Pi}_2\hat{\rho}\} = 0.5$, and $\text{Tr}\{\hat{\Pi}_3\hat{\rho}\} = 0.25$. Note that these outcomes are not mutually exclusive, so we don't need to worry that the three probabilities do not add to 1. Suppose we perform 100 measurements with each projection operator, and we obtain the results $n_1 = 93$, $n_2 = 42$, and $n_3 = 29$. These outcomes are just like the multiple coin tosses used to find the probabilities, and their statistics are governed by the binomial probability distribution. From this data, and using the formula $S_i = (2n_i/N) - 1$, we can estimate the Stokes parameters to be $S_1 = 43/50$, $S_2 = -8/50$, and $S_3 = -21/50$, and thus the measured state is

$$\hat{\rho}_{\text{est}} = \begin{bmatrix} 29/100 & (43 + 8i)/100 \\ (43 - 8i)/100 & 71/100 \end{bmatrix}. \tag{9.19}$$

Since the estimation procedure has introduced uncertainties, the estimated state is not pure. Since we started from the known state $|\varphi_0\rangle = (|0\rangle + \sqrt{3}|1\rangle)/2$, we can calibrate the accuracy of this method by calculating the *fidelity*: $F = \langle\varphi_0|\hat{\rho}_{\text{est}}|\varphi_0\rangle \approx 0.9774$, which implies the measured state is pretty close to the presumed state of the qubit.

9.2.1 Two Qubits

What happens if we want to do quantum state tomography on more than one qubit? For two qubits, the quantum state, again represented by a density matrix, can be expressed using a modified version of Eq. (9.13):

$$\rho = \frac{1}{4}\left(\hat{I} \otimes \hat{I} + \sum_{i=1}^{3} S_{i0}\hat{\sigma}_i \otimes \hat{I} + \sum_{j=1}^{3} S_{0j}\hat{I} \otimes \hat{\sigma}_j + \sum_{i,j=1}^{3} S_{ij}\hat{\sigma}_i \otimes \hat{\sigma}_j\right), \tag{9.20}$$

where S_{i0} are the Stokes parameters for qubit A, treated in isolation; S_{0j} are the Stokes parameters for qubit B; and S_{ij} represent the correlations between the two qubits. If we write the identity matrix as $\hat{I} = \hat{\sigma}_0$, this has a bit more compact form:

$$\rho = \frac{1}{4}\sum_{\mu,\nu=0}^{3} S_{\mu\nu}(\hat{\sigma}_\mu \otimes \hat{\sigma}_\nu). \tag{9.21}$$

Here $S_{\mu\nu}$ is a 4×4 real matrix whose 16 parameters can be considered the two-qubit analogue of the Stokes parameters. Since the Pauli matrices, now four in number, obey the relation $\text{Tr}\{\hat{\sigma}_\mu\hat{\sigma}_{\mu'}\} = 2\delta_{\mu\mu'}$, the analogue of Eq. (9.11) becomes

$$S_{\mu\nu} = \text{Tr}\{\hat{\rho}(\hat{\sigma}_\mu \otimes \hat{\sigma}_\nu)\} \quad (\mu, \nu = 0, 1, 2, 3). \tag{9.22}$$

The normalization of the density matrix implies that $S_{00} = 1$.

These quantities can be estimated by generalizing what we did for single qubits. Using the notation $\hat{U}_1 = \hat{H}, \hat{U}_2 = \hat{H}\hat{S}^\dagger$, and $\hat{U}_3 = \hat{I}$, QC 9.4 allows us to estimate S_{ij} for $i, j \in \{1, 2, 3\}$:

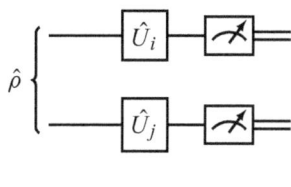

QC 9.4

The probability that *both* detectors register "0" is

$$
\begin{aligned}
p_{ij} &= \langle 00|(\hat{U}_i \otimes \hat{U}_j)\hat{\rho}(\hat{U}_i^\dagger \otimes \hat{U}_j^\dagger)|00\rangle \\
&= \mathrm{Tr}\{(\hat{\Pi}_i \otimes \hat{\Pi}_j)\hat{\rho}\} \\
&= \mathrm{Tr}\left\{\left(\frac{1}{2}(\hat{I} + \hat{\sigma}_i) \otimes \frac{1}{2}(\hat{I} + \hat{\sigma}_j)\right)\hat{\rho}\right\} \\
&= (1 + \mathcal{S}_{i0} + \mathcal{S}_{j0} + \mathcal{S}_{ij})/4.
\end{aligned}
\tag{9.23}
$$

The probability that the lower detector registers "0" regardless of what the other detector gets is

$$
\begin{aligned}
p_{i0} &= \mathrm{Tr}\{(\hat{\Pi}_i \otimes \hat{I})\hat{\rho}\} \\
&= \mathrm{Tr}\left\{\left(\frac{1}{2}(\hat{I} + \hat{\sigma}_i) \otimes \hat{I}\right)\hat{\rho}\right\} \\
&= (1 + \mathcal{S}_{i0})/2.
\end{aligned}
\tag{9.24}
$$

Similarly, the probability that the upper detector registers a "0" regardless of the outcome of the lower is

$$
p_{0j} = (1 + \mathcal{S}_{0j})/2.
\tag{9.25}
$$

Thus, if we can measure the probabilities p_{ij}, p_{i0}, and p_{0j} in the usual manner of counting the number of times you get a particular outcome then dividing by the number of trials, we can determine all 15 Stokes parameters:

$$
\mathcal{S}_{i0} = 2p_{i0} - 1, \quad \mathcal{S}_{0j} = 2p_{0j} - 1, \quad \text{and} \quad \mathcal{S}_{ij} = 4p_{ij} - 2p_{i0} - 2p_{0j} + 1.
\tag{9.26}
$$

Once all of the elements of $\mathcal{S}_{\mu\nu}$ are known, Eq. (9.21) can be used to find a matrix which should be the density matrix.

Let's try it for the singlet state: suppose we can create two qubits in the entangled state $|\beta_2\rangle = \frac{1}{\sqrt{2}}(|01\rangle - |10\rangle)$ (see Eq. (3.41)), and we wish to confirm this by performing tomography. A total of 15 different measurements should be made, and let's suppose we repeated them 100 times. Typical results (we used a random number generator using the appropriate binomial distribution rather than actual pairs of qubits) are as follows:

$$
\begin{bmatrix}
n_{00} & n_{01} & n_{02} & n_{03} \\
n_{10} & n_{11} & n_{12} & n_{13} \\
n_{10} & n_{21} & n_{22} & n_{23} \\
n_{10} & n_{31} & n_{32} & n_{33}
\end{bmatrix}
=
\begin{bmatrix}
100 & 51 & 57 & 54 \\
47 & 0 & 35 & 34 \\
42 & 22 & 0 & 26 \\
51 & 26 & 31 & 0
\end{bmatrix}.
\tag{9.27}
$$

The two-qubit Stokes parameters can then be found using Eq. (9.26):

$$
\begin{bmatrix}
\mathcal{S}_{00} & \mathcal{S}_{01} & \mathcal{S}_{02} & \mathcal{S}_{03} \\
\mathcal{S}_{10} & \mathcal{S}_{11} & \mathcal{S}_{12} & \mathcal{S}_{13} \\
\mathcal{S}_{10} & \mathcal{S}_{21} & \mathcal{S}_{22} & \mathcal{S}_{23} \\
\mathcal{S}_{10} & \mathcal{S}_{31} & \mathcal{S}_{32} & \mathcal{S}_{33}
\end{bmatrix}
=
\frac{1}{50}
\begin{bmatrix}
50 & 1 & 7 & 4 \\
-3 & -48 & 16 & 17 \\
-8 & 1 & -49 & 6 \\
1 & 0 & 4 & -55
\end{bmatrix}.
\tag{9.28}
$$

Using these values in Eq. (9.21), we get the following estimate for the density matrix of the state:

$$\hat{\rho}_{\text{est}} = \frac{1}{200} \begin{bmatrix} 0 & 1-11i & 14+2i & 1-17i \\ 1+11i & 102 & -97+15i & -20+14i \\ 14-2i & -97-15i & 108 & 1-3i \\ 1+17i & -20-14i & 1+3i & -10 \end{bmatrix}. \tag{9.29}$$

This density matrix has trace 1 and is Hermitian, as it should be. Also the fidelity is $F = \langle \beta_2 | \hat{\rho}_{\text{est}} | \beta_2 \rangle = 1.01$, so that there is 101% probability of overlap between $\hat{\rho}_{\text{est}}$ and $|\beta_2\rangle$ (which is *not* as it should be). Another clue that something is rotten in the state of these qubits is $\langle 11 | \hat{\rho}_{\text{est}} | 11 \rangle$ is a negative number, which is not physically possible, since it implies a negative probability of overlap of $\hat{\rho}_{\text{est}}$ and $|11\rangle$ (probabilities are always positive numbers between 0 and 1 inclusive). If further proof of the rottenness of this $\hat{\rho}_{\text{est}}$ is required, its eigenvalues are 1.02878, 0.0958361, 0.0412598, and -0.16588. The negative eigenvalue clinches it: physical density matrices *must* have positive eigenvalues.

This is not an isolated case in which the random variations of the data (i.e., the values of $n_{\mu,\nu}$) resulted in a pathological, non-physical outcome. Rather, such non-positive definite estimates of pure quantum states are what usually happens (which are what is required in quantum technologies). Remember a two-qubit pure state has eigenvalues 1, 0, 0, and 0; a small error induced by the randomness of the measurement outcome can just as easily result in one of the 0 eigenstates being perturbed to a negative value as to a positive value. What can we do? Undergraduates in a teaching laboratory might well be tempted to ignore the problem and hope the grader doesn't notice; but reality is a harsher task-master. There are various strategies to circumvent this difficulty and extract a positive density matrix from the data.

Exercises 9.2.2

1. Write the Stokes parameters for the following states:
   (a) $|0\rangle$.
   (b) $|1\rangle$.
   (c) $(|0\rangle + e^{i\phi}|1\rangle)/\sqrt{2}$.
   (d) The maximally mixed state, $\hat{\rho} = \frac{1}{2}\hat{I}$.
2. Show that the Stokes parameters obey $\sum_{i=1}^{3} S_i^2 \leq 1$. Are Stokes parameters real? Do they have to be positive?
3. Suppose we repeated the one-qubit tomography exercise discussed in the text, only this time we obtain the slightly different results $n_1 = 94$, $n_2 = 43$, and $n_3 = 26$. Find $\hat{\rho}_{\text{est}}$ in this case. Is there anything strange about this density matrix? *Hint: Check the eigenvalues.* What do you think caused this, and what should we do to redress the situation?

9.3 Quantum State Tomography from Method of Least Squares

> ✦ **Mathematics Tips 9.2**
>
> **Linear Regression.** Suppose there are two quantities, x and y, which we think might be related. A simple example might be the temperature on a given day, and the number of ice creams a shop sells. On hotter days, the shop will sell more ice cream, but other variables might also affect ice-cream

sales, so one would not expect a perfect correlation. To investigate further, we have gathered a set of data consisting of the values of two quantities, which we will denote as x_i and y_i, with the subscripts $i = 1, 2, \ldots, N$. Linear regression is a method for modelling the relationship between these two variables by assuming that the relationship between them is given by a linear equation of the form

$$y = mx + b, \tag{9.30}$$

where m is the slope, and b is the y-intercept (the value of y when $x = 0$). If we can determine the best values for m and b, it will allow us to predict, at least approximately, the value of y for a given x (and thus avoid the shop running out of ice cream next time there is a heat wave).

One way to see if our model is any good is to look at the difference between the *predicted* value of y and its *actual* value. We quantify that using a quantity called the *residual* defined for each data point, labelled by i, as

$$\mathcal{R}_i = y_i - Y_i, \tag{9.31}$$

where $\mathcal{R}_i$ is the residual, y_i is the actual value of y, and $Y_i = mx_i + b$ is the predicted value of y. The better the model, the closer $\mathcal{R}_i$ is to 0; i.e., $\mathcal{R}_i \to 0$.

With a set of data points one has to draw the best line that will represent the relationship, so that it minimizes the distance between the line itself and the data points. In other words, the following quantity needs to be minimized:

$$\mathcal{R} = \sum_i \mathcal{R}_i^2. \tag{9.32}$$

Notice the *squared* quantities in (9.32); this is the reason why sometimes the regression line is called the *least squares line*. The idea is to minimize $\mathcal{R}$. What we want is to find the combination of m and b that gives the smallest $\mathcal{R}$. Then, we take the values of this combination to define the relationship between y and x in (9.30). In order to achieve that, we resort to calculus.

Let's write $\mathcal{R}$ explicitly:

$$\mathcal{R}(m, b) = \sum_i^N \left[y_i - (mx_i + b) \right]^2, \tag{9.33}$$

where N is the number of data points. What we need to do is take the derivative of $\mathcal{R}(m, b)$ with respect to m as well as b and then set both derivatives to 0. Using the two independent equations, we can find the values of m and b in terms of the variables y and x.

The equations $\frac{\partial \mathcal{R}}{\partial m} = 0$ and $\frac{\partial \mathcal{R}}{\partial b} = 0$ give the following relations, respectively:

$$\begin{aligned} m\overline{x^2} + b\overline{x} &= \overline{xy}, \\ b &= \overline{y} - m\overline{x}. \end{aligned} \tag{9.34}$$

The overbar on top of each quantity corresponds to its average, e.g., $\overline{x} = \sum_i^N x_i / N$. Combining these equations, we get

$$m = \frac{\overline{xy} - \overline{x}\,\overline{y}}{\overline{x^2} - \overline{x}^2}, \tag{9.35}$$

$$b = \overline{y} - m\overline{x}.$$

The equations in (9.35) applied to (9.30) give us the regression model we are seeking.

As we saw with the coin-tossing during our nocturnal adventure in Section 9.1, $p \neq n/N$; rather, $p \approx n/N$. So it is with the Stokes parameters: $(2n_{i0}/N) - 1$ is not the actual value of S_{i0}, rather it is a measured value, and like all measurements, it can have errors. As we have seen, in the case of estimating a density matrix these can be rather serious, in that the matrix one obtains by evaluating Eq. (9.21) (and Eq. (9.18) for the single-qubit case) might not actually be a density matrix at all. The density matrices obtained from these formulas will be Hermitian and have unit trace, but we also need them to have positive eigenvalues.

There are various methods used to deal with this problem. Most simple is the so-called *quick-and-dirty* method, by which the troublesome negative eigenvalues of $\hat{\rho}_{est}$ are simply discarded, and the remainder of the density matrix is re-normalized. In the example above, the eigenvalues of $\hat{\rho}_{est}$ are (approximately) $\lambda_1 = 1.029$, $\lambda_2 = 0.096$, $\lambda_3 = 0.041$, and $\lambda_4 = -0.166$. We will denote the corresponding orthonormal eigenstates (which can be determined numerically from the $\hat{\rho}_{est}$ matrix), as follows: $|v_1\rangle$, $|v_2\rangle$, $|v_1\rangle$, and $|v_1\rangle$, so that $\hat{\rho}_{est} = \sum_{n=1}^{4} \lambda_n |v_n\rangle\langle v_n|$. We see that $\lambda_4 = -0.166$ is negative, thus the quick-and-dirty estimate of the state is obtained by setting that value to 0 and re-normalizing:

$$\hat{\rho}_{Q\&D} = \frac{1}{\left(\sum_{n=1}^{3} \lambda_n\right)} \sum_{n=1}^{3} \lambda_n |v_n\rangle\langle v_n|. \tag{9.36}$$

Unlike $\hat{\rho}_{est}$, the matrix $\hat{\rho}_{Q\&D}$ will have the key properties of a density matrix: it is Hermitian, it has unit trace, and its eigenvalues are positive or 0. The fidelity is $F = \langle \beta_2|\hat{\rho}_{Q\&D}|\beta_2\rangle = 0.867$, which is a good (but not great) fit.

A more sophisticated technique is to use **regression analysis**, which is a general technique used in data analysis to find the best model that is consistent with noisy experimental data. Perhaps the most well-known example is linear regression based on the least-squares method, widely used to make predictions based on data (see Box 9.2 and Fig. 9.1). For quantum state tomography, we have a set of data – the values of $n_{\mu,\nu}$ – and we assume a mathematical model, i.e., a density matrix. Let the

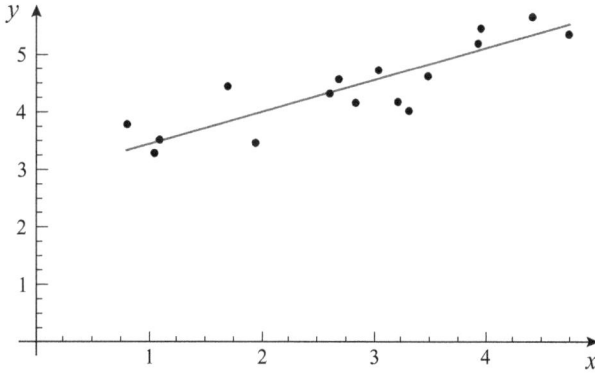

Figure 9.1 Linear regression using the least-squares method. Data pairs $\{x_i, y_i\}$ are plotted as points on the graph. The straight line shows the relation between the two variables and is derived from statistical averages.

symbolic matrix $\hat{T}(u)$ be a function of real-valued variables $u = \{u_1, u_2, \ldots, u_{16}\}$, and defined by the formula

$$\hat{T}(u) = \begin{bmatrix} u_1 & 0 & 0 & 0 \\ u_5 + iu_6 & u_2 & 0 & 0 \\ u_{11} + iu_{12} & u_7 + iu_8 & u_3 & 0 \\ u_{15} + iu_{16} & u_{13} + iu_{14} & u_9 + iu_{10} & u_4 \end{bmatrix}. \tag{9.37}$$

Then the matrix

$$\hat{\rho}(u) = \hat{T}(u)^\dagger \hat{T}(u) / \text{Tr}\{\hat{T}(u)^\dagger \hat{T}(u)\} \tag{9.38}$$

will *always* be **positive definite**, Hermitian, and unit trace, and thus has all the required properties of a density matrix. The relationship is reversible: for any positive definite Hermitian matrix one can find a unique matrix $\hat{T}(u)$ (this is called the *Cholesky decomposition*), so $\hat{\rho}(u) = \hat{T}(u)^\dagger \hat{T}(u) / \text{Tr}\{\hat{T}(u)^\dagger \hat{T}(u)\}$ is our model, akin to the assumption of the linear formula $y = mx + b$ in the linear regression problem. We want to find the "best" values of the 16 variables $\{u_1, u_2, \ldots, u_{16}\}$ which are most consistent with the noisy data $n_{\mu,\nu}$ gathered from repeated experiments on our qubit pairs. How we determine the best is to a certain extent a matter of taste. A simple method widely used in regression analysis is to calculate the *residual*, given by the expression

$$\mathcal{R}_{\mu,\nu}(u) = \frac{n_{\mu,\nu}}{n_{0,0}} - \text{Tr}\{\hat{\rho}(u)(\Pi_\mu \otimes \Pi_\nu)\}. \tag{9.39}$$

The first term on the right is the probability estimate from our data; the second term is the corresponding probability which we would calculate using $\hat{\rho}(u)$. Note that in our set of tomographic bases changes, $\hat{U}_0 = \hat{I}$ so that $n_{0,0}$ is always equal to N, the number of repetitions of the experiment. The sum of the squares of these residuals for all data points is a good means to compare $\hat{\rho}(u)$ with the measured data:

$$\mathcal{R}(u) = \sum_{\mu,\nu=0}^{3} \mathcal{R}_{\mu,\nu}(u)^2. \tag{9.40}$$

The "best" set of u parameters is the one for which $\mathcal{R}(u)$ takes the least value (hence the name "method of least squares"). In the case of the simple linear regression, one can use calculus to find analytic expressions for m and b in terms of the averages of the data. For two-qubit tomography there are 16 variables variables $\{u_1, u_2, \ldots, u_{16}\}$ over which we must optimize, too many to hope for an analytic expression. However, there are a variety of powerful methods of numerical optimization that can find the optimal values of $\{u_1, u_2, \ldots, u_{16}\}$ efficiently, and and can be implemented on a computer (we used the "FindMinimum" function in Mathematica, with initial values for the u-parameters deduced from the Cholesky decomposition of $\rho_{Q\&D}$; but there are plenty of other options). The parameterization of the density operator using the Cholesky decomposition automatically confines our search to the space of positive density matrices. In the above numerical example, the optimal values of u which we found numerically were as follows: $u = \{0.199, 0.657, 0.283, 0.001, -0.067, -0.138, -0.565, 0.059, -0.082, 0.057, 0.238, 0.116, -0.106, 0.052, -0.011, -0.093\}$; substituting these into Eq. (9.38) we obtain the optimal density matrix $\hat{\rho}_{\text{Opt}}$ (see Fig. 9.2). The fidelity $F = \langle \beta_2 | \hat{\rho}_{\text{Opt}} | \beta_2 \rangle = 0.95$.

It turns out that, for a simple single-qubit state, the least-squares approach does not work very well, since the only time the Stokes parameter re-construction given in Eq. (9.18) fails to work is for highly pure states, for which the quick-and-dirty technique is pretty good.

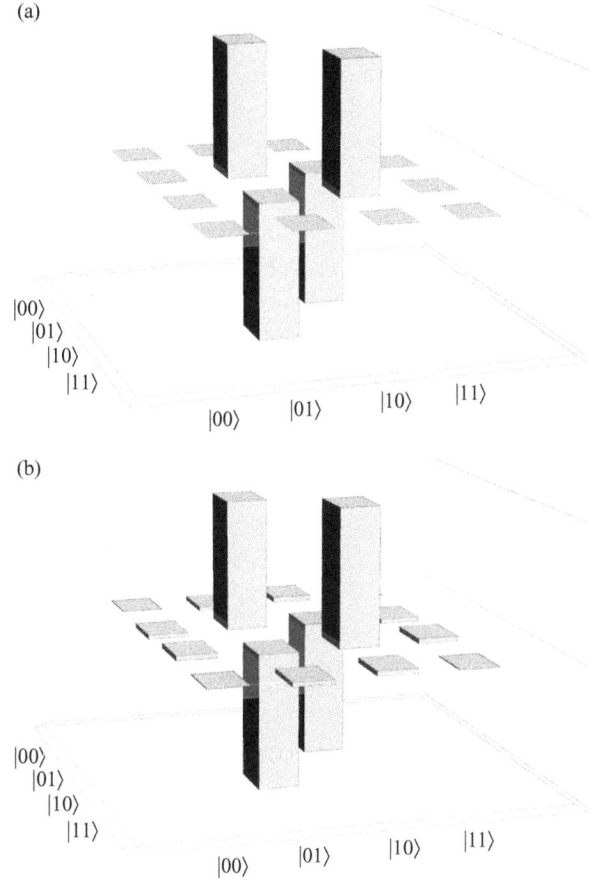

(a)

(b)

Figure 9.2   Results of regression analysis for quantum state tomography. (a) Shows the theoretical density matrix for the singlet state $|\beta_2\rangle$; (b) shows the real part of the density matrix recovered from the simulated data Eq. (9.27) using the least-squares method described in the text. Not shown is the imaginary part of the density matrix (which vanishes in the theoretical case, and is negligible in the least-squares case).

There are refinements to the least-squares method that can improve accuracy; most simply, one can increase the number of trials. The weighted least-squares method involves multiplying the terms in the sum in Eq. (9.40), so that the terms with less uncertainty are given more weight. Also, other means of extracting conclusions from random data, such as Bayesian inference (in which one assumes a particular state at the start, and modifies the assumption based on the outcome of each measurement) have been employed. For some technologies, a different set of measurements used might be useful depending on what unitary operations are easiest to carry out, but the method is readily adaptable. Also, other circumstances might necessitate modification of the procedure: for example, if the number of repetitions N is a random number, or if your quantum detector only works an unknown percentage of the time. But, in general, they follow the basic ideas given here.

Quantum state tomography can be scaled up for multiple qubits. If the power of quantum computation arises from the large dimensionality of Hilbert space, when trying to measure the state of multi-qubit devices, that complexity becomes a headache. Not only are a large number of measurements required (with n qubits, there are 4^n generalized Stokes parameters, which require 3^n distinct quantum circuits to measure), once the data is gathered, the numerical optimization required to process that data into a positive-defined density matrix rapidly requires a search through a 4^n-dimensional parameter space, and so becomes intractable. There are some techniques, such as assuming that the density matrix conforms to some particular model, that can ease this problem somewhat, but there is no practical way of overcoming this exponential scaling: Hilbert space is just too big.

Another generalization of quantum state tomography is the application of similar techniques to perform *quantum process tomography*, by which the effect of a quantum device (or process) can be completely characterized by inputting a variety of initial states and performing state tomography on the outputs. For a process involving n qubits, one must input 4^n distinct inputs, each of which requires 4^n measurements to perform the output state tomography. That makes a total of 4^{2n} measurements; while this is useful for small-scale devices or quantum communication channels, it rapidly becomes unwieldy as the number of qubits increases.

Exercises 9.3.1

1. Starting from Eq. (9.34) prove Eq. (9.35).
2. Suppose one has a pair of three-level systems: how would you formulate tomography in this case?
3. The Stokes parameters for a single qubit were defined in Eq. (9.11). Find the Cholesky decomposition of a single-qubit density matrix in terms of the Stokes parameters.

9.4   Noisy Intermediate-Scale Quantum Devices and Sampling Random Quantum Circuits

Currently, the field of quantum information science has a problem: while the goal is to build mega-qubit devices, right now we have devices with about 100 fully-functioning qubits with controllable decoherence. Tomography, which is the main theme of this chapter, works well for a few qubits, and is an important tool for validating the building blocks of the technology. But tomography can only go so far, since the information that must be gathered experimentally and then processed computationally to yield the quantum state grows as 4^n, where n is the number of qubits: for 10 qubits, one must search through a parameter space with over a million dimensions; for 20 qubits it has become more than a trillion dimensions, and so on. While various work-arounds and assumptions about the state can push the tomography envelope, its ability to fully characterize systems with dozens or hundreds of qubits appears to be problematic.

Thus, to characterize the modern **noisy intermediate-scale quantum (NISQ) devices** with about 100 qubits one must adopt a different approach. A promising and widely adopted method is to use the device to perform some quantum circuits chosen at random and compare the measured probabilities of different outcomes with simulations performed by conventional computers (this, of course, presupposes that such classical simulations are tractable). If the performance matches the conventional simulation, you have validated the device.

The number of qubits involved in the circuit is called the *width*, w, and they are assumed to initially be in the state $|0\rangle$. The quantum circuit consists of a number of layers, each layer consisting of one-qubit gates and two-qubit gates performed in parallel; $w/2$ (or $(w-1)/2$ if w is odd) such gates can be performed simultaneously between different pairs of qubits. The number of layers in the circuit is called the *depth*, d.

To test your quantum device, start with a small number of qubits – say $w = 2$ – and perform a relatively simple quantum circuit – say with two layers, so $d = 2$. For such a circuit calculating the possible outcomes (i.e., the bit-values output when the qubits are measured) and their probabilities should not be too arduous a task. Repeating this procedure for say 100 times should give a reasonable initial estimate of the probabilities associated with each possible outcome, and thus confirm you are implementing the desired unitary. Repeating this for a variety of different unitaries, each chosen at random, confirms that you didn't somehow get lucky with the first circuit.

Now we repeat the procedure for a circuit of width $w = 3$ and depth $d = 3$; if that works as well, we continue to $w = 4$ and $d = 4$ and so on, until a width and a depth is reached for which the outputs are too far from the predicted results for us to have any confidence in the functioning of our quantum device. Let $w_{max} = d_{max}$ be the largest value of w and d for which the quantum computer "works" (in the sense of reproducing the expected output with acceptable accuracy); then the **quantum volume** is defined as

$$\mathcal{V}_Q = 2^{w_{max}}. \tag{9.41}$$

This figure of merit has much to recommend it: all the characteristics that will be needed in a large-scale quantum computer – accuracy of quantum gates and measurements; good inter-connectivity of qubits (so that two-qubit gates can be performed between arbitrary pairs of qubits); a sizeable selection of quantum gates that can be easily performed; good quantum compilers, also called **transpilers** (which combine quantum gates into smaller, less complex combinations of gates which, to a good approximation, have an equivalent effect); all of these will also improve the quantum volume. Current experiments are reporting quantum volumes up to 2^{40}, indicating that the device has 40 qubits which can reliably implement circuits of depth 40.

The random circuits themselves are of a specific type: a random permutation of the qubits (carried out by multiple SWAP gates) followed by two-qubit unitary gates carried out on as many pairs of qubits as are available (i.e., one qubit has to be left out if you have an odd number of qubits in the circuit) constitutes one *layer* of the circuit. Such layers are repeated, each with a different randomly selected permutation followed by different two-qubit gates. The total number of such layers in the circuit is the *depth d* of the circuit. After the final layer is implemented, the qubits are measured, to obtain some w-bit number, x.

⊕ **Mathematics Tips 9.3**

Heavy-Output Generation (HOG) Problem. To illustrate one particular protocol used in NISQ device validation, let's consider the possible outcomes on the simple circuit shown in QC 9.5.

QC 9.5

When the output is measured, one obtains as an output a w-bit integer in the range $\{0, 1, \ldots, 2^w - 1\}$. The probability of each outcome is given by the Born rule as $p(x) = |\langle x | \hat{U} | 0 \rangle|^2$; some outputs will be more likely than others. There will a median probability, easily deduced from the list of probabilities, and 50% of the outputs will be more probable than the median probability: these are called the *heavy-outputs*. For example, we numerically created a random unitary gate (see

Appendix D) which acts on the space of $w = 3$ qubits; the possible outcomes are $\{0, 1, \ldots, 7\}$ and their probabilities (which we calculated using the numerical values of the elements of $\hat{U}$) are $p(0) = 0.081$, $p(1) = 0.031$, $p(2) = 0.137$, $p(3) = 0.047$, $p(4) = 0.057$, $p(5) = 0.156$, $p(6) = 0.380$, and $p(7) = 0.111$. Sorting this list in order of increasing probability, we get $p(1) = 0.031$, $p(3) = 0.047$, $p(4) = 0.057$, $p(0) = 0.081$, $p(7) = 0.111$, $p(2) = 0.137$, $p(5) = 0.156$, and $p(6) = 0.380$. In other words, the median probability is $p_m = 0.096$, and the heavy-outputs are 7, 2, 5, and 6. The total probability of obtaining one of the heavy-outputs is $p_h = p(7) + p(2) + p(5) + p(6) = 0.784$: this quantity depends on $\hat{U}$.

The heavy-output generation problem is, for a given $\hat{U}$, to generate outputs that are heavy two-thirds of the time. It is believed that, classically, solving this problem by working out the probabilities via the Born rule becomes intractable for large systems, while on a quantum computer it's just a question of collecting the outputs; hence, generating a heavy-output is an example of an undeniably quantum process.

Exercises 9.4.1

1. Generate a random unitary matrix for three qubits and find the probability that it will have a heavy-output.
2. Using the method in Appendix D, generate 10,000 Haar-distributed random unitaries, and create a histogram of the heavy-output probabilities.

9.5 CONCEPT CHECK AND EXERCISES

Concept Check

- Quantum states can be measured using methods analogous to measuring probabilities.
- Regression analysis of some sort is usually needed to recover a physical density matrix.
- Characterizing large-scale devices must be accomplished via bench-marking algorithms such as random circuit sampling rather than measuring the quantum state.

Chapter Exercises

1. In Section 9.3 we looked at an example of single-qubit tomography which failed to produce a positive-definite density matrix via the Stokes parameter technique. Use both the quick-and-dirty technique and the method of least squares to deduce a physical density matrix from this data. Which produces the better fidelity, and why?
2. Here is a simulated set of counts data for quantum state tomography of a two-qubit system. Write a computer code to find the optimal density matrix via the least-squares method, and estimate the state of the system.

$$\begin{bmatrix} 1000 & 490 & 485 & 500 \\ 515 & 236 & 248 & 0 \\ 515 & 513 & 265 & 253 \\ 500 & 230 & 486 & 253 \end{bmatrix} \tag{9.42}$$

10 Variations on the DiVincenzo Criteria

Composers of classical music are fond of devising pieces of music that are "variations on a theme," in which a melody from some other composer is subtly altered and improved to produce a different but related composition, which hopefully the fans will still enjoy. So it is with quantum computing architectures: the DiVincenzo criteria, which we discussed at length in Chapter 6, are a composition that provides a very challenging set of requirements for a potential quantum computing technology, and if we can alter and improve on these criteria in subtle ways perhaps we can find easier technological solutions. In this chapter we are going to explore three such *variations on a theme by DiVincenzo.*

Learning Objectives

- Understand the experimental limitations on working with initial qubits that are all in the $|0\rangle$ state, how that affects the accuracy of the results, and the importance of understanding the latter to avoid making incorrect physical claims.
- Understand linear-optical quantum computing in which two-qubit interferometric quantum gates, using beamsplitters, can be implemented non-deterministically.
- Understand the basics of cluster-state quantum computing.

10.1 Pseudo-Pure States

The initialization of qubits requires, in effect, for the qubits to be in the pure state $|0\rangle_1 \otimes |0\rangle_2 \otimes \cdots \otimes |0\rangle_L$. Looking at this in thermodynamics terms, this means the qubits, which are not the abstract perfections of a quantum circuit diagram but are physical systems existing in the real-world, must be at absolute zero temperature, with zero entropy. Unfortunately, at least according to Nerst's formulation of the third law of thermodynamics, this is not possible: cooling a system to absolute zero requires an infinite number of operations. We have to live with a certain amount of heat. Using technologies like laser cooling and trapping, or dilution refrigerators, cooling to nano-kelvin (10^{-9} K) levels can be routinely achieved (although just because it can be done routinely does not imply that it is "easy"); and most quantum technologies require such cooling to function. Small amounts of residual temperatures can be tolerated by employing error correction techniques (which we will describe in the next chapter).

It is a reasonable question to ask, however: are such heroically low temperatures really necessary? Consider, for example, a single qubit in a thermal equilibrium state at temperature T. Using the Boltzmann distribution, its density matrix is given by

$$\hat{\rho}_1 = p_0 |0\rangle\langle 0| + p_1 |1\rangle\langle 1|, \tag{10.1}$$

where

$$p_n = \frac{\exp(-E_n/k_B T)}{\exp(-E_0/k_B T) + \exp(-E_1/k_B T)} \tag{10.2}$$

with E_0 and E_1 being the energies of the two qubit levels, and k_B is Boltzmann's constant, which relates temperature to energy; its value is 1.380649×10^{-23} joules per degree kelvin. Equation 10.1 can be re-written as follows:

$$\hat{\rho}_1 = p_1(|0\rangle\langle 0| + |1\rangle\langle 1|) + (p_0 - p_1)|0\rangle\langle 0|$$
$$= (1 - \varepsilon)\frac{1}{2}\hat{I} + \varepsilon|0\rangle\langle 0|, \tag{10.3}$$

where

$$\varepsilon = p_0 - p_1 \approx \Delta E/2k_B T. \tag{10.4}$$

Such a state as that given in Eq. (10.3) is called a **pseudo-pure state**. It consists of the pure-state projector $|0\rangle\langle 0|$ combined with the maximally mixed state $\frac{1}{2}\hat{I}$. What happens if we try to perform unitary operations on a pseudo-pure state?

$$\hat{\rho}' = \hat{U}\hat{\rho}\hat{U}^\dagger$$
$$= (1 - \varepsilon)\frac{1}{2}\hat{U}\hat{I}\hat{U}^\dagger + \varepsilon\hat{U}|0\rangle\langle 0|\hat{U}^\dagger$$
$$= (1 - \varepsilon)\frac{1}{2}\hat{I} + \varepsilon|\varphi\rangle\langle\varphi|, \tag{10.5}$$

where $|\varphi\rangle = \hat{U}|0\rangle$. In other words, the pure-state projector has been transformed while the maximally mixed component has been left untouched. Since quantum computing is, by and large, about performing unitary operations on a pure state, why not use a pseudo-pure state instead?

One fly in the ointment is that measurement is a bit more involved. If we had a qubit in the pure state $|0\rangle$, then measurement should yield the result "0" with 100% probability. However, measurement of the pseudo-pure state Eq. (10.3) will yield "0" with probability $(1 + \varepsilon)/2$ and "1" (i.e., the wrong answer!) with probability $(1 - \varepsilon)/2$. To avoid things becoming completely hit-or-miss, one must repeat the measurements a large number of times, with multiple qubits each prepared in the same manner. If n is the number of trials, the number of "0" results, m, will be a random variable which conforms to the binomial probability distribution. The average number of correct answers out of n trials will be $n(1 + \varepsilon)/2$, the average number of wrong answers will be $n(1 - \varepsilon)/2$, and the standard deviation of m is $\sigma_m = \sqrt{n(1 - \varepsilon^2)}/2$. In order to achieve a correct result with confidence, we require that the difference between the number of correct results and the number of wrong answers be much greater than the standard deviation, so that the majority result is the correct answer. Thus,

$$n(1 + \varepsilon)/2 - n(1 - \varepsilon)/2 \gg \sqrt{n(1 - \varepsilon^2)}/2$$
$$n\varepsilon \gg \sqrt{n(1 - \varepsilon^2)}/2$$
$$\sqrt{n} \gg \sqrt{(1 - \varepsilon^2)}/2\varepsilon$$
$$n \gg \frac{1 - \varepsilon^2}{4\varepsilon^2} \approx \frac{1}{4\varepsilon^2}. \tag{10.6}$$

Remember: these n separate, uncoupled qubits each in the same pseudo-pure state all represent just *one* qubit of our quantum computer.

Things get a bit more tricky when you want to have more than one qubit involved. For example, two uncoupled qubits in the thermal state have the following density operator:

$$
\begin{aligned}
\hat{\rho}_2 &= \hat{\rho}_1 \otimes \hat{\rho}_1 \\
&= \frac{1}{4}\left\{ (1+\varepsilon)^2|00\rangle\langle00| + (1-\varepsilon^2)|01\rangle\langle01| + (1-\varepsilon^2)|10\rangle\langle10| + (1-\varepsilon)^2|11\rangle\langle11| \right\}.
\end{aligned} \tag{10.7}
$$

This state has three distinct eigenvalues: $(1+\varepsilon)^2$, $(1-\varepsilon)^2$, and $(1-\varepsilon^2)$, the last of which is "doubly degenerate", meaning that it has two linearly independent eigenvectors associated with it. But a two-qubit pseudo-pure state must have the following form:

$$
\hat{\rho}_2^{(\mathrm{PPS})} = \frac{1-\varepsilon'}{4}\hat{I} \otimes \hat{I} + \varepsilon'|\Psi_{12}\rangle\langle\Psi_{12}|, \tag{10.8}
$$

where ε' is a small parameter (distinct from ε). The eigenvalues will be $(1+3\varepsilon')/4$, associated with $|\Psi_{12}\rangle$, and $(1-\varepsilon')/4$, associated with the three possible states orthogonal to $|\Psi_{12}\rangle$. No unitary transforms can alter the state given in Eq. (10.7) into a state of the form given in Eq. (10.8) unitary transforms only alter eigenstates, not eigenvalues.

We can fix this, however: recalling the effect of CNOT gates (and remembering to apply the gates to both the bras and to the kets, i.e., to both sides of the density operator), we find

$$
\begin{aligned}
\mathrm{CNOT}_{1,2}\,\hat{\rho}_2\,\mathrm{CNOT}_{1,2} = \frac{1}{4}\Big\{ &(1+\varepsilon)^2|00\rangle\langle00| + (1-\varepsilon^2)|01\rangle\langle01| \\
&+ (1-\varepsilon^2)|11\rangle\langle11| + (1-\varepsilon)^2|10\rangle\langle10| \Big\},
\end{aligned} \tag{10.9}
$$

$$
\begin{aligned}
\mathrm{CNOT}_{2,1}\,\hat{\rho}_2\,\mathrm{CNOT}_{2,1} = \frac{1}{4}\Big\{ &(1+\varepsilon)^2|00\rangle\langle00| + (1-\varepsilon^2)|11\rangle\langle11| \\
&+ (1-\varepsilon^2)|10\rangle\langle10| + (1-\varepsilon)^2|01\rangle\langle01| \Big\}.
\end{aligned} \tag{10.10}
$$

And so, if we combine $\hat{\rho}_2$ with the two transformed states we get

$$
\begin{aligned}
\hat{\rho}_2' &= \frac{1}{3}\left\{ \hat{\rho}_2 + \mathrm{CNOT}_{1,2}\,\hat{\rho}_2\,\mathrm{CNOT}_{1,2} + \mathrm{CNOT}_{2,1}\,\hat{\rho}_2\,\mathrm{CNOT}_{2,1} \right\} \\
&= \frac{(1+\varepsilon)^2}{4}|00\rangle\langle00| + \frac{3-2\varepsilon-\varepsilon^2}{12}\left\{ |11\rangle\langle11| + |10\rangle\langle10| + |01\rangle\langle01| \right\} \\
&= \frac{1-\varepsilon'}{4}\hat{I} \otimes \hat{I} + \varepsilon'|00\rangle\langle00|,
\end{aligned} \tag{10.11}
$$

where $\varepsilon' = \varepsilon(\varepsilon+2)/3$. Note that the mathematical transform given by Eq. (10.11) is an example of a **completely positive trace-preserving map**, which is a generalization of the idea of a unitary transform of a state such that the entropy of the state can change. How can such a transformation be performed if we can only implement unitary transformations on our qubits? What it is telling us to do physically is to arrange the thermal state qubits into groups consisting of three parts; on one part you do nothing; on the second part you do $\mathrm{CNOT}_{1,2}$; and on the third $\mathrm{CNOT}_{2,1}$. Then one could allow those parts that have had different treatment to mingle together randomly: the resultant state will be the pseudo-pure state Eq. (10.11). This process can readily be scaled up to arbitrary large numbers of qubits.

So, on the whole, this looks a rather promising approach: we have replaced one of the harsher DiVincenzo requirements – initialization of all the qubits in a pure state – by something that is a bit more straightforward: a couple of CNOT gates and some means of averaging over the results. This was the basis of the *liquid-state nuclear magnetic resonance quantum information processor* first proposed in 1997. The qubits were the spins of nuclei in a molecule, each of which is a two-level system with a slightly different resonant frequency, allowing control of individual nuclear qubits to be achieved simply by tuning the external force appropriately. The ensemble of n identical qubits needed for reliable measurement was obtained by simply having a large number of molecules in solution. This technology had been in widespread use in physical chemistry to determine structures of molecules since the 1950s, so it was relatively mature technology ready to go, and thus generated a lot of excitement at the time.

Unfortunately, pseudo-pure state quantum computing has a very serious drawback. If one had an L-qubit system, with a Hamiltonian $\hat{\mathcal{H}}$, in thermal equilibrium at temperature T (we assume that the temperature is high, so that $k_B T \gg |E_{\max}|$, where $E_{\max}$ is the energy eigenvalue with the largest absolute value), the density matrix is going to be

$$\begin{aligned} \hat{\rho}_L &= \frac{\exp(-\hat{\mathcal{H}}/k_B T)}{\mathrm{Tr}\{\exp(-\hat{\mathcal{H}}/k_B T)\}} \\ &\approx \frac{\hat{I}_L - \hat{\mathcal{H}}/k_B T}{\mathrm{Tr}\{\hat{I}_L - \hat{\mathcal{H}}/k_B T\}} \\ &\approx \frac{1}{2^L}\hat{I}_L - \frac{1}{2^L k_B T}(\hat{\mathcal{H}} - \mathrm{Tr}\{\hat{\mathcal{H}}\}/2^L). \end{aligned} \qquad (10.12)$$

Here $\hat{I}_L = \bigotimes_{m=1}^{L} \hat{I}$ is the L-qubit identity operator. We have the beginnings of a pseudo-pure state here: the identity operators to give the maximally mixed part, plus something involving the L-qubit Hamiltonian, which can be transformed into a pure state with a set of unitary operators and averaging, akin to Eq. (10.11). But look at the factor of $1/2^L$ in front of the Hamiltonian term: that factor is not going to go away with averaging, so it is going to be there in the resulting ε for the pseudo-prime state: as the L gets big, pseudo-pure states necessarily get exponentially closer to the maximally mixed state. Now recall Eq. (10.6). for reliable measurements, we need an ensemble of n qubits where $n \gg 1/2\varepsilon^2$. Thus, the ensemble size must have $n \gg 2^{2L-1}$: the size of the ensemble must grow exponentially with the number of qubits. This must negate the potential exponential computational speed-up of calculations with quantum computers: with pseudo-pure states, all you have done is replace an exponentially large set of operations with an exponentially large number of qubits.

Exercises 10.1.1

1. Prove Eq. (10.12). *Hint: You can treat a function of an operator using the Taylor series:* $f(\hat{\mathcal{M}}) = \hat{I}f(0) + \hat{\mathcal{M}}f'(0) + \hat{\mathcal{M}}^2 f''(0)/2 + \cdots$.
2. Prove Eq. (10.11).

10.2 Non-Deterministic Gates and Linear-Optical Quantum Computing

In Section 7.1 we discussed photons, the indivisible, massless "particles" of light, and how they are a very promising candidate to be used as qubits; but in Section 8.1 we saw that the natural photon–photon interaction, due to non-linear polarizability of crystals, was too weak to implement two-qubit quantum logic gates. In this section we will examine how, if we are prepared to accept a quantum gate that works only some of the time, we can perform two-qubit gates using a quantum interference: in other words, one can implement *linear-optical quantum computing* (LOQC). To begin, let's first examine one of the more humble items of optical technology: the **beamsplitter**.

10.2.1 Quantum Mechanics of Beamsplitters

A beamsplitter is a simple optical device, such as a glass block carefully polished and coated, which can either change the propagation direction of a photon by reflecting it, or let it continue in its original direction by transmitting it. The probability amplitudes associated with these two possibilities are r and t, respectively (if the photon is incident from the left or top) or r' and t', respectively (if the photon is incident from the right or bottom), see Fig. 10.1.

Stokes Relations

One of the important principles of optics is that of time reversibility: suppose you shone some optical field onto a system of optical elements that do not absorb or detect light; some output field would result. If you then sent that output field, with all propagation directions reversed, back into the same optical system, then you should recover the original input. Let us apply this principle to the beamsplitter, Fig. 10.2.

In Fig. 10.2(a) a beam of light of amplitude 1 is incident from the left on a beamsplitter; the amplitude of the reflected field is r and the amplitude of the transmitted field is t. In Fig. 10.2(b), the time-reversed situation is depicted: a field of amplitude r is incident on the beamsplitter from above,

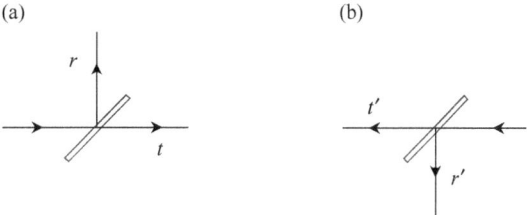

Figure 10.1   Beamsplitter basics: (a) photon incident from the left or from above is transmitted with probability amplitude t and reflected with probability amplitude r; and (b) photon incident from the right or from below is transmitted with probability amplitude t' and reflected with probability amplitude r'.

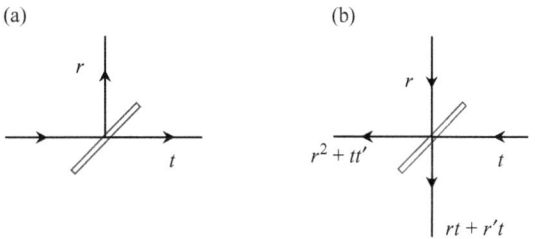

Figure 10.2   (a) Light beam incident on the beamsplitter. (b) Time-reversed situation.

and another field of amplitude t is incident from the right. The field from above results in a reflected field of amplitude r^2 heading towards the left and a transmitted field of strength tr heading downwards. For the field of amplitude t coming in from the right we must use the reflection and transmission amplitudes r' and t'; thus, this field results in a field of amplitude $t't$ heading to the left and a field amplitude $r't$ downwards. Combining these fields, we get a total amplitude $r^2 + t't$ heading towards the left and $tr + r't$ downwards. Thus, if the principle of time reversal holds, we require $r^2 + t't = 1$ and $tr + tr' = 0$. Since $t = 0$ does not represent a beamsplitter (something that only reflects and does not transmit is a mirror!), then the second result implies $r + r' = 0$. Thus we have the *Stokes relations* for the reflection and transmission coefficients of a beamsplitter:

$$\boxed{r^2 + t't = 1, \qquad r = -r'.}\tag{10.13}$$

Beamsplitters and Entanglement

Let us now consider individual photons and beamsplitters. A beamsplitter is in fact a very simple way to create an entangled state. Suppose exactly one photon is incident heading towards the right, and no photon is heading upwards: we write one photon state heading to the right as $|1\rangle_r$ and $|0\rangle_u$ is the "vacuum state" in the mode heading upwards; and, as before, we use the double-struck numbers to distinguish photon states from qubit states. After the beamsplitter, the system will be in the following state:

$$|0\rangle_u \otimes |1\rangle_r \to r|1\rangle_u \otimes |0\rangle_r + t|0\rangle_u \otimes |1\rangle_r,\tag{10.14}$$

which is entangled, *even though only one particle is present*. Unfortunately, this sort of entanglement is of limited utility, simply because there is no way to actually measure a vacuum state, because there is nothing there to measure. Thus, experiments like Bell inequalities or teleportation which rely on correlation measurements or projections onto the Bell basis are not feasible, let alone more complicated circuits required to execute quantum algorithms. In principle one could arrange for the photon modes to interact with atomic qubits after the beamsplitter, and thus this state could create an entangled pair of atoms; the difficulty of ensuring a single photon is absorbed by a particular atom should not be underestimated.

10.2.2 Hong–Ou–Mandel Interference

More impactful is the situation shown in Fig. 10.3, in which one photon is incident from the left and one from below.

Because photons are indivisible, in this case there will be three possible configurations after the beamsplitter:

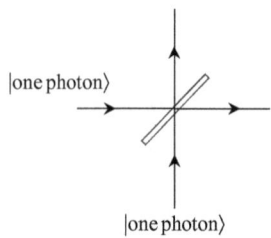

Figure 10.3 Two photons incident on a beamsplitter. We denote this input state by $|1, 1\rangle = |1\rangle_u \otimes |1\rangle_r$.

1. Two photons are heading upwards and zero to the right. This can only happen if the photon incident from the left is reflected (with probability amplitude r) and the photon incident from below is transmitted (with amplitude t'). Thus, this combined state has the probability amplitude $\sqrt{2}rt'$, the extra factor of 2 being required to ensure correct normalization.

2. Two photons are heading right and zero upwards. This can only happen if the photon incident from the left is transmitted (with probability amplitude t) and the photon incident from below is reflected (with amplitude r'), for a combined probability amplitude $\sqrt{2}r't$, again with the extra normalization factor.

3. One photon is heading right and one upwards. There are two possible scenarios that give rise to this outcome: both photons are reflected (with amplitude rr') or both are transmitted (with amplitude tt'). The combined probability amplitude is $tt' + rr'$, which may be re-written using the Stokes relations Eq. (10.13) as $1 - 2r^2$.

Thus, the quantum state of the output photons is

$$|\varphi_{\text{out}}\rangle = \sqrt{2}rt'|2,0\rangle + \sqrt{2}r't|0,2\rangle + (1 - 2r^2)|1,1\rangle. \tag{10.15}$$

We have simplified notation by writing $|\mathbb{n}, \mathbb{m}\rangle = |\mathbb{n}\rangle_u \otimes |\mathbb{m}\rangle_r$: the first double-struck number in the ket refers to the mode heading upwards, while the second refers to the mode travelling to the right. If we use "50-50" beamsplitters, for which $|r| = |t| = |r'| = |t'| = 1/\sqrt{2}$, then the third term disappears, and the photons will either both go up or both go right. This is the famous Hong–Ou–Mandel quantum interference effect, named for the scientists who first observed it in 1987. In a way, we can think of the **Hong–Ou–Mandel effect** as a non-linearity induced on the level of two photons: with only one photon incident, it can go either up or to the right; but with two photons, they bunch together rather than acting as independent particles.

10.2.3 Linear Optics Quantum Computing

How can the Hong–Ou–Mandel effect be exploited to make a quantum gate?

First, we will need to define our qubits: photon polarization is a very nice way to do this. In the above example of quantum interference at a beamsplitter, we implicitly assumed both photons had the same polarization. But, as we saw in Section 7.1, each spatial mode (i.e., "going up" and "going to the right" are our two spatial modes in the above example) can have one of two orthogonal polarization states, horizontal and vertical. Thus, the two-qubit state $|\mathbb{H}\mathbb{V}\rangle$ denotes the state in which the first spatial mode is vertically polarized, and the photon in the second spatial mode is horizontal. An arbitrary two-photon state will be of the following form:

$$|\varphi_{\text{in}}\rangle = \alpha|\mathbb{H}\mathbb{H}\rangle + \beta|\mathbb{H}\mathbb{V}\rangle + \gamma|\mathbb{V}\mathbb{H}\rangle + \delta|\mathbb{V}\mathbb{V}\rangle. \tag{10.16}$$

This can be simply translated into a qubit state by making the replacement $|\mathbb{H}\rangle \rightarrow |0\rangle$ and $|\mathbb{V}\rangle \rightarrow |1\rangle$; the single-struck numbers refer to the qubit states.

Single-qubit gates, as we have seen, can be implemented using wave plates; so what we require is a linear optical device – an interferometer built of prisms, mirrors, lenses or diffraction gratings etc. – which can perform a two-qubit universal operation such as a CZ gate, i.e., it will transform $|\varphi_{\text{in}}\rangle$ into the output state

$$\begin{aligned} |\varphi_{\text{out}}\rangle &= \hat{U}_{CZ}|\varphi_{\text{in}}\rangle \\ &= \alpha|\mathbb{H}\mathbb{H}\rangle + \beta|\mathbb{H}\mathbb{V}\rangle + \gamma|\mathbb{V}\mathbb{H}\rangle - \delta|\mathbb{V}\mathbb{V}\rangle. \end{aligned} \tag{10.17}$$

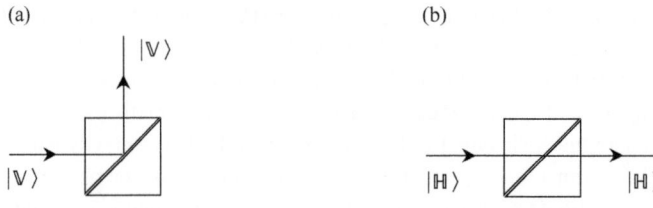

Figure 10.4   A polarizing beamsplitter.

Unfortunately, no-one knows how. However, what can be done is a *non-deterministic* quantum gate, which has the following effect:

$$|\Psi_{\text{out}}\rangle = \hat{U}_{\text{NDG}}\left(|\varphi_{\text{in}}\rangle \otimes |\phi_{\text{in}}\rangle\right)$$
$$= \sqrt{p}|\Psi_{\text{correct}}\rangle + \sqrt{1-p}|\Psi_{\text{wrong}}\rangle, \tag{10.18}$$

where

$$|\Psi_{\text{correct}}\rangle = \left(\hat{U}_{CZ}|\varphi_{\text{in}}\rangle\right) \otimes |\phi_{\text{out}}\rangle. \tag{10.19}$$

What is going on here? In order to make things work, it turns out we need to include some additional auxiliary photon modes, beyond the two modes used for our qubits; the state of these modes is represented by $|\phi\rangle$; the state of the overall multi-photon system (qubits plus auxiliaries) is denoted $|\Psi\rangle$ (we cannot assume this is a separable state in which qubits and auxiliaries are not entangled). Thus, with probability p the non-deterministic gate gives us just what we want: $|\Psi_{\text{correct}}\rangle$, a controlled-Z gate applied to our qubits, and it is not entangled to the auxiliaries. But with probability $1-p$ the output is in the *wrong* state, $|\Psi_{\text{wrong}}\rangle$. If the wrong state is perpendicular to the correct state, i.e.,

$$\langle\Psi_{\text{correct}}|\Psi_{\text{wrong}}\rangle = 0, \tag{10.20}$$

then we can always distinguish whether the gate was performed correctly or not. In some schemes, measurement of the auxiliary photon modes will yield the answer; in others, one must wait until the end of the algorithm since Ψ_{wrong} might involve the qubit modes with more than one photon present (in Ψ_{correct} these modes will only have one photon each).

Here is an example of how such non-deterministic interferometic quantum gates work. We require three beamsplitters (labelled BS1, BS2, and BS3) which have reflection and transmission coefficients r, r', t, and t', just like the beamsplitter illustrated in Fig. 10.1. We assume that these coefficients are independent of the direction of polarization. In optics, the Fresnel coefficients determine the amplitudes of reflection and transmission, and they do depend on the direction of polarization; however, horizontally polarized photons will only interact with BS2, and vertically polarized photons with BS1 or BS3; in order to obtain identical reflection and transmission coefficients for the different polarizations, one must use non-identical beamsplitters.

In addition, we require four **polarizing beamsplitters** (Fig. 10.4), which have the property that vertically polarized light is always reflected, and horizontally polarized light always transmitted.

The layout of the interferometer is shown in Fig. 10.5. The paths taken by two vertically polarized photons are indicated. In this case

$$|\mathbb{VV}\rangle \rightarrow rr'|\mathbb{VV}\rangle + \sqrt{1-r^2r'^2}|\Psi_{\text{wrong}}^{(VV)}\rangle, \tag{10.21}$$

where $|\Psi_{\text{wrong}}^{(VV)}\rangle$ corresponds to either one or both photons being transmitted at BS1 and/or BS3.

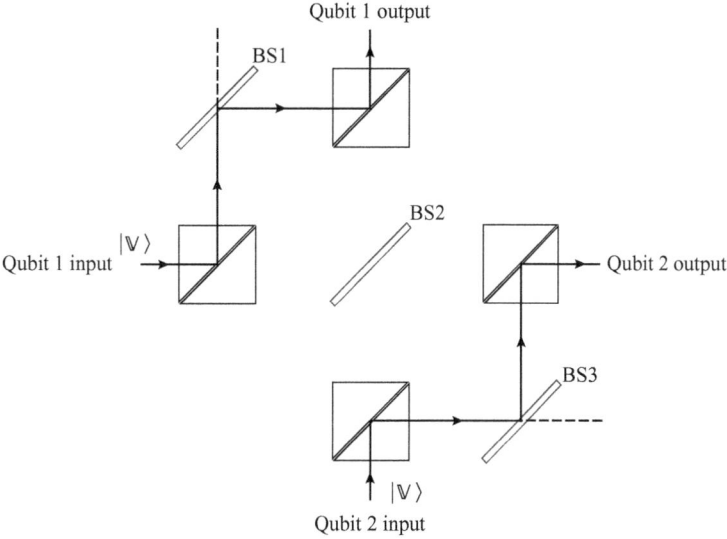

Figure 10.5   The linear-optics quantum computer interferometric gate. The paths of two incident vertically polarized photons are shown; dashed lines indicate error states.

The paths taken when a horizontal and a vertical photon are input are shown in Fig. 10.6. In this case, the output will be

$$|\mathbb{V}\mathbb{H}\rangle \rightarrow r'^2|\mathbb{V}\mathbb{H}\rangle + \sqrt{1 - r'^4}|\Psi_{\text{wrong}}^{(VH)}\rangle. \tag{10.22}$$

In this case $|\Psi_{\text{wrong}}^{(VH)}\rangle$ corresponds to either one or both photons being transmitted at BS1 and/or BS2. In the latter case, it results in *two* photons, one horizontally polarized and one vertically polarized, exiting in the qubit 1 output mode: since this mode should only contain one photon, identifying the error state will be possible when the qubit is measured at the termination of the algorithm.

The output state when $|\mathbb{H}\mathbb{V}\rangle$ is input will be similar, with reflection or transmission occurring at BS2 and BS3 (with amplitudes r and t rather than r' and t'), and $|\Psi_{\text{wrong}}^{(HV)}\rangle$ includes the possibility of two photons exiting in the qubit 2 mode:

$$|\mathbb{V}\mathbb{H}\rangle \rightarrow r^2|\mathbb{H}\mathbb{V}\rangle + \sqrt{1 - r^4}|\Psi_{\text{wrong}}^{(HV)}\rangle. \tag{10.23}$$

Finally, the case in which two horizontally polarized photons are incident is shown in Fig. 10.7. The output state is given by Eq. (10.15), where in this case $|\Psi_{\text{wrong}}^{(HH)}\rangle$ involves either two photons in the qubit 1 mode, or two in the qubit 2 mode:

$$|\mathbb{V}\mathbb{H}\rangle \rightarrow (1 - 2r^2)|\mathbb{H}\mathbb{H}\rangle + 2r\sqrt{1 - r^2}|\Psi_{\text{wrong}}^{(HH)}\rangle. \tag{10.24}$$

Comparing Eqs. (10.21)–(10.24), and if we require $r = 1/\sqrt{3}$ (and not forgetting $r = -r'$, see Eq. (10.13)), then we find

$$\alpha|\mathbb{H}\mathbb{H}\rangle + \beta|\mathbb{H}\mathbb{V}\rangle + \gamma|\mathbb{V}\mathbb{H}\rangle + \delta|\mathbb{V}\mathbb{V}\rangle \rightarrow$$
$$\frac{1}{3}\left(\alpha|\mathbb{H}\mathbb{H}\rangle + \beta|\mathbb{H}\mathbb{V}\rangle + \gamma|\mathbb{V}\mathbb{H}\rangle - \delta|\mathbb{V}\mathbb{V}\rangle\right) + \frac{2\sqrt{2}}{3}|\Psi_{\text{wrong}}\rangle, \tag{10.25}$$

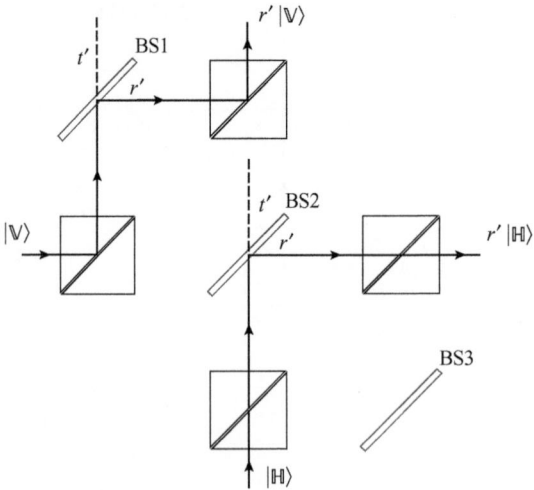

Figure 10.6   The linear-optics quantum computer interferometric gate, with one vertical and one horizontal polarized photon.

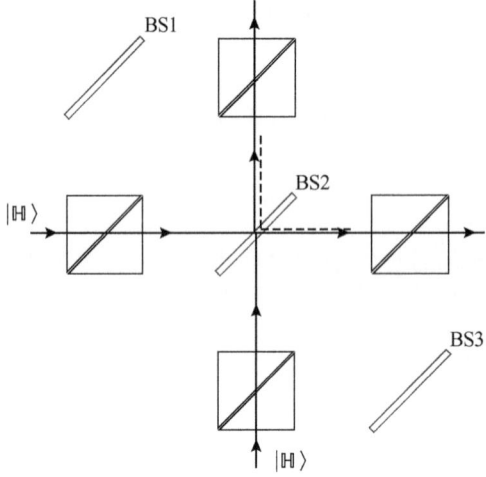

Figure 10.7   The linear-optics quantum computer interferometric gate, with two horizontal polarized photons as input. In this case, quantum interference of the Hong–Ou–Mandel type takes place at BS2.

where $|\Psi_{\mathrm{wrong}}\rangle$ is a complicated state (which involves photons bunching together in the qubit modes), is orthogonal (and therefore distinguishable) from $\alpha|\mathbb{H}\mathbb{H}\rangle + \beta|\mathbb{H}\mathbb{V}\rangle + \gamma|\mathbb{V}\mathbb{H}\rangle \pm \delta|\mathbb{V}\mathbb{V}\rangle$.

At first sight it might appear that each time you perform a gate, your probability of success drops by a factor of p, so that if your algorithm involves N quantum gates, the probability of successfully completing it will be p^N; and one must repeat the algorithm $(1/p)^N$ times in order to have a chance of getting the correct outcome. However, photons are relatively easy to create, and this sort of linear-optics quantum computer (LOQC) non-deterministic gate could be used, for example, to create a large reservoir of entangled photons which could be used as a resource in a variety of schemes to gain quantum advantage.

1. In what sense is a 50-50 beamsplitter equivalent to a Hadamard gate?
2. Suppose two classical fields, both of equal strength (1, say) are incident one on each side of a 50-50 beamsplitter, in the arrangement of the Hong–Ou–Mandel effect. What is the output field in that case? Compare it with the situation in which single photons are incident: what is the fundamental difference?
3. Suppose two pairs of photons are simultaneously incident, one pair on each side of the beamsplitter. What will be the output state?

10.3 Cluster-State Quantum Computing

Arguably the most irksome technological bottleneck to the development of quantum computers is the means by which two-qubit entangling gates such the the CNOT gate can be implemented. As we have seen, they require either a physical qubit–qubit interaction, or the use of a quantum information bus involving coupling the qubits to some other degree of freedom: in both cases, the interaction is not strong, and therefore the whole process will take a relatively long time to complete, thousands of qubit oscillation times $(2\pi/\omega)$; and a long duration means there is more chance for things to go wrong. Thus, it would be very nice if there were some viable alternative.

Cluster-state quantum computing, also called *measurement-based* quantum computing, is just such an alternative. The idea is that if we can prepare a large number of qubits in some standard massively entangled state, called a **cluster state**, which can be created efficiently at the beginning of our quantum algorithm, then the whole algorithm can be implemented just by performing one-qubit unitary operations and measurements. The results of the measurements on one set of qubits are fed-forward to determine what local unitaries should be performed on the next set of qubits prior to their measurement.

Consider the simple circuit in QC 10.1:

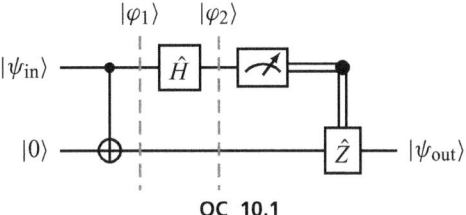

QC 10.1

With qubit #2 in the arbitrary single-qubit state $|\psi_{\text{in}}\rangle = \alpha|0\rangle + \beta|1\rangle$, the overall two-qubit initial state is $|\varphi_0\rangle = |\psi_{\text{in}}\rangle \otimes |0\rangle = \alpha|00\rangle + \beta|10\rangle$. Applying the CNOT gate thus results in the following entangled state:

$$|\varphi_1\rangle = \alpha|00\rangle + \beta|11\rangle. \tag{10.26}$$

The Hadamard gate applied to qubit #2 results in the state:

$$\begin{aligned}
|\varphi_2\rangle &= \frac{1}{\sqrt{2}} \left(\alpha\{|00\rangle + |10\rangle\} + \beta\{|01\rangle - |11\rangle\} \right) \\
&= \frac{1}{\sqrt{2}} \left(|0\rangle \otimes \{\alpha|0\rangle + \beta|1\rangle\} + |1\rangle \otimes \{\alpha|0\rangle - \beta|1\rangle\} \right) \\
&= \frac{1}{\sqrt{2}} \left(|0\rangle \otimes |\psi_{\text{in}}\rangle + |1\rangle \otimes \hat{Z}|\psi_{\text{in}}\rangle \right). \tag{10.27}
\end{aligned}$$

Thus, after the measurement of qubit #2 and the (classically) controlled application of the $\hat{Z}$ gate to qubit #1, the final output state is

$$|\psi_{\text{out}}\rangle = |\psi_{\text{in}}\rangle. \tag{10.28}$$

This is a two-qubit version of teleportation, in which the receiving qubit interacts directly, via a CNOT gate, with the input qubit.

Using it as our foundation, and using the identity $\hat{H}^2 = \hat{I}$, the two-qubit teleportation circuit implies (QC 10.2)

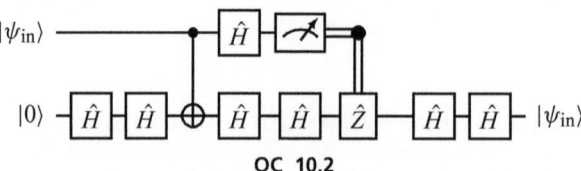

QC 10.2

Using QC 3.6 and denoting $|+\rangle = \hat{H}|0\rangle = \{|0\rangle + |1\rangle\}/\sqrt{2}$ this becomes QC 10.3

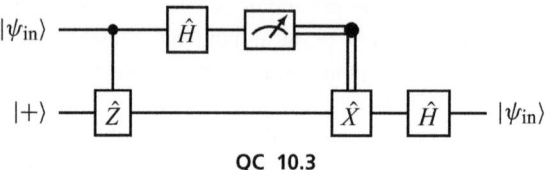

QC 10.3

where we have used the identity $\hat{H}\hat{Z}\hat{H} = \hat{X}$. Let us now write $|\psi_{\text{in}}\rangle = \hat{U}_z(\phi)|\chi\rangle$, where $\hat{U}_z(\phi)$ is the arbitrary phase-shift gate:

$$\hat{U}_z(\phi) = \cos\phi/2\,\hat{I} - i\sin\phi/2\,\hat{Z} = \begin{bmatrix} e^{-i\phi/2} & 0 \\ 0 & e^{i\phi/2} \end{bmatrix}$$

(see Section 2.2, Exercise 5). Thus we get the following circuit, which constitutes the building block for understanding the functioning of cluster states.

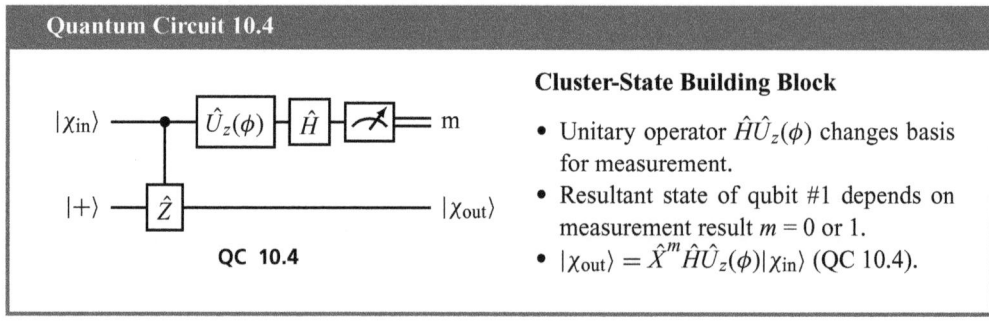

Quantum Circuit 10.4

QC 10.4

Cluster-State Building Block

- Unitary operator $\hat{H}\hat{U}_z(\phi)$ changes basis for measurement.
- Resultant state of qubit #1 depends on measurement result $m = 0$ or 1.
- $|\chi_{\text{out}}\rangle = \hat{X}^m\hat{H}\hat{U}_z(\phi)|\chi_{\text{in}}\rangle$ (QC 10.4).

In other words, starting with a pair of qubits prepared in an entangled state (by the controlled-$\hat{Z}$) one can perform a measurement in some particular basis (i.e., the $\hat{H}\hat{U}_z(\phi)$ applied to qubit #2 is a change of basis; effectively, the measurement basis is now $\{(|0\rangle + \exp(i\phi)|1\rangle)/\sqrt{2}, (|0\rangle - \exp(i\phi)|1\rangle)/\sqrt{2}\}$); the resultant state of qubit #1 is the input state of qubit #2 transformed by the basis change $\hat{H}\hat{U}_z(\phi)$ and, dependent on the outcome of the measurement, an $\hat{X}$-gate as well.

These building blocks can be cascaded. Consider QC 10.5:

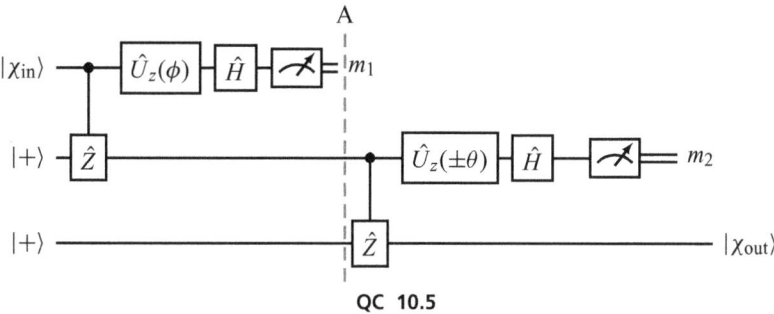

QC 10.5

At A, the quantum state of qubit #2 is $\hat{X}^{m_1}\hat{H}\hat{U}_z(\phi)|\chi_{in}\rangle$; i.e., the output of the first block forms the input for the second. For the second block, the sign of the angle θ depends on the value of m_1: specifically, if $m_1 = 0$ the sign is positive; if $m_1 = 1$ it is negative. If we make this choice, the final state $|\chi_{out}\rangle$ is given by

$$
\begin{aligned}
|\chi_{out}\rangle &= \left\{\hat{X}^{m_2}\hat{H}\hat{U}_z\left(\{-1\}^{m_2}\theta\right)\right\}\left\{\hat{X}^{m_1}\hat{H}\hat{U}_z(\phi)\right\}|\chi_{in}\rangle \\
&= \hat{X}^{m_2}\hat{H}\hat{X}^{m_1}\hat{U}_z(\theta)\hat{H}\hat{U}_z(\phi)|\chi_{in}\rangle \\
&= \hat{X}^{m_2}\hat{H}\hat{X}^{m_1}\hat{H}\hat{H}\hat{U}_z(\theta)\hat{H}\hat{U}_z(\phi)|\chi_{in}\rangle \\
&= \hat{X}^{m_2}\hat{Z}^{m_1}\hat{U}_x(\theta)\hat{U}_z(\phi)|\chi_{in}\rangle, \quad (10.29)
\end{aligned}
$$

where we have used $\hat{U}_z(-\theta)\hat{X} = \hat{X}\hat{U}_z(\theta)$ and $\hat{H}\hat{H} = \hat{I}$.

Now let's re-draw the circuit for the particular case that $|\chi_{in}\rangle = |+\rangle$, and we will start from the initial state with all qubits being in $|0\rangle$, so there will be some extra Hadamard gates used to create the $|+\rangle$ states. We have grouped the two controlled-$\hat{Z}$ gates together at the very beginning, and the local unitary operations in QC 10.6:

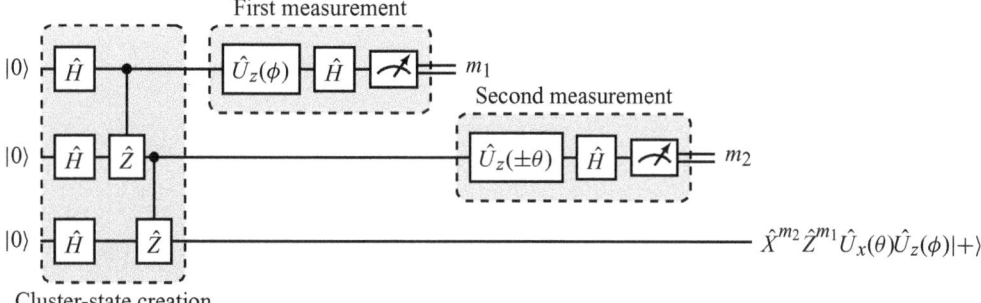

QC 10.6

This is rather remarkable: what we have done is perform an (admittedly very simple) quantum algorithm, namely creation of an arbitrary single-qubit state, by an initial step of creating a multi-qubit entangled state ("cluster-state creation") followed only by measurements made in carefully chosen bases. The measurements are performed in a specific order: the choice of sign of the angle θ in the second measurement depends on the outcome of the first measurement.

This same basic scheme can be scaled up to perform arbitrary complex quantum computations. Suppose that L **logical qubits** are required for the desired quantum algorithm (i.e., there are L rows in the quantum circuit diagram one needs to implement), and there are a total of M steps in the circuit – the local unitary operations or the multi-qubit gates each constitute a step in this context, although performing as many operations as possible in parallel reduces M to a minimum. Then the total number of physical qubits required to implement the measurement-based scheme is $L \times M$: a considerably larger number than required in the circuit approach. Think of these $L \times M$ qubits arranged in a lattice: L rows each with M qubits; then the initial cluster-state creation step involves first applying a Hadamard gate to each of these qubits (depending on the control mechanism, these might all be done simultaneously); second, controlled-$\hat{Z}$ gates must be applied between each qubit's nearest neighbours, both in the left–right and in the up–down directions. These controlled-$\hat{Z}$ gates also can be done in parallel, so the whole $L \times M$ cluster state can be created in the time taken to perform just four controlled-$\hat{Z}$ gates. Only nearest-neighbour interactions are required. Once the cluster state is created, the execution of the algorithm proceeds by a series of basis changes followed by measurements.

Exercises 10.3.1

1. Write down the three-qubit cluster state shown in the cluster-state creation section of the circuit (QC 10.6).
2. Write out the cluster state in a bipartite form, with qubit #3 being one of the sub-systems, and qubits #2 and #1 forming the other part. Is the system entangled?
3. Use the Griffiths–Niu theorem and standard circuit identities to show that the two-qubit teleportation circuit (QC 10.1) shown at the start of this section is equivalent to the SWAP gate.

10.4 CONCEPT CHECK AND EXERCISES

Concept Check

- Pseudo-pure states make initialization of the quantum computer easier, at the cost of signal strength declining exponentially with the number of qubits.
- Linear-optical quantum computing exploits quantum interference to perform quantum gates on photon qubits, with at most a 1/9 chance of success.
- Cluster-state quantum computing envisions the creation of a massively entangled register of qubits, which, through measurement alone, can execute quantum circuits.

Chapter Exercises

1. A Mach–Zehnder interferometer (see Fig. 10.8) consists of two beamsplitters and two mirrors.

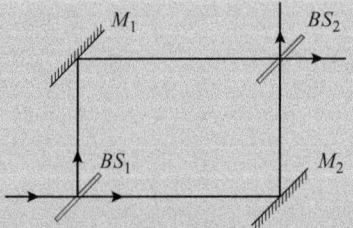

Figure 10.8   The Mach–Zehnder interferometer. Assume both beamsplitters are identical, with $|t| = |r| = 1/\sqrt{2}$.

(a) Suppose that a single photon is incident in the bottom left, as indicated. Assuming both mirrors have 100% reflectivity, what is the output state?

(b) If the the mirror M_1 had reflectivity r, what would the output state be?

(c) If an opaque object, which absorbs any photon that interacts with it, is inserted into the upper arm of the interferometer, what would be the output state? Can a single photon detect the presence of the object without actually interacting with it?

2. Calculate the value of $\langle \hat{a} \otimes \hat{c} \rangle$ for a pseudo-pure singlet state $\rho = (1 - \epsilon)\hat{I} \otimes \hat{I} + \epsilon|\beta_2\rangle\langle\beta_2|$ (cf. Eq. (3.53)). What is the smallest value of ϵ for which one might observe a violation of Bell's inequalities?

11 Decoherence and How to Defeat It

Quantum computers are not perfect. Indeed, the strange phenomena due to quantum entanglement raise the question: why don't these things happen in everyday life? The answer is generally ascribed to one thing: *decoherence*. As we have observed already, quantum states are like a rare, delicate orchid that withers and dies whenever it interacts with its environment; and, worse, the larger the system, the more quickly this happens.

In quantum computing, we are trying to build just such a large-scale device in which the quantum state is shielded from decoherence. There are three basic strategies to achieve this. First, the qubits themselves are chosen to be long-lived: for example, in atoms or ions, we choose the levels $|0\rangle$ and $|1\rangle$ which do not rapidly decay due to spontaneous emission. Second, we can combine two or more "physical" qubits to create a single "logical" qubit which is immune to decoherence: two or more noisy qubits have been sacrificed to make a more stable single qubit; generically, this approach is called the method of **decoherence-free subspaces**. And finally, and most shockingly, **quantum error correction** again uses multiple "physical" qubits to create a single "logical" qubit, one can actually measure and correct an error, without disturbing the quantum information stored in the device.

In this chapter, we give a brief introduction to this topic, which remains an active and exciting area of research.

Learning Objectives

- Understand the causes of decoherence and its effects on the density matrix of the system.
- Learn how to build an effective (or logical) qubit using several physical qubits such that the subspace of the logical qubit is decoherence free.
- Learn about possible quantum circuit arrangements that can correct the effect of decoherence, thereby reducing the errors in a quantum algorithm.

11.1 Random Errors in Gates and Qubits

So far we have usually assumed that qubits are nice well-behaved entities over which we have complete control. So, if we create a state such as $\alpha|0\rangle + \beta|1\rangle$, then leave it alone for a time while we do some other tasks associated with our quantum algorithm, then the qubit will remain in exactly the state as when we left it. Unfortunately, things are not that simple.

As an example of what can go wrong, we saw in Section 6.2 that physical qubits usually have an energy difference between the two qubit states $|0\rangle$ and $|1\rangle$, so that the state is actually evolving in time according to Eq. (6.8), i.e.,

$$|\varphi(t)\rangle = A(0)\exp(i\omega t/2)|0\rangle + B(0)\exp(-i\omega t/2)|1\rangle. \tag{11.1}$$

As we saw in Section 6.3, in order to perform single-qubit gates we have to apply a time-varying force to the qubit, oscillating on-resonance at frequency ω. Thus, we have two physical systems – the qubit itself, and the force that we apply to it – which must be oscillating at precisely the same frequency. Maintenance of such a perfect resonance for ever is just not possible: there will always be something that will cause either the qubit or the force to drift off-resonance over time.

For example, suppose the applied force is due to a laser. Laser frequencies are stabilized by precise control of the separation of the cavity mirrors, which can be done using a feedback signal generated by measuring the interference between the laser and light from some reference frequency standard. (The inventors of this technique, John L. Hall and Theodor W. Hänsch, were awarded the 2005 Nobel Prize in Physics.) It works very well and can stabilize a laser frequency for as long as *seconds*, or 10^{15} cycles of the oscillations – but a few seconds is not for ever. Other control mechanisms for other types of qubit can be frequency stabilized by other techniques, but none of them is truly perfect. Similarly, the energy difference between the qubit levels might be susceptible to changes in the magnetic field at the qubit's location: again, this can be stabilized by sophisticated experimental techniques using Lorentz coils to null out any ambient magnetic field; but again, it is not perfect.

Let us therefore assume that the frequency ω is time varying:

$$\omega \rightarrow \omega(t) = \bar{\omega} + \delta\omega(t), \tag{11.2}$$

where $\bar{\omega}$ is the average frequency and $\delta\omega(t)$ is a *random* time-varying component of the frequency (whose average value is zero).

We have encountered random *variables* already; random *functions* are a simple generalization: a function whose values are random. An everyday example is the outdoor temperature where you live: every second there is a temperature which you can measure using a thermometer, so temperature is a continuous function of time. The temperature varies during each 24-hour period, usually it is warmer at noon than it is at midnight. Also, temperature changes through the year, being cold in winter and warm in summer. But, on top of these predictable variations, there is a random variation of temperature – usually a few degrees only, but sometimes more dramatic – which it is virtually impossible to predict for more than a few days ahead. Generally speaking, tomorrow's temperature is likely to be not too different from today's; but in a week or so it is much more uncertain what the temperature will be. This time period over which the random function is likely to be more-or-less constant is (at least in physical science) called the **coherence time** of the function.

Incorporating the random variation of frequency into the oscillation terms in Eq. (11.1) we find

$$\exp(\pm i\omega t/2) \rightarrow \exp(\pm i\{\bar{\omega}t + \zeta(t)\}/2), \tag{11.3}$$

where $\zeta(t)$ is a random variable defined by

$$\zeta(t) = \int_0^t \delta\omega(t')dt'. \tag{11.4}$$

Since the average value of $\delta\omega$ is zero, so is the average of $\zeta(t)$. Thus, the variance of $\zeta(t)$ is given by

$$\sigma^2 = \overline{\zeta(t)^2}$$
$$= \iint_0^t \overline{\delta\omega(t')\delta\omega(t'')}dt'\,dt'', \tag{11.5}$$

where the overbar indicates an average over the random function $\delta\omega(t)$. Let us assume that the random function is highly uncorrelated, in the sense that its value at time t' is completely unconnected with its value a short time later at t''. For example, if $\delta\omega(t') = 1$ s^{-1}, then its value at $\delta\omega(t'')$ could be 1 s^{-1},

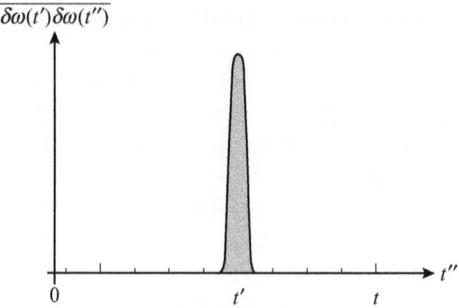

Figure 11.1   Sketch of the integrand appearing in Eq. (11.5).

or, equally likely, it could be -1 s^{-1}; or indeed any value in between, not to mention larger positive or more highly negative values. If the two values are unconnected, then the average $\overline{\delta\omega(t')\delta\omega(t'')}$, called an **auto-correlation function**, is just equal to 0. Thus, the integrand in Eq. (11.5) is as sketched in Fig. 11.1.

The height of the peak is the variance of $\delta\omega$, $\overline{\delta\omega(t')^2}$, which is assumed to be independent of time. (Random functions whose statistical properties are independent of time are called **stationary**; this is a common simplifying assumption for random functions.) Similarly, the width and shape of the peak are independent of t', and hence

$$\int_0^t \overline{\delta\omega(t')\delta\omega(t'')} dt'' = 2/T, \tag{11.6}$$

where T is a constant (with units of time); the factor of 2 is for later convenience. Thus,

$$\sigma^2/2 = t/T. \tag{11.7}$$

The state $|\varphi(t)\rangle$ now has the sort of additional "classical" randomness we discussed in Section 2.6, and so the state is now described by a density matrix:

$$|\varphi(t)\rangle\langle\varphi(t)| \rightarrow \hat{\rho} = \overline{|\varphi(t)\rangle\langle\varphi(t)|}$$

$$= \begin{bmatrix} |A(0)|^2 & A(0)B(0)^* e^{-i\bar{\omega}t}\overline{e^{i\zeta}} \\ A(0)^*B(0)e^{-i\bar{\omega}t}\overline{e^{-i\zeta}} & |B(0)|^2 \end{bmatrix}. \tag{11.8}$$

The effect of the random frequency variations is all contained in the $\overline{e^{\pm i\zeta}}$ factors which appear in the off-diagonal terms.

To proceed, we have to find a way to perform the average. For this, we need to know the **probability density function** of the random variable ζ. If we assume that the random nature of ζ arises from a combination of multiple, independent, yet identically distributed random events – for example, collisions with some thermal cloud of impurities interacting with the qubit; or multiple small voltage spikes in the electromagnetic coils stabilizing the magnetic field – then the combined random effect will have a *normal distribution*, so that the probability of its value falling between ζ and $\zeta + d\zeta$ is $P(\zeta)d\zeta$, where

$$P(\zeta) = \frac{1}{\sqrt{2\pi}\sigma} \exp\left(-\frac{\zeta^2}{2\sigma^2}\right). \tag{11.9}$$

Thus, the average value will be given by the following integral:

$$\overline{e^{\pm i\zeta}} = \int_{-\infty}^{\infty} P(\zeta) \exp(\pm i\zeta) d\zeta$$

$$= \frac{1}{\sqrt{2\pi}\sigma} \int_{-\infty}^{\infty} \exp\left(-\frac{\zeta^2}{2\sigma^2} \pm i\zeta\right) d\zeta$$

$$= \frac{1}{\sqrt{2\pi}\sigma} \int_{-\infty}^{\infty} \exp\left(-\frac{1}{2\sigma^2}\left\{(\zeta \mp i\sigma^2)^2 + \sigma^4\right\}\right) d\zeta$$

$$= \exp(-\sigma^2/2) \int_{-\infty}^{\infty} P(\zeta \mp i\sigma^2) d\zeta. \tag{11.10}$$

If we change the variable of integration to $\zeta' = \zeta \mp i\sigma^2$, and use the fact that the probability density function is normalized (strictly speaking, we are evaluating the analytic function $P(z)$ over a rectangular contour in the complex plane; the outcome is the same, however), $\int_{-\infty}^{\infty} P(\zeta') d\zeta' = 1$. Thus, we find

$$\overline{e^{\pm i\zeta}} = \exp(-\sigma^2/2)$$

$$= \exp(-t/T). \tag{11.11}$$

Thus, writing $\alpha = A(0)\exp(-i\bar{\omega}t/2)$ and $\beta = B(0)\exp(i\bar{\omega}t/2)$, we get the following expression for the qubit state after it is subjected to random phase fluctuations or *dephasing*:

$$\hat{\rho} = \begin{bmatrix} |\alpha|^2 & \alpha\beta^* e^{-t/T} \\ \alpha^*\beta e^{-t/T} & |\beta|^2 \end{bmatrix}. \tag{11.12}$$

The dephasing time T (often called T_2 in the literature, T_1 being the timescale on which bit-flip errors can occur, which tends to be much longer) is an empirical constant dependent on the physical circumstances of the qubit. It is a burden shared by all quantum device engineers to maximize T.

The effect of the decay and disappearance of the off-diagonal elements of the single-qubit density matrix means that our qubit reverts to its classical counterpart, namely the coin we discussed in Chapter 1. Probability amplitudes have disappeared, and ordinary probabilities ($p_0 = |\alpha|^2$ and $p_1 = |\beta|^2$) now specify the system; the whole notion of quantum superposition is gone, and with it the basic building blocks of our quantum computer. Or, to be concise: decoherence is bad for quantum computers.

In the next two sections we will describe the two strategies that have been introduced to defeat decoherence.

Exercises 11.1.1

1. Calculate the purity of a one-qubit state undergoing dephasing.
2. Show that the density matrix undergoing dephasing obeys the following equation:

$$\frac{d}{dt}\hat{\rho}(t) = \frac{1}{4T}\left([\hat{Z}\hat{\rho}(t), \hat{Z}] + [\hat{Z}, \hat{\rho}(t)\hat{Z}]\right).$$

Note: This is a special case of the Lindblad **master equation** which describes the dynamics of quantum density operators. A more general form is

$$\frac{d}{dt}\hat{\rho}(t) = \frac{1}{i\hbar}[\hat{\mathcal{H}},\hat{\rho}(t)] + \sum_i \frac{1}{4T_i}\left([\hat{\Gamma}_i\hat{\rho}(t),\hat{\Gamma}_i^\dagger] + [\hat{\Gamma}_i,\hat{\rho}(t)\hat{\Gamma}_i^\dagger]\right),$$

where $\hat{\Gamma}_i$ is an operator that describes one of the decay processes of the system, and T_i is the characteristic time over which the decay process happens.

11.2   Decoherence-Free Spaces

11.2.1   Logical and Physical Qubits

A common feature of methods used to mitigate decoherence in quantum computers is to use multiple physical qubits (i.e., individual two-level ions or superconducting circuits) to represent a single *logical qubit* (i.e., the idealized two-level quantum systems we used as the building-blocks of a quantum computer algorithm). The physical qubits existing in a laboratory are thus susceptible to the effects of decoherence. Working in combination, two or more qubits can be used to not only store our quantum information, but can also be used to cancel out the physical perturbations that give rise to decoherence. When the qubit needs to lie dormant while other computation tasks involving different qubits take place, it makes sense to "store" the quantum information distributed across several qubits. This is rather similar to the dynamic random access memory (DRAM) cell used for quick logic operations and the static random access memory (SRAM) cell used for more permanent memory in modern conventional computers.

11.2.2   Collective Decoherence

The first strategy that can be deployed to defeat decoherence is based on the physics of the device in question. Often the error that happens to one qubit might also be happening to the adjacent qubits. For example, if the qubits are trapped ions lying a few micrometres apart, both qubits will be equally susceptible to any drift in the magnetic field, since magnetic fields generally do not vary much on those sorts of length scales; thus, both qubits undergo the same random phase shifts. Or, if a common frequency reference is being used for the control of both qubits, then again a random drift in that reference will apply to both qubits. We can take advantage of this *collective* dephasing to mitigate its effect.

Let ϕ be the random phase shift common to two qubits; thus, for one qubit:

$$\hat{U}_z(\phi)|0\rangle = e^{-i\phi/2}|0\rangle, \quad \hat{U}_z(\phi)|1\rangle = e^{i\phi/2}|1\rangle. \tag{11.13}$$

As we have seen, when we average over these phase fluctuations it has the effect of washing out the quantum nature of the superposition state.

What about two qubits?

$$\begin{aligned}
\hat{U}_z(\phi) \otimes \hat{U}_z(\phi)|00\rangle &= e^{-i\phi}|00\rangle, \\
\hat{U}_z(\phi) \otimes \hat{U}_z(\phi)|01\rangle &= |01\rangle, \\
\hat{U}_z(\phi) \otimes \hat{U}_z(\phi)|10\rangle &= |10\rangle, \\
\hat{U}_z(\phi) \otimes \hat{U}_z(\phi)|11\rangle &= e^{i\phi}|11\rangle.
\end{aligned} \tag{11.14}$$

Thus, the dephasing has no effect on the states $|01\rangle$ and $|10\rangle$. These two states form a two-dimensional subspace of the four-dimensional space spanned by the two qubits. It is a simple example of a *decoherence-free subspace*.

Let's use the immunity to dephasing of this decoherence-free space of two physical qubits to create a single logical qubit. Suppose we have a single physical qubit state $|\varphi\rangle = \alpha|0\rangle + \beta|1\rangle$; encoding it in the decoherence-free subspace is a relatively simple circuit (QC 11.1):

QC 11.1

One can also devise multi-qubit operations that perform quantum gates on the logical basis without straying from the decoherence-free subspace. For example, the logical $\hat{X}$-gate is

$$\hat{X}_L \left(\alpha|0\rangle_L + \beta|1\rangle_L \right) = \alpha|1\rangle_L + \beta|0\rangle_L \tag{11.15}$$

(where the subscript L stands for "logical"), which is equivalent to

$$\hat{X} \otimes \hat{X} \left(\alpha|01\rangle + \beta|10\rangle \right) = \alpha|10\rangle + \beta|01\rangle. \tag{11.16}$$

Decoherence-free spaces are not infallible, and ultimately they will decohere, since the correlation between the phases of the individual qubits is not perfect. However, the method does allow a dramatic improvement in dephasing, with coherence times of several seconds being reported in some experiments.

Exercises 11.2.3

1. Suppose two physical qubits have perfectly *anti-correlated* dephasing: what would be a good decoherence-free subspace to use in this case?
2. Find the equivalent operations on the physical qubits for the following logical gates:
   - **(a)** $\hat{Y}_L$.
   - **(b)** $\hat{Z}_L$.
   - **(c)** $\hat{H}_L$.
   - **(d)** CZ.

11.3 Error-Correcting Codes

The decoherence-free subspace is a passive, physics-based approach to reducing the effects of errors in a quantum computer. There is also an active, algorithmic approach called *quantum error correction*.

It will be easier to consider how quantum error correction will detect and nullify a *bit-flip error*, in which a single physical qubit state $\alpha|0\rangle + \beta|1\rangle$ is transformed into $\alpha|1\rangle + \beta|0\rangle$; as we will see, the techniques involved will readily be adaptable to phase errors as well.

At first sight, correcting such an error *without* measuring the state, and thereby destroying the superposition, seems impossible. Anything you do to detect an error should involve some sort of revelation about the state, right? That this is not the case is one of the truly remarkable discoveries in quantum information.

To begin our discussion, let's think about a simple way to do error correction classically by *majority voting*. Suppose we send a single bit down some noisy channel where there is some chance p of the bit getting flipped (or, more precisely, of getting an odd number of bit flips, since an even number of such flips produces no effect). If we send just one copy, the probability of getting the wrong message is just p. But instead, we can send the message three times, and take the "majority" vote, thus the receiver will interpret "111", "110", "101", and "011" as "1"; and "000", "001", "010", and "100" as "0". Now the probability of getting the wrong answer is the probability of two or three of the bits getting flipped: $p^3 + 3p^2(1 - p) = p^2(3 - 2p)$, which is always less than p provided the channel is not very noisy, and is a lot less than p if p is small; for example, if $p = 0.1$, implying there is a one in 10 chance of a bit flip, then the chance of majority voting getting the wrong answer is 0.028, or one chance in 36 or so; if we did majority voting with five bits, the probability of error is reduced to 0.00856, or one chance in 116.

Let us try to carry this strategy over into the quantum realm. Using three physical qubits, we will define the "Logical 0" to be $|000\rangle$, and the "Logical 1" to be $|111\rangle$, so that our one logical qubit state is $\alpha|000\rangle + \beta|111\rangle$. Such a state can be created from a physical qubit state by a relatively simple circuit (QC 11.2):

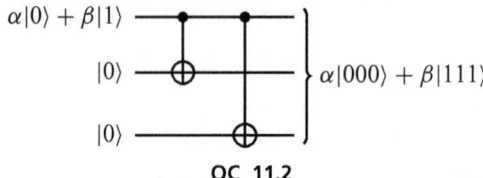

QC 11.2

So, what happens if we transfer our logical qubit to three-qubit encoding, then leave it to wait for a bit-flip error to occur. Suppose the bit flip occurs on qubit #1, so the state is now $\alpha|001\rangle + \beta|110\rangle$: Does this help? At first sight, we cannot figure out which qubit has flipped without measuring them, and thus destroying the superposition.

But on closer examination, we can employ quantum logic gates, together with three properly initialized ancilla qubits, to diagnose and correct the errors. In QC 11.3, the upper three *physical* qubits form the single *logical* qubit of our memory. Step I is to encode our physical qubit into the logical qubit, as above. The dotted lines indicate that these three qubits are going to be sitting around for a while, during which period a bit-flip error might have occurred to one of the qubits.

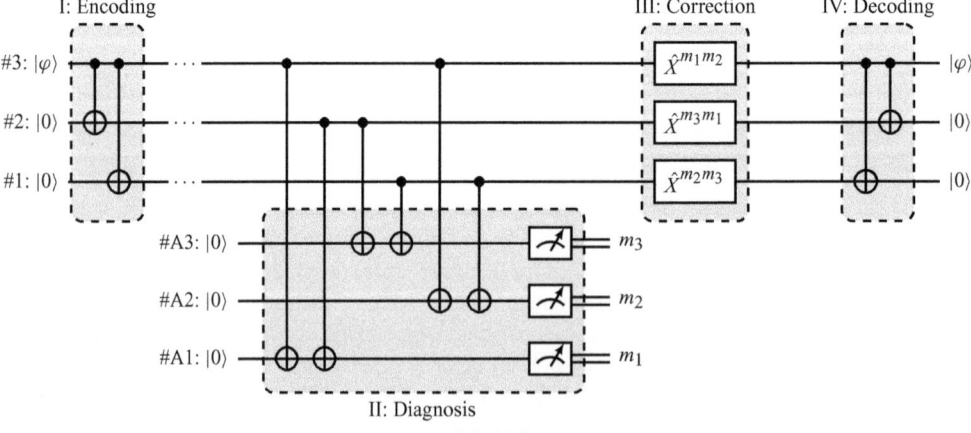

QC 11.3

Table 11.1 Results.

Error	State of logical qubits	State of ancillae after diagnosis
No error	$\alpha|000\rangle + \beta|111\rangle$	$|000\rangle$
Bit flip on #1	$\alpha|001\rangle + \beta|110\rangle$	$|110\rangle$
Bit flip on #2	$\alpha|010\rangle + \beta|101\rangle$	$|101\rangle$
Bit flip on #3	$\alpha|100\rangle + \beta|011\rangle$	$|011\rangle$

Step II is diagnosis of which qubit, if any, has been flipped. We perform a series of six CNOT gates between the physical qubits constituting the logical qubit, and the three ancillae, all in state $|0\rangle$ as shown. Let us consider the following four cases separately:

- *No bit flip has occurred.* The three upper qubits will all either be in state $|000\rangle$, so none of the CNOTs have any effect; or they will all be in state $|111\rangle$, in which case each of the ancillae undergoes two CNOT gates, leaving them in the state they started in.
- *Bit flip has occurred on qubit #1.* In this case qubits #2 and #3 will be in the same state, and so the first pair of CNOT gates with the ancillae will have no effect: qubit #A1 remains in state $|0\rangle$. However, qubits #1 and #2 and qubits #1 and #3 will be in opposite states, and so one, and only one, of the CNOTs acting on #A2 and #A3 will have an effect: thus, both #A2 and #A3 will end up in state $|1\rangle$.
- *Bit flip has occurred on qubit #2.* Now it is qubits #1 and #3 which will be in the same state, and so the last pair of CNOT gates with the ancillae will have no effect: qubit #A2 remains in state $|0\rangle$. The first two pairs of CNOTs will have an effect: and #A1 and #A3 will end up in state $|1\rangle$.
- *Bit flip has occurred on qubit #3.* Now it is the middle pair of CNOT gates which will have no effect, leaving qubit #A3 in state $|0\rangle$, while #A1 and #A2 will end up in state $|1\rangle$.

These results are summarized in Table 11.1

In step III, we carry out the actual reversal of the bit-flip error. Measurement of the ancillae qubits in the computational basis yields m_1, m_2, and m_3, each of which has value 0 or 1. As shown in the circuit, the operation $\hat{X}^{m_2 m_3} \otimes \hat{X}^{m_3 m_1} \otimes \hat{X}^{m_1 m_2}$ is applied to the three logical qubits, $m_i m_j$ just being the ordinary product of the two outcomes. Thus, if $m_1 = 0$, $m_2 = 1$, and $m_3 = 1$ we get $m_1 m_2 = 0$, $m_2 m_3 = 1$, and $m_3 m_1 = 0$, so the required correction is $\hat{X} \otimes \hat{I} \otimes \hat{I}$, i.e., a reverse bit flip on qubit #3.

Step IV is the reverse of step I, reverting to using a single physical qubit to be the logical qubit.

11.3.1 Quantum Error Correction without Measurement

One can replace the measurement at the end of the diagnosis and the resultant correction operations by three-qubit *Toffoli gates* (QC 11.4):

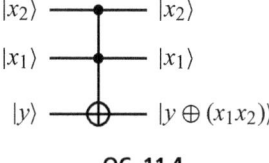

QC 11.4

In effect, the Toffoli gate is a bit flip applied to qubit #1 conditional on *both* qubit #2 and qubit #3 being in state $|1\rangle$; if either of them are in state $|0\rangle$ then nothing happens. Toffoli gates can be implemented using CNOT gates and local operations (QC 11.5):

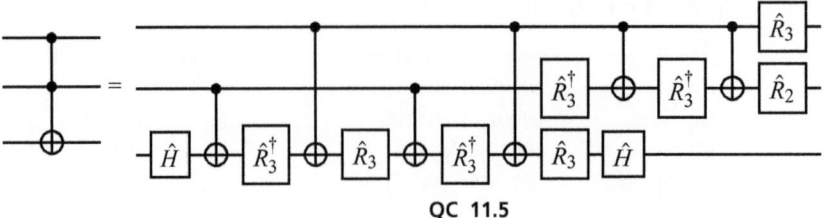

QC 11.5

where, as before (see Eq. (5.17))

$$\hat{R}_k = \begin{bmatrix} 1 & 0 \\ 0 & \exp(2\pi i/2^k) \end{bmatrix}.$$
(11.17)

Using the Toffoli gate, the error correction ciruit looks like QC 11.6:

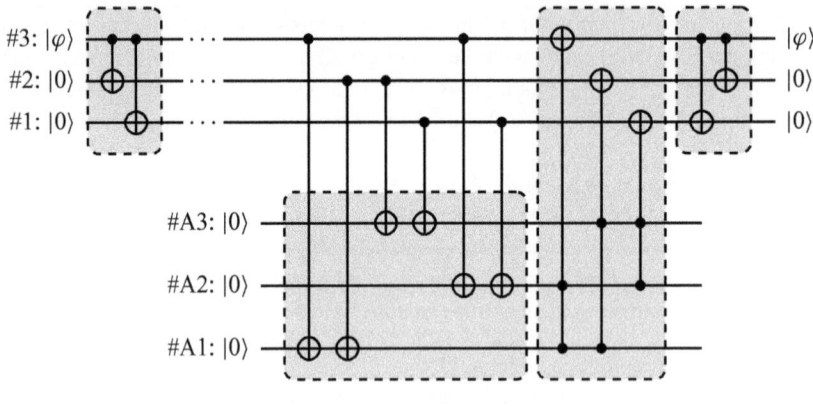

QC 11.6

What this circuit does in essence is filter out the random component of the logical qubit's state, and deposit it in the ancilla qubits, which are left in a mixed state at the end. In the approximation that only single bit flips occur, the logical qubit's final state factors out, so there is no entanglement between the ancillae and the logical qubits at the end (which is a good thing).

11.3.2 Phase Errors

Thus far we have seen that quantum error-correcting codes can provide a method to protect quantum information from spontaneous bit flips, or, mathematically, a random application of an $\hat{X}$-gate on one of the three qubits that constitutes our logical qubit. But things other than bit flips can go wrong: indeed, in the preceding sections we discussed phase errors in detail; these are in fact the most pernicious sort of errors in many technologies. Can quantum error correction work in this case as well? As before, the logical qubit is in state $|\Psi\rangle = \alpha|000\rangle + \beta|111\rangle$, and let us suppose that a random phase flip is applied to qubit #1; $\left(\hat{Z} \otimes \hat{I} \otimes \hat{I}\right)|\Psi\rangle = \alpha|000\rangle - \beta|111\rangle$. But you get exactly the same state if the phase flip occurred on qubit #2 or qubit #3, so there is no way any quantum algorithm using this encoding can identify which qubit got phase flipped.

However, we can always convert a phase flip into a bit flip, simply by applying two Hadamard gates: $\hat{H}\hat{Z}\hat{H} = \hat{X}$. Thus, if we add three Hadamard gates immediately after the initial encoding stage, and three more immediately prior to the start of the diagnostic stage (see QC 11.7), any phase-flip error will look just like a bit flip.

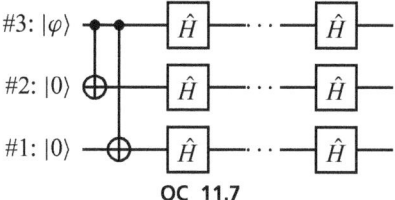

QC 11.7

11.3.3   Concatenation and Fault-Tolerant Quantum Computing

We have seen a recipe for nullifying either bit-flip or phase errors; but what happens if there is a possibility of both happening at the same time? There is nothing stopping us from converting each of the three physical qubits in the bit-flip error-correction circuit into logical qubits, each consisting of three physical qubits, and performing the phase-flip error correction on each of them: thus, nine physical qubits can, with some overhead in terms of ancilla qubits, correct both types of error. Thus, using nine physical qubits to form one compound logical qubit we can correct for both types of error simultaneously; one can in fact devise alternative encoding schemes with fewer qubits that correct all errors in this manner, with different circuits.

All of our analysis so far is based on an assumption that at most only one of the physical qubits undergoes a random error; if more errors occur, the error-correcting code still produces a final state that has an error, although, since that would require two random events, it would have a much smaller probability. In other words, error-correcting codes do not eliminate errors, they just reduce their likelihood. But there is a price to pay: we need more qubits, and more quantum logic gates in order to perform the error-correcting circuit; this will maybe increase the probability of error once again. There is a threshold (variously estimated at 10^{-6} to 10^{-4} for error probabilities) when performing quantum gates which, if it can be met, ensures the error correction does improve things. In principle, we could perform quantum logic operations with accuracy that exceeds this threshold; we can keep adding layers of **concatenation** (i.e., each of the nine qubits in the error-correcting code is replaced by nine qubits; then each of those gets replaced by nine more, and so on) so that errors in the ultimate output state are reduced to negligible probability. This ultimate goal of **fault-tolerant quantum computing**.

Exercises 11.3.4

1. A noisy classical communications channel has a probability $p = 1/3$ of randomly flipping a bit an odd number of times during transmission. In order to send messages reliably, multiple copies of each bit are sent. How many copies of each bit should be sent to ensure that the probability of a bit-flip error is less than 0.1?
2. Let p be the probability that any one qubit has undergone an odd number of bit flips. Write down the full density matrix you would get for three qubits.
3. What effect does the error-correcting circuit have on a state that has undergone two bit flips on different qubits? Suppose you have a single logical qubit you wish to protect against bit-flip error. The probability of such an error on a single qubit is p; assume that the bit-flipped state is orthogonal to the non-bit-flipped state. You use a three-qubit error-correction code.

(a) What is the fidelity if no error correction was employed?
(b) What effect does the error-correcting circuit have on the uncoded state if two different qubits flipped while it was encoded?
(c) What is the maximum factor by which state fidelity is enhanced using a three-qubit code?

11.4 CONCEPT CHECK AND EXERCISES

Concept Check

- Decoherence is a pervasive source of error in quantum computation, and large-scale devices will be impossible if it is not defeated.
- Passive error correction exploits the physics of devices and the correlations in error mechanisms to identify and exploit decoherence-free sub-spaces in which quantum computation can be realized.
- Quantum error correction actively seeks to trap, localize, and nullify individual qubit errors.

Chapter Exercises

1. An arbitrary quantum operation that preserves the basic properties of a density matrix can be written in the *operator sum form* as follows:

$$\hat{\rho}' = \sum_m \hat{K}_m^\dagger \hat{\rho} \hat{K}_m,$$

where the $\hat{K}_m$ are called **Kraus operators**.
   (a) Show that, if $\mathrm{Tr}\{\hat{\rho}'\} = 1$ for all $\hat{\rho}$, then $\sum_m \hat{K}_m \hat{K}_m^\dagger = \hat{I}$.
   (b) Show that Eq. (11.12) can be re-written in the operator-sum form, and give explicit expressions for the Kraus operators.
2. Construct a circuit to protect against dephasing errors using a decoherence-free subspace while correcting for bit-flip error. You may assume that the ancillae are not susceptible to dephasing.
3. Is it possible to do error correction with just two rather than three qubits?
4. The Lindblad master equation for a qubit undergoing spontaneous decay from $|1\rangle$ to $|0\rangle$ is

$$\frac{d}{dt}\hat{\rho}(t) = \frac{1}{2T_1}\left(2|0\rangle\langle 1|\hat{\rho}(t)|1\rangle\langle 0| - |1\rangle\langle 1|\hat{\rho}(t) - \hat{\rho}(t)|1\rangle\langle 1|\right).$$

Find the solution for $\hat{\rho}(t)$ if the initial state is $\hat{\rho}(0)$. Express the result as a **CPTP map**.

12 Quantum Technologies

In this chapter, we will learn about other technologies that exploit the unique properties of quantum mechanics. We will learn how it can be used to make message sending more secure, to improve the measurement of phase shifts, to detect objects more reliably, and to model and study the physics of complicated quantum systems using simpler quantum circuits.

Learning Objectives

- Understand how the one-time pad can be shared more securely using quantum key distribution.
- Learn, through a quantum metrology example, about the power of entanglement in improving the precision of measurements.
- Get a flavour of how entanglement can be exploited in target detection.
- Learn about the basics of quantum simulation and how it simplifies the study of complicated quantum systems.

12.1 Secure Communications: Quantum Key Distribution

In Chapter 5, we saw how using quantum power, namely the entanglement in Shor's algorithm, threatens the security of the (classical) RSA encryption method. In this section, we will see the opposite effect of quantum power, where it can be used to make a (classical) encryption method even more secure.

Let us say that Alice and Bob agree beforehand that the messages they send to each other will be four characters long. This means that each message will be $4 \times 8 = 32$ bits long. Say Alice wants to send Bob the message: *Read*

In the binary number system, the ASCII code for each of the characters is given by an 8-bit integer as follows:

$$
\begin{aligned}
R &= 01010010, \\
e &= 01100101, \\
a &= 01100001, \\
d &= 01100100.
\end{aligned}
\tag{12.1}
$$

Therefore, Alice wants to send Bob the following bit sequence:

$$01010010011001010110000101100100. \tag{12.2}$$

However, she should not send the message as is because if someone intercepts it, then it is easy to convert the binary digits back to the characters they represent and read the message.

Instead, Alice should encrypt the message and then send it to Bob, and Bob should have a way to decrypt it. What Alice uses is a sequence of 32 bits (i.e., the agreed upon length of messages) that she only uses once. Hence, it is given the name *one-time pad*. She chooses to use this one-time pad:

$$00011101010111100100001110000011. \tag{12.3}$$

Now, she adds her message to the one-time pad using bitwise addition modulo 2 (see Eq. (4.6)) to get

$$01001111001110110010001011100111 \tag{12.4}$$

and send it to Bob, using conventional (and not necessarily secure) channels. In addition, Alice enlists the help of a very trusty and secretive friend to hand Bob the sequence of the one-time pad she used as given in (12.3). It is not a big deal if someone intercepts the message she sent (Eq. (12.4)) because it is encrypted. In fact, if they try to convert the binary digits sequence to the character representation, they will get: O;"ç.

These characters are not only meaningless, but they give no indication of what the original message is.

Bob receives the confidential one-time pad sequence and adds it to the message he received from Alice, using bitwise addition modulo 2, to obtain

$$01010010011001010110000101100100 \tag{12.5}$$

which is the message Alice wanted him to get.

After retrieving Alice's message, Bob destroys the one-time pad so that it is not used again in future message sending. That is because even if the one-time pad remains safe and secret, if it is used more than once, then by looking at the encrypted messages, one can deduce patterns and start (correctly) guessing what the encoded messages say.

This looks like a very nice and secure way to send messages. However, there is a vulnerability in it: What if someone intercepts the trusted friend and copies the one-time pad without her knowing? Worse still, what if someone bribes the friend to see the secret sequence? This means that that person can easily decrypt the message that is being sent. If Alice can send Bob this sequence in a *guaranteed* secure way, that would remove this vulnerability in message sending. This is where the power of quantum physics comes to the rescue.

In order to share the one-time pad securely in a way that will also tell Alice and Bob whether someone has been eavesdropping during the process, Alice sends Bob a series of qubits, one at a time, prepared in a way that will allow them to construct a series of 0s and 1s to represent the one-time pad securely. The qubits need to be easy to transmit long distances, and the photon-based qubits we discussed in Section 7.1 are ideal for this application.

These are the steps for this process:

1. First, Alice comes up with a random string, of length n, made of a combination of two different symbols, say a and b.
2. Alice prepares a qubit for transmission to Bob. If the next symbol in her list is a, she prepares it in either $|0\rangle$ or $|1\rangle$; if the symbol is b, then she prepares it in either $|+\rangle$ or $|-\rangle$ (see Eq. (1.16)).
3. As Bob receives the qubits one by one, he needs to figure out what are their states. All he knows is that there are four different possibilities $\{|0\rangle, |1\rangle, |+\rangle, |-\rangle\}$; he does not know whether the qubit was prepared in the $\{|0\rangle, |1\rangle\}$ basis or in the $\{|+\rangle, |-\rangle\}$ basis. If he did know, he could just apply a Hadamard as appropriate and then measure. So Bob decides randomly to apply $\hat{H}$ on the qubit

before measurement. In other words, sometimes he applies $\hat{H}$ and sometimes not. Because this is done randomly, sometimes he will correctly figure out the state of the qubit that Alice sent and sometimes he will get it wrong. The latter occurs when he applies $\hat{H}$ when he should not or when he does not apply it when he should.

4. Now Alice gets in touch with Bob and tells him the correct basis for each qubit she sent (but not the actual state: only the basis). Bob then compares Alice's list with the measurements he actually carried out: if the bases match, then he records the measurement outcome (0 or 1) as one bit of the one-time pad he will use to communicate with Alice; if they do not match, the measurement is discarded. Repeating this with all the qubits in sequence allows construction of the one-time pad, bit by bit. He will also tell Alice which of the qubits he is ignoring, so she can do the same.

5. Now, on both ends, Alice and Bob share a one-time pad, generated as required, without the need of a trustworthy courier. Alice can write a message, convert it to binary, encipher it with her one-time pad, then send the result to Bob who can use his identical one-time pad to decipher Alice's message.

One of the chief potential vulnerabilities of this system is the "man-in-the-middle" attack: an eavesdropper intercepts the qubits Alice has sent and measures them, guessing at which measurement basis to use; she then prepares a qubit in the state she just measured and sends it on to Bob, who will receive it unaware of what has happened. But the eavesdropper cannot prepare a perfect copy of the qubit she measured, since she did not know the basis, so Bob will not get the correct state that Alice sent him. In order to discover whether this kind of interception occurred, Alice and Bob, after deciding which qubits to incorporate into their one-time pad, can randomly select a few (you never know who might be listening) of the ones left and compare notes on what states they both have. If a large percentage do not match, then they know someone has been intercepting the qubits. Then, they have to make the decision whether the estimated number of intercepted qubits is big enough or not to compromise the security of using this particular one-time pad. If it is, they repeat everything again until they are pleased that it is secure enough.

The aforementioned way to share the one-time pad securely is known as **BB84**, named after its inventors (Bennett and Brassard) as well as the year (1984) they developed it. The naming of **quantum key distribution** comes from the fact that the laws of quantum mechanics are used to send the key (one-time pad) securely.

 Interesting Facts 12.1

Cryptography in History. Encrypting messages to hide them from certain individuals and decrypting them to discover what they say, even when one is not the intended recipient, are acts resulting from human nature; some want to hide secrets, and some want to know them. Concealing messages or writing them in unintelligible (except for the recipient) forms, whether for amusement or for the necessity of secrecy, has been going on for thousands of years. The earliest evidence we have is from ancient Egypt roughly four thousand years ago. Moreover, Julius Caesar, who lived more than two thousand years ago, was fond of encrypting his personal correspondence. The *Caesar cipher*, which refers to encryption by substituting a letter for the one a given number of positions down the alphabet, is named after him. There is also evidence in old Indian texts from the fourth century CE for other methods of encryption by substitution. The strength of these methods came from the fact that to guess the *key* to break the code, one would have to go through a very impractical number of possibilities. That is why they persisted for many centuries.

This happy situation of using these contemporary encryption techniques securely came to an end when the Arabs, during the Islamic Golden Age, invented *cryptanalysis*, the science of breaking encrypted messages. They developed and applied statistical techniques to encrypted messages to decrypt them without having to figure out the key, using methods such as *frequency analysis*. The oldest known book on the subject of cryptanalysis, *Treatise on Deciphering Cryptographic Messages*, was written in the ninth century by the polymath Yaqub bin Ishaq Al-Kindi.

Fast-forward several centuries, monks interested in examples of cryptography in biblical texts ignited Europe's interest in the subject. As progress was made in this field, cryptanalysis experts specialized in frequency analysis became so prevalent that it rendered it ridiculous (and at times dangerous) to depend solely on the known encryption techniques of that time. Something had to be done to make these techniques not so vulnerable to such analyses. In the sixteenth century, during the Renaissance, standing on the works of many intellectuals before him, the French cryptographer Blaise de Vigenère put together a powerful new encryption method, where letters in a message were encoded using different keys. This is known as the *Vigenère cipher*, and it is immune to decryption using frequency analysis. However, since it can be time consuming and impractical at times, other encryption methods were also developed and used, such as ones that are hybrids of the earlier encryption methods and those developed by Vigenère's contemporaries as well as the *Great Cipher* developed in the seventeenth century by the Rossignol father-and-son duo. Then, just as frequency analysis proved a breakthrough in decryption a thousand years earlier, in the nineteenth century, the *unbreakable* Vigenère cipher was defeated by the British polymath Charles Babbage (father of the computer) and the German cryptanalyst Friedrich Kasiski.

It seems that every time a *clever* encryption idea is used, another *clever* cryptanalyst figures out how to decrypt the message by exploiting some weakness. Then, in the late nineteenth century and early twentieth century, came the idea of the one-time pad that we discussed in this section. This, as we saw, can also be insecure when being shared. That is where the BB84 comes to the rescue. Do you think the latter also has a weakness that can be exploited?

Exercises 12.1.1

1. In this exercise, we will get a flavour of why using one-time pads more than once is not a good idea if we want our messages to be secure. We will simplify the message sending to 1, 2, and 3-bit messages.

   (a) Let us say that Alice wants to send Bob 1-bit messages, using the one-time pad 0. Calculate the encrypted message for the two messages $m_1 = 0$ and $m_2 = 1$ to obtain the corresponding encrypted messages d_1 and d_2. What can an eavesdropper conclude about m_1 and m_2 by looking at d_1 and d_2?

   (b) Now, let us say that Alice wants to exchange 2-bit messages with Bob by using the one-time pad 11 to encrypt the messages. Let the two messages she wants to send be $m_1 = 00$ and $m_2 = 10$. Find d_1 and d_2 and explain what an eavesdropper can conclude about m_1 and m_2 by looking at them.

   (c) Finally, if Alice wants to send Bob the 3-bit messages $m_1 = 010$ and $m_2 = 101$ using the one-time pad 101, find d_1 and d_2 and explain what an eavesdropper can conclude about m_1 and m_2 by analyzing them.

2. We discussed how Alice sent Bob the message: *Read*. In this exercise, work out how Bob can encrypt the response: *Okay*, send it to Alice, and how Alice can decrypt the response. Note: Come up with your own one-time pad; i.e., do not reuse the one Alice used to send the message to Bob.

3. What happens when the following qubit states are measured in the $\{|0\rangle, |1\rangle\}$ basis? (a) $|0\rangle$, (b) $|1\rangle$, (c) $|+\rangle$, (d) $|-\rangle$.

4. What happens when the following qubit states have a Hadamard gate applied on them before they are measured in the $\{|0\rangle, |1\rangle\}$ basis? (a) $|0\rangle$, (b) $|1\rangle$, (c) $|+\rangle$, (d) $|-\rangle$.

5. Suppose you know that a qubit is prepared in one of two possible states, as shown below. What is the best circuit to distinguish between them with a single measurement in the $|0\rangle$, $|1\rangle$ basis? What is the probability of successfully distinguishing them?

   (a) $|0\rangle$ or $|1\rangle$?

   (b) $|+\rangle$ or $|-\rangle$?

   (c) $|0\rangle$ or $|+\rangle$?

   (d) $|0\rangle$ or $|-\rangle$?

6. Suppose Alice sends Bob the state $|+\rangle$, and an adversary, Inez (who is aware of the protocol and knows the qubit must be in one of the four possible states $|0\rangle$, $|1\rangle$, $|+\rangle$ or $|-\rangle$), intercepts the qubit, measures it in the $\{|0\rangle, |1\rangle\}$ basis, prepares a new qubit in state corresponding to the outcome of her measurement, and, pretending to be Alice, transmits the new qubit to Bob. When he receives the qubit, Bob measures in the $\{|+\rangle, |-\rangle\}$ basis (if he did not, the case would be moot anyway, since that bit would be discarded when Alice and Bob compare keys). What are the outcomes and probabilities of Bob's measurement? Does Inez intercept any useful information about the key?

7. How would the analysis of Exercise 6 be different if Inez used the $\{|+\rangle, |-\rangle\}$ basis?

8. When Alice tells Bob that she sent him a qubit in an eigenstate of $\hat{X}$, what is the likelihood Bob will discard the result? In the cases in which Bob and Alice agree on the basis, what are the chances that Bob receives the correct state that Alice sent? What are the chances he disagrees with Alice if there is an adversary intercepting the qubits? What is the maximum probability that the adversary has acquired the correct bit value?

12.2   Quantum Metrology

Metrology is the science of measurement (not to be confused with *meteorology*, which is the study of weather, or *meteoritics*, which is the study of meteors; there does not appear to be a word for the study of the Parisian underground railway). It is very important to be able to measure quantities as accurately and precisely as possible, although how far we want to go depends on the situation. For example, finding that someone is 149 kg in weight (technically we are talking about mass, of course) is probably sufficient for a dietician devising a diet and exercise regime: they do not need to know that the person weighs 149.00034569 kg; nor does a lumberjack care that the diameter of a tree trunk is 50.00803256 cm: 0.5 m should be sufficient precision for what he needs to do. However, there are cases where the finer measurement scales are vitally important, such as in precision manufacture of nano-electronics or the detection of the slight polarization rotation of scattered light that would indicate protein-folding has occurred; we need a finer ruler for such applications.

This section is about how the ideas of quantum information can be used to improve precision of certain measurements. We will use a simple generic example of measurement of a phase shift using abstract qubits. To utilize these results in a practical setting, one would have to apply them to one of the qubit technologies – which technology depends on the application one has in mind. Usually in many sensing applications, light waves have pride of place for distance measurements, and our qubits will be photons; but one could envision situations in which precise measurements of a frequency standard using trapped ions or fine measurement of magnetic field variation using electron or nuclear spin is called for.

First let's consider a simple method for measurement of a phase shift ϕ using single qubit probes, which corresponds to the standard classical approach to the problem (QC 12.1).

QC 12.1

We prepare a qubit in state $|0\rangle$, apply a Hadamard gate, then allow it to pass through the phase-shift gate given by

$$\hat{U}_\phi = \begin{bmatrix} \exp(i\phi/2) & 0 \\ 0 & \exp(-i\phi/2) \end{bmatrix} = \cos(\phi/2)\hat{I} + i\sin(\phi/2)\hat{Z}. \tag{12.6}$$

Our goal is to deduce the value of ϕ from repeated measurements. After passing through the phase shifter, a second Hadamard is followed by measurement. The probability of obtaining the outcome "0" in the measurement is

$$\begin{aligned} p_0 &= |\langle 0|\hat{H}\hat{U}_\phi\hat{H}|0\rangle|^2 \\ &= |\langle 0|\hat{H}\left(\cos(\phi/2)\hat{I} + i\sin(\phi/2)\hat{Z}\right)\hat{H}|0\rangle|^2 \\ &= |\langle 0|\left(\cos(\phi/2)\hat{I} + i\sin(\phi/2)\hat{X}\right)|0\rangle|^2 \\ &= \cos^2(\phi/2), \end{aligned} \tag{12.7}$$

where we have used $\hat{H}\hat{H} = \hat{I}$ and $\hat{H}\hat{Z}\hat{H} = \hat{X}$. Let us repeat this circuit N times, with the number of "0" results being n. The expected value of n is $\bar{n} = N\cos^2(\phi/2)$, so our estimate of the value of ϕ will be

$$\phi \approx 2\arccos\sqrt{n/N}. \tag{12.8}$$

The standard deviation $\sigma_{\phi,0}$ is most important to assess the precision of the measurement: we know from Section 9.1 that the standard deviation of n is $\sigma_n = \sqrt{Np_0(1-p_0)} = \sqrt{\bar{n}(1-\bar{n}/N)}$. The standard deviation of ϕ will be

$$\sigma_{\phi,0} = \sigma_n \left|\frac{d\phi}{d\bar{n}}\right|. \tag{12.9}$$

To find $d\phi/d\bar{n}$, take the derivative of $\bar{n} = N\cos^2(\phi/2)$:

$$\begin{aligned} \frac{d\bar{n}}{d\bar{n}} &= \frac{d\phi}{d\bar{n}}\frac{d}{d\phi}\left(N\cos^2(\phi/2)\right) \\ 1 &= N\frac{d\phi}{d\bar{n}}\left(-\cos(\phi/2)\sin(\phi/2)\right) = -\frac{d\phi}{d\bar{n}}\sqrt{N\bar{n}(1-\bar{n}/N)}, \end{aligned} \tag{12.10}$$

and hence

$$\frac{d\phi}{d\bar{n}} = -\frac{1}{\sqrt{N\bar{n}(1-\bar{n}/N)}}. \tag{12.11}$$

Substituting into Eq. (12.9) we get

$$\sigma_{\phi,0} = \frac{\sqrt{\bar{n}(1-\bar{n}/N)}}{\sqrt{N\bar{n}(1-\bar{n}/N)}} = \frac{1}{\sqrt{N}}. \tag{12.12}$$

This is referred to as the **standard quantum limit** on the precision of measurements.

Using quantum entanglement we can improve on this result. Consider QC 12.2:

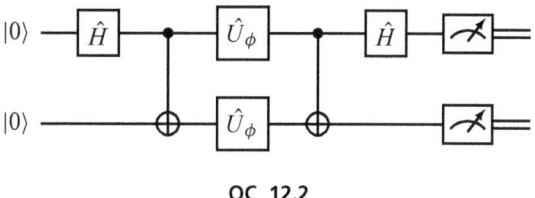

QC 12.2

In this case, the probability that both qubits yield a result "0" will be

$$p_2 = |\langle 00| \left(\hat{H} \otimes \hat{I} \right) \text{CNOT} \left(\hat{U}_\phi \otimes \hat{U}_\phi \right) \text{CNOT} \left(\hat{H} \otimes \hat{I} \right) |00\rangle|^2. \tag{12.13}$$

(The notation is a bit idiosyncratic: it is easier to write p_2 than p_{00}, and p_m instead of $p_{00\ldots 0}$; they both refer to the probability of obtaining "all-0" measurement outcomes with the subscript denoting the number of entangled qubits.) The state $\text{CNOT} \left(\hat{H} \otimes \hat{I} \right) |00\rangle = \frac{1}{\sqrt{2}}(|00\rangle + |11\rangle)$ is one of the maximally entangled Bell-basis states we encountered in Chapter 3; it is also an example of the "GHZ" state, named for Daniel Greenberger, Michael Horne, and Anton Zeilinger, who introduced such multi-qubit entangled states in 1989. We will use the notation:

$$|\text{GHZ}_m\rangle = \frac{1}{\sqrt{2}} \left(\underbrace{|00\ldots 0\rangle}_{m \text{ qubits}} + \underbrace{|11\ldots 1\rangle}_{m \text{ qubits}} \right). \tag{12.14}$$

Such a state can be created using multiple CNOT gates (QC 12.3):

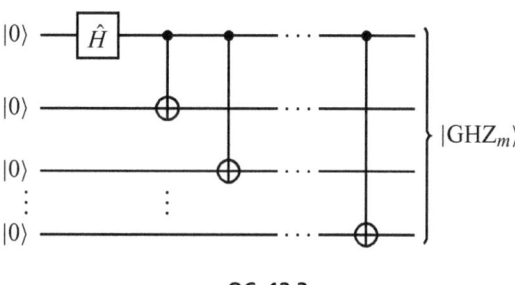

QC 12.3

Using this notation, Eq. (12.13) becomes

$$\begin{aligned}
p_2 &= |\langle \text{GHZ}_2 | \left(\hat{U}_\phi \otimes \hat{U}_\phi \right) |\text{GHZ}_2\rangle|^2 \\
&= \frac{1}{4} \left| (\langle 0|\hat{U}_\phi|0\rangle)^2 + (\langle 1|\hat{U}_\phi|1\rangle)^2 \right|^2 \\
&= \frac{1}{4} |\exp(i\phi) + \exp(-i\phi)|^2 \\
&= \cos^2(\phi).
\end{aligned} \tag{12.15}$$

If we scaled up the circuit to have an m-qubit GHZ state, with each qubit interacting with the phase shift operator, the circuit would look like QC 12.4:

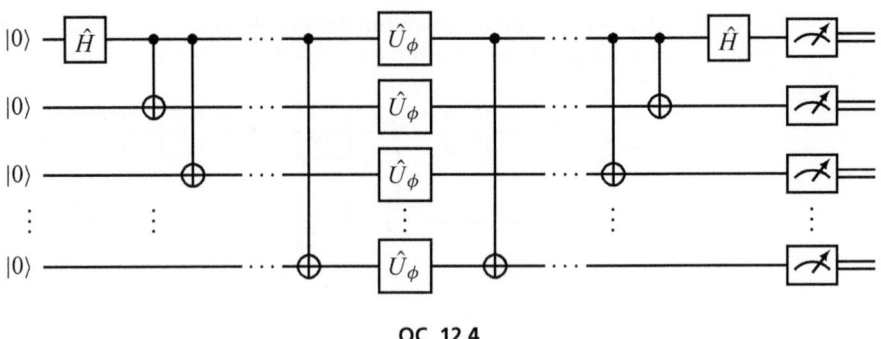

QC 12.4

In this case, the probability of obtaining "0" in all m detectors will be

$$
\begin{aligned}
p_m &= |\langle \mathrm{GHZ}_m|(\hat{U}_\phi)^{\otimes m}|\mathrm{GHZ}_m\rangle|^2 \\
&= \frac{1}{4}\left|(\langle 0|\hat{U}_\phi|0\rangle)^m + (\langle 1|\hat{U}_\phi|1\rangle)^m\right|^2 \\
&= \cos^2(m\phi/2).
\end{aligned}
\tag{12.16}
$$

Thus, in order to estimate ϕ in either the two-qubit or the m-qubit cases, we repeat the circuit multiple times and observe the number of occurrences of each detector registering "0". In order to have a fair comparison of the standard single-qubit approach with the m-qubit entangled approach, we will require the total number of times the $\hat{U}_\phi$ operation is carried out to be always equal to N. Thus, if we use a circuit with m entangled qubits, we should repeat the circuit N/m times (we'll assume that N/m is an integer). Thus, the expected number of "all-0" results with an m-qubit entangled circuit will be

$$
\bar{n}_m = \frac{N}{m}p_m = \frac{N}{m}\cos^2(m\phi/2),
\tag{12.17}
$$

and the standard deviation will be

$$
\sigma_n^{(m)} = \sqrt{\frac{N}{m}p_m(1-p_m)} = \sqrt{\frac{\bar{n}_m(N-m\bar{n})}{N}}.
\tag{12.18}
$$

Thus, our estimate for ϕ based on N/m repetitions of an m-qubit entangled circuit will be

$$
\phi \approx (2/m)\arccos\sqrt{mn_m/N}.
\tag{12.19}
$$

If we take the derivative of Eq. (12.17) with respect to $\bar{n}_m$ we obtain

$$
\frac{d\phi}{d\bar{n}_m} = -\frac{1}{\sqrt{m\bar{n}_n(N-m\bar{n}_m)}},
\tag{12.20}
$$

and hence the standard deviation of the estimated ϕ based on N/m repetitions of an m-qubit entangled circuit will be

$$
\sigma_{\phi,m} = \frac{\sqrt{\bar{n}_m(N-m\bar{n})/N}}{\sqrt{m\bar{n}_n(N-m\bar{n}_m)}} = \frac{1}{\sqrt{Nm}} = \frac{\sigma_{\phi,0}}{\sqrt{m}}.
\tag{12.21}
$$

Thus, by using the m-qubit entangled circuit instead of the simple circuit we originally envisioned, we have obtained a factor of $\sqrt{m}$ improvement in the precision of the experiment, while keeping the number of calls on the phase-shift operator constant.

Imagine, for example, we wanted to probe the thickness of some delicate organism using an optical probe – in other words, photon qubits. Each time a photon passes through the sample, there is a chance of damaging it; a large number of photons will surely destroy it. But using **quantum metrology** one can perform a measurement of equal precision using a factor of $1/\sqrt{m}$ fewer photons and lower concomitant risk of damage. It is applications of this sort that may be much closer to practical realization and deployment than the mega-qubit quantum computers needed to execute truly impactful quantum algorithms.

Exercises 12.2.1

1. Suppose that, instead of estimating a phase shift, one had to detect the possibility of a bit-flip: how would you modify the single-qubit measurement protcol?
2. Instead of counting the "all-0" outcomes of the two-qubit entangled circuit, suppose one were to count the total number of "0" outcomes for each repetition of the circuit; thus the outcome "00" would be recorded as 0, "01" and "10" would both be recorded as 1, and "11" as 2, and so on. Calculate the expected value of this counting method. What is its standard deviation? Does it offer significant advantages over simply recording the number of "all-0" outcomes?

12.3   Quantum Detection

In this section, we will learn about the advantage that quantum power, in the form of quantum entanglement, gives us when we try to detect objects. Before we delve into this, let us revisit the consequences of entanglement on measurement results by analyzing the situation with the following entangled state:

$$|\psi\rangle = a|00\rangle + b|11\rangle. \tag{12.22}$$

If we measure the first qubit (qubit A) and obtain $|0\rangle$, then we know measurement of the second qubit (qubit B) will yield $|0\rangle$ as well. What if the measurement of qubit B yielded $|1\rangle$ instead? This could mean that we measured the wrong qubit (qubit C), one that is not entangled to qubit A as described by the state in (12.22). This is one way we could tell if we had the *right* qubit from the pair of entangled qubits.

One way we can use quantum power to detect an object or a target is to use entangled qubit pairs by taking advantage of our ability to tell that a qubit is from the pair. In the next example, we will consider the situation where the qubits are photons of light.

Say we are all set in our *detection headquarters*; we have entangled pairs of photons as well as a measurement setup that allows us to test whether a given photon is from a pair that we have. The first step is to send one of the photons from an entangled pair (we keep the second one from that pair in the headquarters) in the direction that we suspect there is an object (Fig. 12.1).

In the best case scenario, the photon we send will reach the target, and will reflect back into the headquarters, where we are ready to measure it to verify whether it is the one entangled with the pair we kept with us (Fig. 12.2).

Sometimes, after sending a photon, we do not get any back. With no photon reflected back, one can conclude that perhaps there is no object or maybe the photon was not sent in the right direction (Fig. 12.3).

Figure 12.1   Photons sent towards suspect object.

Figure 12.2   Photon reflected from object.

Figure 12.3   No photon reflected back.

Even if we do get a photon, it does not mean it is the one we sent. It could be a random photon from the environment that found its way into the detection headquarters (Fig. 12.4).

We can use the photon that we kept with us to determine whether the one that comes into the detection headquarters is the other one of the pair or an unrelated one from the environment, using the same logic that we introduced at the beginning of this section.

Figure 12.4   Photons from the environment detected.

Exercises 12.3.1

1. If we have the following state:

$$|\psi\rangle = a|01\rangle + b|10\rangle, \tag{12.23}$$

and we are told that when the first qubit is measured first, we find that it is in state $|1\rangle$, and then immediately after that, when the second qubit is measured, we find that it is in state $|1\rangle$. Give two possible explanations for such an observation.

2. We assume that the state in (12.23) is expressed in the computational basis; i.e., the basis made of the eigenvectors of $\hat{Z}$. Say that the first qubit is measured in the appropriate (computational) basis with $|1\rangle$ obtained as the measurement result, but then immediately after that the second qubit is measured in the basis made of the eigenvectors of $\hat{X}$.

   (a) What are the possible measurement results?

   (b) What can be concluded about whether the second qubit is the actual qubit from the entangled pair or not?

3. We can send uncorrelated photons towards a suspected target, and if they bounce back, then we know we detected something. Why are we bothered about sending photons from entangled pairs to detect the object then?

4. Let us assume that we are in the situation where there is an object in the direction in which we are sending a photon from the entangled pair. In other words, the photon we send will reach the object and bounce back. Say that, for unforeseen environmental obstacles, the chance that this photon bounces back to the headquarters is 10%. If we send 10 such photons, all of which are independent of each other, what is the probability of the following?

   (a) We detect 1 of the photons.

   (c) We detect 5 of the photons.

   (b) We detect all 10 photons.

12.4 Quantum Simulations

If we are studying a quantum system made of a few qubits, then the mathematical description of this system will not be very complicated, and we might be able to study it analytically (i.e., by solving equations and without resorting to numerical techniques such as in writing computer programs), using its complete description, on a piece of paper, making all sorts of predictions about how it will evolve in time depending on the conditions around it. However, if we study very large systems made of many quantum particles, such as is common in condensed matter systems, then, except for a few special models, we can wave goodbye to analytic results. One might think that this is not a big deal; we can describe these huge quantum systems using the computational power that many classical computers are able to provide. This is a good point, and yes, we can do that, but as the number of quantum particles, especially if they are entangled, increases, the resources needed to study them classically increase very very quickly, reaching a stage where another computational power has to take over. What better than quantum computation?!

We can, for example, choose a simpler and more controlled quantum system as a model to represent a more complicated quantum system whose degrees of freedom are not very accessible. For instance, we can design and build an optical lattice, where we have all these nicely arranged mini potentials that have an atom trapped in each, where the whole lattice arrangement can be a representation for some other complicated condensed matter system. The advantage of the optical lattice is that we can tweak its various properties, including these individual potential wells, to make it as acceptable a model as we need to. Moreover, because it is a quantum system, we can also investigate quantum properties in the model itself to get insight into quantum phenomena that could manifest themselves in the system of interest. Another approach is to model the unitary evolution of the system with a combination of quantum gates. It is this latter approach we will look at in more detail in this section, but for demonstration purposes, we will focus on systems made of few qubits.

Recall that the Schrödinger equation, which describes the dynamics of a closed quantum system represented by the Hamiltonian $\hat{\mathcal{H}}$, is given by the following first-order differential equation:

$$i\hbar\frac{d}{dt}|\psi(t)\rangle = \hat{\mathcal{H}}|\psi(t)\rangle. \tag{12.24}$$

Also recall the solution of this equation:

$$|\psi(t)\rangle = \exp(-i\hat{\mathcal{H}}t/\hbar)|\psi(0)\rangle. \tag{12.25}$$

In other words,

$$|\psi(0)\rangle \longrightarrow \exp(-i\hat{\mathcal{H}}t/\hbar)|\psi(0)\rangle. \tag{12.26}$$

If we can find a way to just apply the exponential operator $\exp(-i\hat{\mathcal{H}}t/\hbar)$ directly as a gate, that would be great. Note though that if $\hat{\mathcal{H}}$ describes more than one qubit, then there is a problem on the practical level; i.e., during the physical implementation of the setup: Applying multi-qubit gates is not easy. However, if we can write this exponential as a product of exponentials, each corresponding to one qubit, so we have to deal with single-qubit gates, which are easier to implement, that would solve the problem. Unfortunately, in general, we cannot do that. However, there are situations in which it is possible. In particular, if the operators in the exponential commute, then the exponential can be factored into a product of exponentials involving the individual operators. Let's say that $[\hat{A}, \hat{B}] = 0$; i.e., $\hat{A}$ and $\hat{B}$ commute, then $\exp(\hat{A} + \hat{B}) = \exp(\hat{A})\exp(\hat{B})$. If the Hamiltonian equals the sum of the operators $\hat{A}$ and $\hat{B}$, that would simplify $\exp(-i\hat{\mathcal{H}}t/\hbar)$ so that it can be written as $\exp(-i\hat{A}t/\hbar)\exp(-i\hat{B}t/\hbar)$. This will allow us to apply two single-qubit phase gates at the same time to correspond to that evolution. Unfortunately, often the situation in which the operators do not commute arises, and technically we cannot separate the operators into single-qubit operators and get away with it. However, there is an acceptable approximation that one can perform using the Trotter formula in which the exponential can be broken down into a product of exponentials, each corresponding to one system. This makes the experiment much easier to implement. When we discuss our next example, we will use this approximation, but before we do that, we need to comment on another issue, namely, why physicists choose to ignore the individual independent evolution of subsystems, especially when studying large quantum systems.

⊕ Mathematics Tips 12.2

The Trotter Formula and Simplifying Quantum Modelling. For any (not necessarily commuting) operators, $\hat{A}$ and $\hat{B}$, the Trotter product formula states that

$$\exp(\hat{A} + \hat{B}) = \lim_{n \to \infty} \left(\exp(\hat{A}/n) \exp(\hat{B}/n) \right)^n. \tag{12.27}$$

When this approximation is applied in constructing the quantum simulation circuits we discuss in this section, the accuracy of this approximation is dependent on how *small* the argument in the exponential is. In the physical implementation of the quantum gates, this is done by making the time the field (representing the quantum gate) interacts with the qubit as short as possible; i.e., $\Delta t \to 0$.

When we are able to apply the Trotter approximation with very good accuracy, this makes modelling quantum systems much easier, as it allows breaking the evolution part of the state into several simpler gates. Otherwise, we would have to work with one big gate that represents all the terms in the Hamiltonian, which is much harder to implement.

When modelling large quantum systems, physicists usually ignore the evolution of the individual subsystems and do not include any terms in the Hamiltonian to correspond to them. Instead they focus on writing out terms that correspond to the interaction of the subsystems together. This is also referred to in physics as studying the system in the *interaction picture*. The interesting dynamics is usually a result of these interactions, while the individual evolution could just lead to boring global phases. In the next example, we will study the situation where the individual qubits have special gates applied to them individually, so we will see terms that correspond to individual qubits. However, keep in mind that these terms do not correspond to the evolution of the individual qubits if they were left all alone with no interactions, but they are a result of *interacting* with an external field. In other words, we are still working in the interaction picture.

Recall that, since the (2×2) identity operator along with the three Pauli matrices form a complete basis, any 2×2 matrix can be written as a linear combination of these four matrices: $a\hat{I} + b\hat{X} + c\hat{Y} + d\hat{Z}$ (a, b, c, and d are constants). This means that any operation on a single qubit can be expressed in terms of these matrices, and any operation on two qubits can be expressed as a linear combination of tensor products of these matrices: $\sum_{ij} c_{ij} \hat{\sigma}_i \otimes \hat{\sigma}_j$ (c_{ij} represents constants). However, to simplify calculations and predictions, and since it is possible to design experimental setups (by rotation, for example) to accommodate this, we can choose operations that are made of the actual Pauli matrices, not a linear combination of them. We will, of course, not worry about operations described solely by the identity, as they do nothing and are, therefore, boring. For example, in a system of two qubits, if we want to flip the state of one of the qubits only (for example, through an external field), we apply the operator $\hat{X} \otimes \hat{I}$ if we want to flip the state of the first qubit, and we apply the operator $\hat{I} \otimes \hat{X}$ instead if we want to flip the second qubit. Similarly with the two-qubit operation, we can select a specific axis of interaction so that the same Pauli operator is applied on each qubit such as in the following: $\hat{X} \otimes \hat{X}$, $\hat{Y} \otimes \hat{Y}$, $\hat{Z} \otimes \hat{Z}$.

Consider the following Hamiltonian (in the interaction picture):

$$\hat{\mathcal{H}} = \frac{\hbar\Delta}{2}\{\hat{Z}\otimes\hat{I} + \hat{I}\otimes\hat{Z}\} + \hbar J\{\hat{X}\otimes\hat{X}\}, \tag{12.28}$$

where Δ and J are constants. The two parts $\{\hat{Z}\otimes\hat{I} + \hat{I}\otimes\hat{Z}\}$ and $\{\hat{X}\otimes\hat{X}\}$ do not commute, hence we cannot simply factor the time evolution operator. However, using the Trotter approximation, we can factor the unitary operator representing the evolution of the system as follows:

$$\exp(-i\hat{\mathcal{H}}t/\hbar) \approx \left[\left(\underbrace{\exp(-i\hat{Z}\Delta t/2n)}_{\hat{O}(t/n)}\otimes\underbrace{\exp(-i\hat{Z}\Delta t/2n)}_{\hat{O}(t/n)}\right)\underbrace{\exp(-i\{\hat{X}\otimes\hat{X}\}Jt/n)}_{\hat{Q}(t/n)}\right]^n. \tag{12.29}$$

Notice that the term involving $\{\hat{Z}\otimes\hat{I} + \hat{I}\otimes\hat{Z}\}$ has resulted in a pair of local operations (which are relatively easy to perform) plus $\hat{Q}$, which corresponds to a two-qubit non-local entangling gate, which although harder to implement is still achievable. The formula is approximate, and only becomes exact in the limit that $n \to \infty$; however, pretty accurate results can be obtained with relatively few steps.

The quantum circuit representing these gates is given in QC 12.5: there are n repetitions of the basic three-gate subroutine.

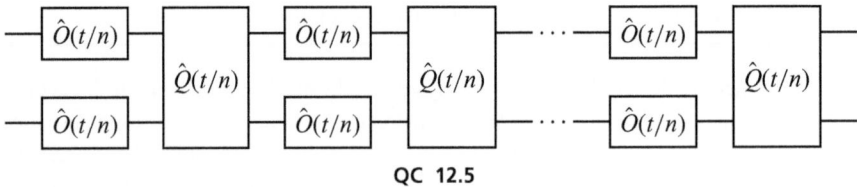

QC 12.5

Exercises 12.4.1

1. **A Consequence of the Baker–Campbell–Hausdorff Formula.**
   (a) If $[\hat{A}_1,\hat{A}_2] = 0$, show that $\exp(\hat{A}_1 + \hat{A}_2) = \exp(\hat{A}_1)\exp(\hat{A}_2)$. *Hint: Consider the Taylor expansion in powers of λ of $\exp[\lambda(\hat{A}_1 + \hat{A}_2)]$.*
   (b) Given $\hat{A} = \sum_i \hat{A}_i$, where $0 \le i \le n$, prove that if, for all i and j, $[\hat{A}_i,\hat{A}_j] = 0$, then $e^{\hat{A}} = e^{\hat{A}_1}e^{\hat{A}_2}e^{\hat{A}_3}\ldots e^{\hat{A}_n}$. *Hint: Use mathematical induction.*

2. Work out the following commutators of two-qubit operators:
   (a) $\left[\hat{X}\otimes\hat{X}, \hat{Y}\otimes\hat{Y}\right]$,
   (b) $\left[\hat{I}\otimes\hat{Z} + \hat{Z}\otimes\hat{I}, \hat{X}\otimes\hat{X}\right]$,
   (c) $\left[\hat{X}\otimes\hat{Y}, \hat{Y}\otimes\hat{X}\right]$.

3. Suppose we have a two-qubit system, initially in the state $|00\rangle$ and whose dynamics are described by the Hamiltonian (12.28). By solving Schrödinger's equation exactly, show that the probability the system remains in the state $|00\rangle$ is given by

$$P_{00}(t) = 1 - \frac{J^2}{\Delta^2 + J^2}\sin^2\left(\sqrt{\Delta^2 + J^2}t\right). \tag{12.30}$$

4. Use the approximate Trotter formula with $n = 2, 5$, and 10 to work out numerically the time evolution of $P_{00}(t)$, and compare your results with Eq. (12.30). Assume $J = 2\Delta$.

12.5 CONCEPT CHECK AND EXERCISES

Concept Check

- Quantum key distribution provides a secure way to send messages using one-time pads by taking advantage of quantum superposition states and logical reasoning.
- Quantum metrology allows a significant increase in the precision of fine measurements by use of multiple entangled probes.
- Sending photons to detect an object by being reflected back is one way to detect objects. However, adding quantum power by sending photons from entangled pairs that one possesses gives higher confidence in the findings.
- Complicated quantum systems can be modelled using simpler more controlled quantum systems or carefully designed quantum circuits.
- By using quantum circuits and providing the right conditions to use the Trotter approximation, one can break down the complicated evolution of a given quantum system into a collection of gates, many of which are single-qubit gates that are easier to implement and control.

Chapter Exercises

1. Alice sends Bob the following 4-bit messages: $m_1 = 0100$ and $m_2 = 1111$ by using the same one-time pad 1111 in both cases. Find the encrypted messages d_1 and d_2. By looking at d_1 and d_2, what conclusions can an outsider make about m_1 and m_2 without knowing what the one-time pad is?

2. Using the same one-time pad, Alice sends Bob the following two encrypted messages: $d_1 = 010110001001001111011101$ and $d_2 = 010110001001110111000000$.

   (a) By comparing the two messages, what can you conclude about m_1 and m_2?

   (b) If you were told that the messages were $m_1 = $ Bad, and $m_2 = $ Boy, what is the one-time pad?

3. Suppose, due to some technical glitch in the implementation of the BB84 protocol, the adversary has a chance $p = 0.01$ of intercepting each bit of the quantum key. What is the probability that a 20-bit key is completely secure? What steps could Alice and Bob take to use the key they have to produce a smaller, yet more secure key?

4. Say we have 100 entangled photon pairs that we want to use to detect an object. Let us also assume that there is an object in the direction in which we are sending the photons. Assume that the chance that any photon we send bounces back to us is 5%, and also assume that whatever happens to each photon is independent.

   (a) What is the probability that one photon will bounce back?

   (b) What is the probability that at least one photon will bounce back?

   (c) What is the probability that no more than 10 photons will bounce back?

5. **Quantum Simulations Using CNOT Gates.** In this exercise, we will investigate another strategy for simulating quantum systems using CNOT gates and a single-qubit phase gate. In the setup discussed in this exercise, the number of CNOT gates needed equals the number of control qubits, making it harder and harder to implement as the number of qubits increases.

   States evolve according to the formula $|\psi(t)\rangle = \exp(-i\hat{\mathcal{H}}t/\hbar)|\psi(0)\rangle$. The exponential operator looks like a phase term (although it is not a number but an operator). To model such a quantum system by a series of quantum gates, we need to make sure that the ket representing the state is the same at the end of the quantum circuit and also that the appropriate phase term is added somewhere during the process.

We have enough knowledge and experience with quantum circuits by now to construct ones, whether boring or interesting, that satisfy these conditions. Let us first focus on this combination of CNOT gates, where the state of the three-qubit system is represented by $|a\rangle|b\rangle|c\rangle$ and corresponds to the *controls*, while the *target* qubit is given by $|x\rangle$ (QC 12.6):

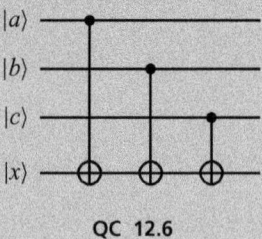

QC 12.6

(a) Find the resultant state at the far right.

(b) Now we will add three more CNOT gates in the opposite order (QC 12.7):

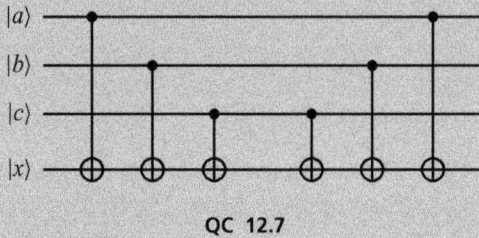

QC 12.7

What is the state at the far right?

(c) Now try the same circuit, but for a CZ combination instead, as shown in QC 12.8:

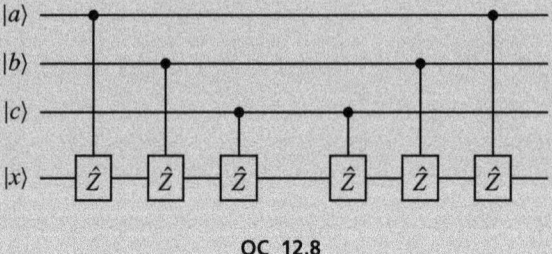

QC 12.8

What is the final state at the end?

(d) To the previous quantum circuit, let us add the single-qubit phase gate $e^{i\Delta t\hat{Z}}$, where Δt is a parameter that can be controlled by controlling the strength of the interaction between the external field applied and that single qubit, and $\hat{Z}$ is the Pauli-Z operator; this parameter corresponds to the time degree of freedom in the evolution of the system. QC 12.9 shows the quantum circuit representation:

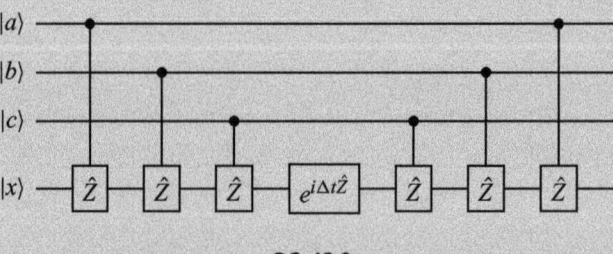

QC 12.9

(e) What is the state at the end of the circuit? What does the phase term depend on? Explain what kind of phase this is.

(f) What is the Hamiltonian of the state that corresponds to the previous circuit? Describe what kind of evolution it represents.

(g) Let us add the phase gate to the CNOT combination circuit instead, so we have QC 12.10:

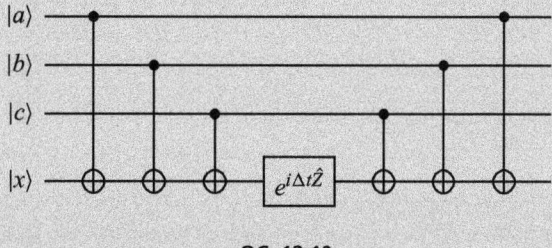

QC 12.10

What does the state look like at the end of the circuit? What does the phase term depend on? Explain what type of phase it is.

(h) Knowing that the possible values for each of a, b, and c in the previous circuit are 0 and 1, write out all the possible values for the sum $a \oplus b \oplus c \oplus$.

(i) Find the Hamiltonian that represents the last circuit. What operator represents this evolution?

Afterword: Whither Quantum Computing?

Quantum computing technology was born in the 1970s and 1980s when a handful of visionary thinkers such as Paul Benioff, Richard Feynman, and David Deutsch first speculated about how the precepts of quantum mechanics might impact computer science. In 1984 Gilles Brassard, a computer scientist and cryptographer, and Charles Bennett, a specialist in physics and information theory, devised a practical application for quantum mechanics in the field of secure communication. The following year David Deutsch showed that some computational tasks can be done better using quantum mechanics. These early developments culminated in 1995 – the *annus mirabilis* of quantum computing – when Peter Shor published his famous algorithm for factor-finding; Ignacio Cirac and Peter Zoller together proposed a plausible architecture for building a quantum computer using trapped ions; and the practicality of quantum information processing was demonstrated in an experiment carried out by David Wineland and co-workers. Since then, this new science has seen rapid development. By the early years of the new millennium, experiments involving two or three qubits were at the cutting edge; by the mid-teens of the twenty-first century, apparatus with a dozen or so qubits were being demonstrated. Today, devices with registers of 100 or more qubits on which dozens of quantum gates can be performed with low (but not zero) chance of error have been demonstrated by corporations such as Google and IBM (both using superconducting qubits) or by Quantinuum or IonQ using trapped ions; and many other start-up companies, as well research institutions and university laboratories, are in hot pursuit with a variety of technologies. Very loosely speaking, capabilities seem to be improving by an order of magnitude every decade.

Nearly 30 years after its discovery, Shor's algorithm arguably remains the *killer app* for quantum computing, combining as it does an exponential speed-up over classical computation with a compelling and impactful application. A recent study by Craig Gidney and Martin Ekerå, using realistic noise modelling and a variety of ingenious algorithmic tricks, suggests that for a quantum computer to implement Shor's algorithm to factor a 2048-bit (617-digit) integer in 8 hours would require approximately 2×10^7 qubits: five orders of magnitude more than are available today. In terms of mountaineering, if our goal were to conquer Mount Everest (8849 metres above sea-level), we have so far reached a height of only about 10 centimetres. If the trend continues (and enthusiasm and resources are sustained), we might expect to have such a device sometime around 2075. Of course, some as yet undreamed of advance or discovery may confound such a pessimistic forecast: The desirability of such an advance was part of our rationale for discussing paradigm shifts in quantum computing technologies in Chapter 10 of this book. *Treat this as a challenge, gentle reader!*

Solutions to Select Exercises

Chapter 1

1.1 Exercises

1.

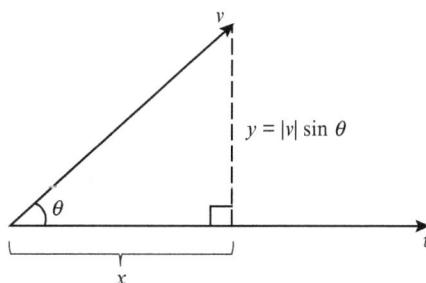

Figure S.1 Exercise 1.1.1.

(a) $\cos\theta = \frac{x}{|v|}$ gives $x = |v|\cos\theta$. Therefore, $\text{Proj}_u v = |v|\cos\theta\,\hat{\mathbf{u}}$.

(b) $|v|\hat{\mathbf{v}} - |v|\cos\theta\,\hat{\mathbf{u}} = |v|\sin\theta\,\hat{\mathbf{u}}_\perp$.

(c) $|v|\sin\theta\,\hat{\mathbf{u}}_\perp.\hat{\mathbf{u}} = 0$ (since $\hat{\mathbf{u}}_\perp.\hat{\mathbf{u}} = 0$). Therefore, the projection of $\hat{\mathbf{v}}$ on $\hat{\mathbf{u}}$ subtracted from $\hat{\mathbf{v}}$ gives a vector that is perpendicular to $\hat{\mathbf{v}}$.

3. (a) $\mathbf{w}_1 = \mathbf{v}_1 = \mathbf{e}_x$, $\mathbf{w}_2 = \mathbf{v}_2 - |\mathbf{v}_2|\cos(\pi/4)\mathbf{e}_y = 1/\sqrt{2}\mathbf{e}_y$.

(b) $\mathbf{w}_1 = \mathbf{v}_1 = \frac{1}{\sqrt{2}}(\mathbf{e}_x - \mathbf{e}_y)$, $\mathbf{w}_2 = \mathbf{v}_2 - \mathbf{w}_1(\mathbf{w}_1.\mathbf{v}_2)/\sqrt{\mathbf{w}_1.\mathbf{w}_1} = \frac{\sqrt{3}}{2}(\mathbf{e}_x + \mathbf{e}_y)$.

(c) $\mathbf{w}_1 = \mathbf{v}_1 = \frac{1}{\sqrt{2}}(\mathbf{e}_x - \mathbf{e}_y)$, $\mathbf{w}_2 = \mathbf{v}_2 - \mathbf{w}_1(\mathbf{w}_1.\mathbf{v}_2)/\sqrt{\mathbf{w}_1.\mathbf{w}_1} = \mathbf{v}_2 = \frac{1}{\sqrt{2}}(\mathbf{e}_x + \mathbf{e}_y)$.

5. (a) $\begin{bmatrix} 1 \\ 0 \end{bmatrix}$. (b) $\begin{bmatrix} 0 \\ 1 \end{bmatrix}$. (c) $\frac{1}{5}\begin{bmatrix} 4 \\ 3 \end{bmatrix}$. (d) $\frac{1}{\sqrt{3}}\begin{bmatrix} 1 \\ -i\sqrt{2} \end{bmatrix}$.

7. (a), (c), and (d).

9. (a) $|\phi\rangle$ is normalized already, so we set $|v_1\rangle = |\phi\rangle$. $|v_2\rangle = (\hat{I} - |v_1\rangle\langle v_1|)|\psi\rangle/N$, where N is the normalization constant. This gives $|v_2\rangle = \frac{1}{\sqrt{2}}(|H\rangle + |T\rangle)$.

(b) $|v_1\rangle = |\phi\rangle = |T\rangle$. $|v_2\rangle = (\hat{I} - |v_1\rangle\langle v_1|)|\psi\rangle/N$, which gives $|v_2\rangle = |H\rangle$.

11. $|+\rangle = \frac{1}{\sqrt{2}}(|H\rangle + |T\rangle)$ and $|-\rangle = \frac{1}{\sqrt{2}}(|H\rangle - |T\rangle)$ give $|H\rangle = \frac{1}{\sqrt{2}}(|+\rangle + |-\rangle)$ and $|T\rangle = \frac{1}{\sqrt{2}}(|+\rangle - |-\rangle)$. This means that $\frac{1}{\sqrt{2}}(|HH\rangle + |TT\rangle) = \frac{1}{\sqrt{2}}\left(\frac{|+\rangle+|-\rangle}{\sqrt{2}} \otimes \frac{|+\rangle+|-\rangle}{\sqrt{2}} + \frac{|+\rangle-|-\rangle}{\sqrt{2}} \otimes \frac{|+\rangle-|-\rangle}{\sqrt{2}}\right) = \frac{1}{2\sqrt{2}}(|++\rangle + |+-\rangle + |-+\rangle + |--\rangle + |++\rangle - |+-\rangle - |-+\rangle + |--\rangle) = \frac{1}{\sqrt{2}}(|++\rangle + |--\rangle)$. ∎

13. The normalization constant N has the property that $N^2 = 2^2 + 5^2$. The state $|\psi\rangle$ after normalization would be $\frac{1}{N}|\psi\rangle = \frac{2}{\sqrt{29}}|H\rangle + \frac{5}{\sqrt{29}}|T\rangle$. Prob($|H\rangle$) = $\frac{4}{29}$, Prob($|T\rangle$) = $\frac{25}{29}$.

15. $\langle\phi|\psi\rangle = 1/\sqrt{2}$.

17. $|\varphi\rangle = \frac{1}{\sqrt{3}}(\sqrt{2}|HH\rangle + |TT\rangle)$.

19. Nothing can be said about the second coin in this case.

1.2 Exercises

1. Possible digits in the binary number system = 0, 1; i.e., two possibilities. $\therefore$ $2 \times 2 \times 2 = 8$ is the number of whole numbers that can be represented by three digits. These numbers with their binary representation are $0 \to 000$, $1 \to 001$, $2 \to 010$, $3 \to 011$, $4 \to 100$, $5 \to 101$, $6 \to 110$, $7 \to 111$.

3. 19 characters, each = 1 byte = 8 bits. $\therefore$ #bytes = 19 and #bits = $19 \times 8 = 152$.

5. A general two-(quantum coin) state is given by $a|HH\rangle + b|HT\rangle + c|TH\rangle + d|TT\rangle$, where $|a|^2 + |b|^2 + |c|^2 + |d|^2 = 1$. Since the coefficients are, in general, complex, they can be expressed in terms of their real and imaginary parts as follows: $a = a_1 + ia_2$, $b = b_1 + ib_2$, $c = c_1 + ic_2$, $d = d_1 + id_2$. From this it looks like there are eight parameters: $a_1, a_2, b_1, b_2, c_1, c_2, d_1, d_2$. However, normalization implies that $a_1^2 + a_2^2 + b_1^2 + b_2^2 + c_1^2 + c_2^2 + d_1^2 + d_2^2 = 1$. Using this, we can get rid of the dependency of one of the parameters, say a_1. Then we can use the global phase freedom to get rid of another dependency; without loss of generality, we can do so by setting $a_2 = 0$. Therefore, six real parameters specify the state.

7. 100 times on 100 qubits, once on a general superposition state, 100 times on a non-superposition state.

Chapter 1 Exercises

1. $\mathbf{e}_1 = \mathbf{a}/\sqrt{\mathbf{a}.\mathbf{a}}$, $\mathbf{e}_2 = (\mathbf{b} - \mathbf{e}_1(\mathbf{e}_1.\mathbf{b}))/\sqrt{\mathbf{b}.\mathbf{b} - (\mathbf{e}_1.\mathbf{b})^2}$, $\mathbf{e}_3 = \frac{\mathbf{c} - \mathbf{e}_1(\mathbf{e}_1.\mathbf{c}) - \mathbf{e}_2(\mathbf{e}_2.\mathbf{c})}{\sqrt{\mathbf{c}.\mathbf{c} - (\mathbf{e}_1.\mathbf{c})^2 - (\mathbf{e}_2.\mathbf{c})^2}}$.

3. (a) $|\psi\rangle \to \begin{bmatrix} a \\ b \end{bmatrix}$, $|\phi\rangle \to \begin{bmatrix} c \\ d \end{bmatrix}$.

   (b) $\langle\psi|\psi\rangle = \begin{bmatrix} a^* & b^* \end{bmatrix} \begin{bmatrix} a \\ b \end{bmatrix} = |a|^2 + |b|^2$. Similarly, $\langle\phi|\phi\rangle = |c|^2 + |d|^2$, $\langle\psi|\phi\rangle = a^*c + b^*d$, and $\langle\phi|\psi\rangle = c^*a + d^*b$.

   (c) $|\psi\rangle\langle\psi| = \begin{bmatrix} a \\ b \end{bmatrix} \begin{bmatrix} a^* & b^* \end{bmatrix} = \begin{bmatrix} |a|^2 & ab^* \\ ba^* & |b|^2 \end{bmatrix}$. Similarly, $|\phi\rangle\langle\phi| = \begin{bmatrix} |c|^2 & cd^* \\ dc^* & |d|^2 \end{bmatrix}$, $|\psi\rangle\langle\phi| = \begin{bmatrix} ac^* & ad^* \\ bc^* & bd^* \end{bmatrix}$, and $|\phi\rangle\langle\psi| = \begin{bmatrix} ca^* & cb^* \\ da^* & db^* \end{bmatrix}$.

5. If we apply $\hat{X}$ on the two states: $|H\rangle = \begin{bmatrix} 1 \\ 0 \end{bmatrix}$ and $|T\rangle = \begin{bmatrix} 0 \\ 1 \end{bmatrix}$, we find that $\hat{X}|H\rangle = |T\rangle$ and $\hat{X}|T\rangle = |H\rangle$. This means that $\hat{X}$ flips one state into the other. A two-(quantum coin) state is given most generally by $a|HH\rangle + b|HT\rangle + c|TH\rangle + d|TT\rangle$. We need to find the matrix representation that renders that state to $a|TT\rangle + b|TH\rangle + c|HT\rangle + d|HH\rangle = d|HH\rangle + c|HT\rangle + b|TH\rangle + a|TT\rangle$.

If we write the two-(quantum coin) state in matrix form before, $\begin{bmatrix} a \\ b \\ c \\ d \end{bmatrix}$, and after, $\begin{bmatrix} d \\ c \\ b \\ a \end{bmatrix}$, we find

that the following matrix transforms the first into the second: $\begin{bmatrix} 0 & 0 & 0 & 1 \\ 0 & 0 & 1 & 0 \\ 0 & 1 & 0 & 0 \\ 1 & 0 & 0 & 0 \end{bmatrix}$, which means it

is the matrix representation for $\hat{X} \otimes \hat{X}$.

7. $1000000000 = 2^9 = 512$.

9. Each character is represented by 1 byte $= 8$ bits. $\therefore$ we need $100 \times 8 = 800$ bits. Since $2^6 < 100 < 2^7$, we only need seven quantum coins for the same purpose.

11. (a) $p_H = 0.5$.
    (b) The binomial distribution is given by $p(x) = \binom{n}{x} p^x q^{n-x}$, where p and q correspond to the probability of success and failure, respectively, and x corresponds to the number of trials. In this exercise, $p = q = 1/2$.

    For $n = 2$, $p(0) = \binom{2}{0}\left(\frac{1}{2}\right)^2 = \frac{1}{4}$, $p(1) = \binom{2}{1}\left(\frac{1}{2}\right)^1\left(\frac{1}{2}\right)^1 = \frac{1}{2}$, and $p(2) = \binom{2}{2}\left(\frac{1}{2}\right)^2\left(\frac{1}{2}\right)^0 = \frac{1}{4}$.

    (c) Similarly, for $n = 3$, $p(0) = \frac{1}{8}$, $p(1) = \frac{3}{8}$, $p(2) = \frac{3}{8}$, $p(3) = \frac{1}{8}$.
    (d) In this case, the two-(quantum coin) state is $\frac{1}{2}(\underbrace{|HH\rangle}_{p(2)} + \underbrace{|HT\rangle}_{p(1)} + \underbrace{|TH\rangle}_{p(1)} + \underbrace{|TT\rangle}_{p(0)})$. Using the coefficients that multiply each ket, $p(0) = \frac{1}{4}$, $p(1) = \frac{1}{4} + \frac{1}{4} = \frac{1}{2}$, and $p(0) = \frac{1}{2}$. These are the same numbers seen in in the two classical coins case in part (b).

    (e) Here, the state is $\frac{1}{2\sqrt{2}}(|HHH\rangle + |HHT\rangle + |HTH\rangle + |HTT\rangle + |THH\rangle + |THT\rangle + |TTH\rangle + |TTT\rangle)$. Using the same analysis as in part (d), $p(0) = \frac{1}{8}$, $p(1) = \frac{3}{8}$, $p(2) = \frac{3}{8}$, $p(3) = \frac{1}{8}$. These are the same results as in the three classical coins case in part (c).

Chapter 2

2.1 Exercises

1. (a) 7, (b) 1, (c) $\begin{bmatrix} 3 & 6 \\ 9 & 12 \end{bmatrix}$, (d) $\begin{bmatrix} -4 & -4 \\ -4 & -4 \end{bmatrix}$, (e) $\begin{bmatrix} 19 & 22 \\ 43 & 50 \end{bmatrix}$, (f) $\begin{bmatrix} 23 & 34 \\ 31 & 46 \end{bmatrix}$, (g) $\left[(-i + 6)(-2i + 12)\right]$,

(h) $\begin{bmatrix} 5i + 18 \\ 7i + 24 \end{bmatrix}$, (i) 10, (j) $\begin{bmatrix} 1 & 3i \\ -3i & 9 \end{bmatrix}$, (k) $\begin{bmatrix} i \\ 3 \end{bmatrix}$, (l) $\begin{bmatrix} 1 & 2 \\ 3 & 4 \end{bmatrix}$, (m) $\frac{-1}{2}\begin{bmatrix} 4 & -2 \\ -3 & 1 \end{bmatrix}$.

3. (a) For rotation angle $\pi/4 = 45°$, the rotation matrix is $\frac{1}{\sqrt{2}} \begin{bmatrix} 1 & -1 \\ 1 & 1 \end{bmatrix}$. Applying that on the

given state gives $\begin{bmatrix} 0 \\ \sqrt{2} \end{bmatrix}$.

(b) Initial length − final length = $(1^2 + 1^2) - (0^2 + (\sqrt{2})^2) = 0$.

(c) $R^{-1}(\theta) = \begin{bmatrix} \cos\theta & \sin\theta \\ -\sin\theta & \cos\theta \end{bmatrix} = \frac{1}{\sqrt{2}} \begin{bmatrix} 1 & 1 \\ -1 & 1 \end{bmatrix}$.

5. (a) Here, the matrix $\hat{M}$ is equal to $\hat{M}^T$, which means $\begin{bmatrix} a & b \\ c & d \end{bmatrix} = \begin{bmatrix} a & c \\ b & d \end{bmatrix} \implies b = c \implies \hat{M} =$

$\begin{bmatrix} a & b \\ b & d \end{bmatrix}$.

(b) $\hat{M} = \hat{M}^\dagger \implies \begin{bmatrix} a & b \\ c & d \end{bmatrix} = \begin{bmatrix} a^* & c^* \\ b^* & d^* \end{bmatrix} \implies a, d \in \mathbb{R} \ \& \ b = c^* \implies \hat{M} = \begin{bmatrix} a & b \\ b^* & d \end{bmatrix}$. Note $\mathbb{R}$

represents the set of *real* numbers.

(c) See Section 2.1.1.

(d) See Section 2.1.1.

7. We need to show that $\hat{\mathscr{H}}|\mu_-\rangle = \mu_-|\mu_-\rangle$, where $|\mu_-\rangle = \frac{1}{\sqrt{2R(R-S)}} \begin{bmatrix} R - S \\ -2b^* \end{bmatrix}$, and $\mu_- =$

$(T - R)/2$, where $S = a - d$, $T = a + d$, and $R = \sqrt{S^2 + 4|b|^2}$, as given in the text. First,
apply $\hat{\mathscr{H}}$ on the column matrix part of $|\mu_-\rangle$, without the $\frac{1}{\sqrt{}}$ term. You will be able to show

that $\begin{bmatrix} a & b \\ b^* & d \end{bmatrix} \begin{bmatrix} R - S \\ -2b^* \end{bmatrix} = \begin{bmatrix} a(R - S) - 2|b|^2 \\ (R - S)b^* - 2db^* \end{bmatrix}$, which with some algebraic manipulation can

be written as $\begin{bmatrix} a & b \\ b^* & d \end{bmatrix} \begin{bmatrix} R - S \\ -2b^* \end{bmatrix} = \mu_- \begin{bmatrix} R - S \\ -2b^* \end{bmatrix}$. To confirm the normalization factor, take the

inner product of the column vector with itself: $\begin{bmatrix} R - S & -2b \end{bmatrix} \begin{bmatrix} R - S \\ -2b^* \end{bmatrix} = \cdots = 2R(R - S)$.

$\therefore |\mu_-\rangle = \frac{1}{\sqrt{2R(R-S)}} \begin{bmatrix} R - S \\ -2b^* \end{bmatrix}$ is a normalized eigenvector. ∎

9. Here, we have $\hat{\Pi}_+ = \frac{-\mu_-\hat{I}+\hat{M}}{\mu_+-\mu_-}$ and $\hat{\Pi}_- = \frac{\mu_+\hat{I}-\hat{M}}{\mu_+-\mu_-}$. $\hat{\Pi}_+^2 = \frac{(-\mu_-\hat{I}+\hat{M})(-\mu_-\hat{I}+\hat{M})}{(\mu_+-\mu_-)^2} = \frac{\mu_-^2\hat{I}-2\mu_-\hat{M}+\hat{M}^2}{(\mu_+-\mu_-)^2}$.
This can be simplified further if we use this consequence of the Cayley–Hamilton formula:
$\hat{M}^2 = (\mu_- + \mu_+)\hat{M} + \mu_-\mu_+\hat{I}$. Substituting this expression for $\hat{M}^2$ back into the last step in the cal-
culations, we get $\frac{\mu_-^2\hat{I}-2\mu_-\hat{M}+(\mu_-+\mu_+)\hat{M}-\mu_+\mu_-\hat{I}}{(\mu_+-\mu_-)^2} = \frac{\mu_-(\mu_--\mu_+)\hat{I}+(\mu_+-\mu_-)\hat{M}}{(\mu_+-\mu_-)^2} = \frac{\hat{M}-\mu_-\hat{I}}{\mu_+-\mu_-} = \hat{\Pi}_+$. ∎
The Cayley–Hamilton formula can be used again to simplify the other four expressions proving
that $\hat{\Pi}_-^2 = \hat{\Pi}_-$ and $\hat{\Pi}_-\hat{\Pi}_+ = \hat{\Pi}_+\hat{\Pi}_- = 0$ for the general matrix $\hat{M}$.

11. $\hat{\mathscr{H}}$ has eigenvalues $4 + 65$ and $4 - 65$. $\lambda = 69 : \hat{\Pi}_+ = \frac{1}{10} \begin{bmatrix} 9 & 3e^{i\phi} \\ 3^{-i\phi} & 1 \end{bmatrix}$, $\lambda = -61 : \hat{\Pi}_- =$

$\frac{1}{10} \begin{bmatrix} 1 & -3e^{i\phi} \\ -3^{-i\phi} & 9 \end{bmatrix}$. $\therefore f(\hat{\mathscr{H}}) = \frac{f(69)}{10} \begin{bmatrix} 9 & 3e^{i\phi} \\ 3^{-i\phi} & 1 \end{bmatrix} + \frac{f(-61)}{10} \begin{bmatrix} 1 & -3e^{i\phi} \\ -3^{-i\phi} & 9 \end{bmatrix}$.

(a) $\hat{\mathscr{H}}^2 = \begin{bmatrix} 4657 & 312e^{i\phi} \\ 312e^{-i\phi} & 3825 \end{bmatrix}$.

(b) $\sqrt{\hat{\mathscr{H}}} = \frac{\sqrt{69}}{10} \begin{bmatrix} 9 & 3e^{i\phi} \\ 3^{-i\phi} & 1 \end{bmatrix} + \frac{i\sqrt{61}}{10} \begin{bmatrix} 1 & -3e^{i\phi} \\ -3^{-i\phi} & 9 \end{bmatrix}$.

(c) $\hat{\mathscr{H}}^{-1} = \frac{1}{69} \begin{bmatrix} 9 & 3e^{i\phi} \\ 3^{-i\phi} & 1 \end{bmatrix} - \frac{1}{61} \begin{bmatrix} 1 & -3e^{i\phi} \\ -3^{-i\phi} & 9 \end{bmatrix} = \frac{1}{4209} \begin{bmatrix} 48 & 15+36i \\ 15-36i & -56 \end{bmatrix}$.

2.2 Exercises

1. $\hat{M} = \begin{bmatrix} 0 & 1 \\ 1 & 0 \end{bmatrix} \implies \hat{M} = \begin{bmatrix} 0 & 1 \\ 1 & 0 \end{bmatrix} \implies \hat{X}\hat{X}^{\dagger} = \hat{I} = \begin{bmatrix} 1 & 0 \\ 0 & 1 \end{bmatrix}$ and $\hat{X}^{\dagger}\hat{X} = \hat{I} = \begin{bmatrix} 1 & 0 \\ 0 & 1 \end{bmatrix}$.

   Similarly, $\hat{Y} = \hat{Y}^{\dagger}$ and $\hat{Y}\hat{Y}^{\dagger} = \hat{Y}^{\dagger}\hat{Y} = \hat{I}$, and $\hat{Z} = \hat{Z}^{\dagger}$ and $\hat{Z}\hat{Z}^{\dagger} = \hat{Z}^{\dagger}\hat{Z} = \hat{I}$.

3. (a) $\hat{Z}\{a|0\rangle + b|1\rangle\} = a|0\rangle - b|1\rangle$,
      $\hat{X}\hat{Z}\{a|0\rangle + b|1\rangle\} = a|1\rangle - b|0\rangle = |\psi_1\rangle$.

   (b) $\hat{X}\{a|0\rangle + b|1\rangle\} = a|1\rangle + b|0\rangle$,
      $\hat{Z}\hat{X}\{a|0\rangle + b|1\rangle\} = -a|1\rangle + b|0\rangle = |\psi_2\rangle$.

   (c) $|\psi_1\rangle = -|\psi_2\rangle$. The minus sign does not matter because it amounts to the phase factor $e^{i\pi}$, where π is the global phase, and it will, therefore, have no effect on the probability amplitudes of the state.

5. (a) This is straightforward. Substitute in the matrix representation of $\hat{I}$ and $\hat{Z}$ on the right side of the equation, and simplify.

   (b) Substitute in the matrix representation of all the matrices on the right side of the equation. Multiply them, then use the trigonometric relations that express the sin and cos functions in terms of exponentials to simplify. You will eventually end up with this matrix:
   $\begin{bmatrix} e^{-i(\phi+\chi)/2}\cos\theta/2 & -ie^{-i(\phi-\chi)/2}\sin\theta/2 \\ -ie^{i(\phi-\chi)/2} & e^{i(\phi+\chi)/2}\cos\theta/2 \end{bmatrix}$. Analyzing this matrix will reveal that it has the

   form $\begin{bmatrix} a & b \\ -b^* & a \end{bmatrix}$, which is the form of a general unitary matrix, as given in the section.

7. If we multiply the matrices in (a) and (b), each by their Hermitian adjoint, we find that the product

   equals the identity matrix $\hat{I} = \begin{bmatrix} 1 & 0 \\ 0 & 1 \end{bmatrix}$, so in these two cases, yes, the given matrices are unitary.

   On the other hand, when we mulitiply the matrix in (c) by its Hermitian adjoint, we end up with
   $\begin{bmatrix} 1 & 2i\cos\theta\sin\theta \\ -2i\cos\theta\sin\theta & 1 \end{bmatrix} \neq \hat{I}$. Therefore, the matrix in (c) is not unitary.

   Multiplying the expression in (d) with its Hermitian adjoint yields $\hat{I} + 2\cos\theta\sin\theta\hat{X} \neq \hat{I}$. Therefore, it's not unitary either.

9. (a) $|0'\rangle = \frac{1}{\sqrt{2}}|0\rangle + \frac{1}{\sqrt{2}}|1\rangle \rightarrow \frac{1}{\sqrt{2}}\begin{bmatrix} 1 \\ 1 \end{bmatrix}$, $|1'\rangle = \frac{1}{\sqrt{2}}|0\rangle - \frac{1}{\sqrt{2}}|1\rangle \rightarrow \frac{1}{\sqrt{2}}\begin{bmatrix} 1 \\ -1 \end{bmatrix}$. $|b_1\rangle = \hat{U}|0'\rangle =$

   $\frac{1}{\sqrt{2}}\begin{bmatrix} a & b \\ -b^* & a^* \end{bmatrix}\begin{bmatrix} 1 \\ 1 \end{bmatrix} = \frac{1}{\sqrt{2}}\begin{bmatrix} a+b \\ -b^*+a^* \end{bmatrix} = \frac{1}{\sqrt{2}}\{(a+b)|0\rangle + (a^*-b^*)|1\rangle\}$, $|b_2\rangle = \hat{U}|1'\rangle =$

   $\frac{1}{\sqrt{2}}\begin{bmatrix} a & b \\ -b^* & a^* \end{bmatrix}\begin{bmatrix} 1 \\ -1 \end{bmatrix} = \frac{1}{\sqrt{2}}\begin{bmatrix} a-b \\ -b^*-a^* \end{bmatrix} = \frac{1}{\sqrt{2}}\{(a-b)|0\rangle - (a^*+b^*)|1\rangle\}$.

(b) $|0'\rangle = \frac{1}{\sqrt{2}}|0\rangle + \frac{1}{\sqrt{2}}|1\rangle \rightarrow \frac{1}{\sqrt{2}}\begin{bmatrix} 1 \\ 1 \end{bmatrix}$, $|1'\rangle = \frac{1}{\sqrt{2}}|0\rangle + \frac{i}{\sqrt{2}}|1\rangle \rightarrow \frac{1}{\sqrt{2}}\begin{bmatrix} 1 \\ i \end{bmatrix}$. $|b_1\rangle = \hat{U}|0'\rangle =$

$\frac{1}{\sqrt{2}}\begin{bmatrix} a & b \\ -b^* & a^* \end{bmatrix}\begin{bmatrix} 1 \\ 1 \end{bmatrix} = \frac{1}{\sqrt{2}}\begin{bmatrix} a+b \\ -b^*+a^* \end{bmatrix} = \frac{1}{\sqrt{2}}\{(a+b)|0\rangle + (a^*-b^*)|1\rangle)\}$, $|b_2\rangle = \hat{U}|1'\rangle =$

$\frac{1}{\sqrt{2}}\begin{bmatrix} a & b \\ -b^* & a^* \end{bmatrix}\begin{bmatrix} 1 \\ i \end{bmatrix} = \frac{1}{\sqrt{2}}\begin{bmatrix} a+ib \\ -b^*+ia^* \end{bmatrix} = \frac{1}{\sqrt{2}}\{(a+ib)|0\rangle + (-b^*+ia^*)|1\rangle)\}$.

(c) Unitary transformations will preserve the inner product between two vectors, both of which have been transformed. Thus, $\langle\phi|\psi\rangle = \langle\phi'|\psi'\rangle$. This includes the norm $\langle\phi|\phi\rangle = \langle\phi'|\phi'\rangle$. Unitary transforms do NOT preserve the component of a vector with respect to a fixed basis; thus, $\langle 0|\psi'\rangle \neq \langle 0|\psi\rangle$.

2.3 Exercises

1. Let $\{v_i\}$ and $\{|v_i\rangle\}$ be the eigenvalues and eigenvectors of $\hat{M}$, respectively.

(a) $\hat{M}|v_i\rangle = v_i|v_i\rangle \implies \langle v_i|\hat{M}|v_i\rangle = v_i\langle v_i|v_i\rangle$. $\therefore \langle v_i|\hat{M}^\dagger = \langle v_i|v_i^* \implies \langle v_i|\hat{M}^\dagger|v_i\rangle = v_i^*\langle v_i|v_i\rangle$, $\hat{M}^\dagger = \hat{M} \implies v_i = v_i^* \implies \{v_i\} \in \mathbb{R}$.

(b) $\hat{M}|v_i\rangle = v_i|v_i\rangle$, $\hat{M}|v_j\rangle = v_j|v_j\rangle \implies \langle v_j|\hat{M}\dagger = v_j^*\langle v_j|$, and using $\hat{M} = \hat{M}^\dagger$ the results from (a), $\implies \langle v_j|\hat{M} = v_j\langle v_j|$. $\langle v_j|\hat{M}|v_i\rangle = v_i\langle v_j|v_i\rangle = v_j\langle v_j|v_i\rangle$, $\therefore (v_i - v_j)\langle v_j|v_i\rangle = 0$. Either $v_i - v_j = 0$ or $\langle v_j|v_i\rangle = 0$ (orthogonality). If $v_i = v_j$, one can use Gram–Schmidt to orthogonalize the vectors.

3. (a) $\lambda_0 = \hbar\omega_0$ $|v_1\rangle = \begin{bmatrix} 1 \\ 0 \end{bmatrix} \implies \begin{bmatrix} 1 \\ 0 \end{bmatrix}\begin{bmatrix} 1 & 0 \end{bmatrix} = \begin{bmatrix} 1 & 0 \\ 0 & 0 \end{bmatrix}$, $\lambda_1 = \hbar\omega_1$ $|v_2\rangle = \begin{bmatrix} 0 \\ 1 \end{bmatrix} \implies$

$\begin{bmatrix} 0 \\ 1 \end{bmatrix}\begin{bmatrix} 0 & 1 \end{bmatrix} = \begin{bmatrix} 0 & 0 \\ 0 & 1 \end{bmatrix}$

$e^{-i\mathcal{H}t/\hbar} \rightarrow e^{-i\omega_0 t}\begin{bmatrix} 1 & 0 \\ 0 & 0 \end{bmatrix} + e^{-i\omega_1 t}\begin{bmatrix} 0 & 0 \\ 0 & 1 \end{bmatrix} = \begin{bmatrix} e^{-i\omega_0 t} & 0 \\ 0 & e^{-i\omega_1 t} \end{bmatrix}$

$|\psi(t)\rangle = \begin{bmatrix} e^{-i\omega_0 t} \\ 0 \end{bmatrix} = e^{-i\omega_0 t}|0\rangle$.

(b) $|\psi(t)\rangle = \frac{1}{\sqrt{2}}\begin{bmatrix} e^{-i\omega_0 t} & 0 \\ 0 & e^{-i\omega_1 t} \end{bmatrix}\begin{bmatrix} 1 \\ 1 \end{bmatrix} = \frac{1}{\sqrt{2}}\{e^{-i\omega_0 t}|0\rangle + e^{-i\omega_1 t}|1\rangle)\}$.

(c) Here, using these tips: (1) the general form for finding the function of a matrix $\hat{M}$, (2) the identity matrix can be expressed in terms of the complete set of eigenvectors as the sum of their outer products (for a 2×2 matrix with eigenvectors $|m_1\rangle$ and $|m_2\rangle$, $\hat{I} = |m_1\rangle\langle m_1| + |m_2\rangle\langle m_2|$), and (3) the eigenvalues of a 2×2 matrix with trace $= 0$, as is the case in this exercise, are negatives of each other (so if one is m, the other is $-m$), this general formula can be derived: $f(\hat{M}) = \hat{I}\left(\frac{f(m)+f(-m)}{2}\right) + \hat{M}\left(\frac{f(m)-f(-m)}{2m}\right)$. The eigenvalues for the Hamiltonian $\mathcal{H}$ here are $\pm\sqrt{\omega^2 + \Omega^2}\hbar/2$, $\therefore e^{-i\mathcal{H}t/\hbar} = \hat{I}\cos(\sqrt{\omega^2+\Omega^2}t/2) - i\frac{\sin(\sqrt{\omega^2+\Omega^2}t/2)}{\sqrt{\omega^2+\Omega^2}}\begin{bmatrix} -\omega & \Omega \\ \Omega & \omega \end{bmatrix}$. Applying this on the initial state $|\psi(0)\rangle = 0$, gives $|\psi(t)\rangle = \cos(\sqrt{\omega^2+\Omega^2}t/2)|0\rangle - i\frac{\sin(\sqrt{\omega^2+\Omega^2}t/2)}{\sqrt{\omega^2+\Omega^2}}\{-\omega|0\rangle + \Omega|1\rangle)\}$.

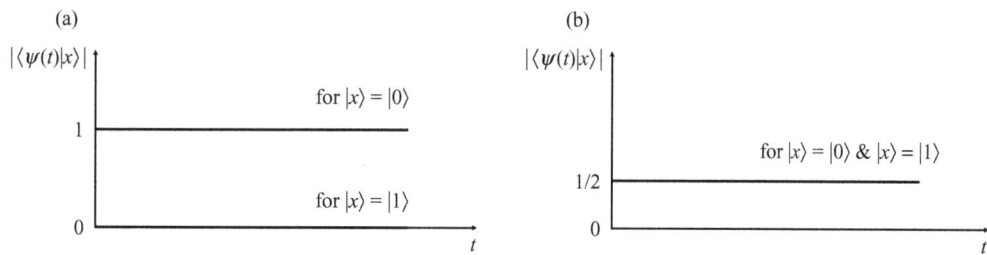

Figure S.2 Exercise 2.3.3 Probability amplitudes.

2.4 Exercises

1. False.

3. To the left.

5. (a) $\hat{D}\hat{C}\hat{B}\hat{A}|0\rangle$.

(b)

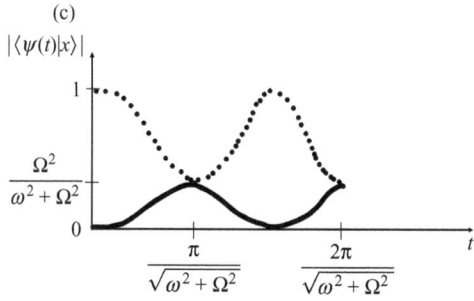

QC S.1 Exercise 2.4.5(b).

7. $\hat{X}\hat{Y}\hat{Z} = \begin{bmatrix} 0 & 1 \\ 1 & 0 \end{bmatrix} \begin{bmatrix} 0 & -i \\ i & 0 \end{bmatrix} \begin{bmatrix} 1 & 0 \\ 0 & -1 \end{bmatrix} = \begin{bmatrix} i & 0 \\ 0 & i \end{bmatrix} = i\hat{I}$

$$\boxed{\hat{Z}}-\boxed{\hat{Y}}-\boxed{\hat{X}} \quad = \quad \rule{2cm}{0.4pt}$$

QC S.2 Exercise 2.4.7.

Note that we ignored the global phase represented by $i = e^{i\pi/2}$.

2.5 Exercises

1. "1".

3. $\hat{\Omega}' = \hat{H}\hat{\Omega}\hat{H} = \frac{1}{\sqrt{2}} \begin{bmatrix} 1 & 1 \\ 1 & -1 \end{bmatrix} \begin{bmatrix} 0 & 0 \\ 0 & 1 \end{bmatrix} \frac{1}{\sqrt{2}} \begin{bmatrix} 1 & 1 \\ 1 & -1 \end{bmatrix} = \frac{1}{2} \begin{bmatrix} 0 & 1 \\ 0 & -1 \end{bmatrix} \begin{bmatrix} 1 & 1 \\ 1 & -1 \end{bmatrix} = \frac{1}{2} \begin{bmatrix} 1 & -1 \\ -1 & 1 \end{bmatrix} =$

$\frac{1}{2}(\hat{I} - \hat{X})$.

$\hat{\Omega}|\varphi\rangle = \lambda|\varphi\rangle \implies (\hat{H}\hat{\Omega}\hat{H})\hat{H}|\varphi\rangle = \lambda|\varphi\rangle \implies \hat{\Omega}'|\varphi'\rangle = \lambda|\varphi'\rangle, |\varphi'\rangle = \hat{H}|\varphi\rangle$.

$$\lambda = 1 \quad |\varphi_1'\rangle = \tfrac{1}{\sqrt{2}}(|0\rangle - |1\rangle),$$
$$\lambda = 0 \quad |\varphi_0'\rangle = \tfrac{1}{\sqrt{2}}(|0\rangle + |1\rangle).$$

(a) Possible outcomes: 0 or 1.

(b) Probabilities: $p_0 = |\langle \varphi_0'|\psi\rangle|^2 = |\tfrac{1}{\sqrt{2}}(0.8+0.6)|^2 = 0.98, p_1 = |\langle \varphi_1'|\psi\rangle|^2 = |\tfrac{1}{\sqrt{2}}(0.8-0.6)|^2 = 0.02$.

(c) $|\varphi_0'\rangle = \tfrac{1}{\sqrt{2}}(|0\rangle + |1\rangle), \quad |\varphi_1'\rangle = \tfrac{1}{\sqrt{2}}(|0\rangle - |1\rangle)$.

(d) $\overline{\Omega} = p_0\lambda_0 + p_1\lambda_1 = p_0(0) + p_1(1) = p_1 = 0.02$.

2.6 Exercises

1. (a) $\hat{\rho}_1 = |\psi_1\rangle\langle\psi_1| = \begin{bmatrix} 1 & 0 \\ 0 & 0 \end{bmatrix}, \hat{\rho}_2 = |\psi_2\rangle\langle\psi_2| = \tfrac{1}{2}\begin{bmatrix} 1 & -i \\ i & 1 \end{bmatrix}, \hat{\rho}_3 = |\psi_3\rangle\langle\psi_3| = \begin{bmatrix} |\alpha|^2 & \alpha\beta^* \\ \alpha^*\beta & |\beta|^2 \end{bmatrix}.$

3. The density matrices in (a) and (d) are pure, and have one pure-state decomposition, defined by the kets in Exercise 2. On the other hand, the other density matrices have more than one possible pure-state decomposition, as they are mixed.

5. A general density matrix $\hat{\rho}$ can be written as $\hat{\rho} = \tfrac{1}{2}(\hat{I} + \overline{X}\hat{X} + \overline{Y}\hat{Y} + \overline{Z}\hat{Z})$.

(a) $\hat{\rho} = |+\rangle\langle+| = \tfrac{1}{2}\begin{bmatrix} 1 & 1 \\ 1 & 1 \end{bmatrix} = \tfrac{1}{2}(\hat{I} + \hat{X})$. This implies that $\overline{X} = 1, \overline{Y} = 0, \overline{Z} = 0$.

(b) Here $\hat{\rho} = \tfrac{1}{2}\hat{I} \therefore \overline{X} = 0, \quad \overline{Y} = 0, \quad \overline{Z} = 0$.

(c) Here $\hat{\rho} = \tfrac{1}{4}(2\hat{I} - \hat{Y}) \therefore \overline{X} = 0, \quad \overline{Y} = -1/2, \quad \overline{Z} = 0$.

Note that the averages of the Pauli matrices are components of the Bloch vector.

Chapter 2 Exercises

1. Let $\{|u_n\rangle\}$ and $\{|v_n\rangle\}$ be two orthonormal bases. $\hat{U} = \sum_n |v_n\rangle\langle u_n|$ is the unitary that transforms one basis to the other.

Let $|u_1\rangle = (4|0\rangle + 3i|1\rangle)/5, |u_2\rangle = (3i|0\rangle + 4|1\rangle)/5 \implies \langle u_1|u_2\rangle = 0$.

$|v_1\rangle = (|0\rangle - |1\rangle)/\sqrt{2}, |v_2\rangle = (|0\rangle + |1\rangle)/\sqrt{2} \implies \langle v_1|v_2\rangle = 0$.

$\hat{U} = |v_1\rangle\langle u_1| + |v_2\rangle\langle u_2| = \tfrac{1}{5\sqrt{2}}\left\{ \begin{bmatrix} 1 \\ -1 \end{bmatrix}\begin{bmatrix} 4 & -3i \end{bmatrix} + \begin{bmatrix} 1 \\ 1 \end{bmatrix}\begin{bmatrix} -3i & 4 \end{bmatrix} = \tfrac{1}{5\sqrt{2}}\begin{bmatrix} 4-3i & 4-3i \\ -4-3i & 4+3i \end{bmatrix} \right\}.$

3. (a) Changes relative phase of α and β.

(b) $\hat{T} = e^{i\pi/8}\begin{bmatrix} e^{-i\pi/8} & 0 \\ 0 & e^{i\pi/8} \end{bmatrix}$, called a $\pi/8$ gate since that's the amount it shifts phases by.

(c) Writing $e^{i\theta} = \cos\theta + \sin\theta$, $\hat{T} = \underbrace{e^{i\pi/8}}_{\text{global phase}}\begin{bmatrix} \cos(\pi/8) - i\sin\pi/8 & 0 \\ 0 & \cos(\pi/8) + i\sin\pi/8 \end{bmatrix}$

$= e^{i\pi/8}\{\cos(\pi/8)\hat{I} - i\sin(\pi/8)\hat{Z}\}$.

5. "Equivalent" implies $|\psi\rangle = e^{i\phi}|\chi\rangle$. $|\psi_1\rangle, |\psi_3\rangle$ are equivalent. $|\psi_2\rangle, |\psi_4\rangle, |\psi_5\rangle$ are equivalent.

7. $\hat{U} = \exp(-i\hat{\mathcal{H}}t/\hbar) = \exp\left[\frac{-i}{2}\underbrace{\begin{bmatrix} -\omega & \Omega \\ \Omega & \omega \end{bmatrix}}_{\hat{M}}t\right]$,

$\hat{M}^2 = \begin{bmatrix} -\omega & \Omega \\ \Omega & \omega \end{bmatrix}\begin{bmatrix} -\omega & \Omega \\ \Omega & \omega \end{bmatrix} = \begin{bmatrix} \omega^2 + \Omega^2 & 0 \\ 0 & \omega^2 + \Omega^2 \end{bmatrix} = (\omega^2 + \Omega^2)\hat{I} \therefore \hat{M}^{2m} = (\omega^2 + \Omega^2)^m\hat{I}, \hat{M}^{2m+1} =$

$(\omega^2 + \Omega^2)^m\hat{M} \therefore \exp\left[\frac{-i}{2}t\hat{M}\right] = \hat{I} + \sum_{n=1}^{\infty}\frac{1}{n!}(\frac{-it}{2})^n\hat{M}^n = \hat{I} + \sum_{m=1}^{\infty}\frac{1}{(2m)!}(\frac{-it}{2})^{2m}(\omega^2 + \Omega^2)^m\hat{I} +$

$\sum_{m=0}^{\infty}\frac{1}{(2m+1)!}(\frac{-it}{2})^{2m+1}(\omega^2 + \Omega^2)^{m+1/2}\frac{\hat{M}}{\sqrt{\omega^2+\Omega^2}}$,

$\hat{U}(t) = \cos(\sqrt{\omega^2 + \Omega^2}t/2)\hat{I} - i\sin(\sqrt{\omega^2 + \Omega^2}t/2)\frac{\hat{M}}{\sqrt{\omega^2+\Omega^2}}$.

9. (a) Yes, (b) not Hermitian $\implies$ No, (c) trace $= 0$, diagonal elements < 0 $\implies$ No, (d) determinant $< 0 \implies$ eigenvalues $< 0 \implies$ No.

11.

QC S.3 Chapter Exercise 2.11.

13. $(5 - \mu)(4 - \mu) - 2 = 0 \implies \mu^2 - 9\mu + 18 = 0 \implies \mu = \frac{9\pm\sqrt{81-4(18)}}{2} = 6$ and $3 \implies \mu_1 = 3, \mu_2 = 6. f(\hat{M}) = f(\mu_1)|r_1\rangle\langle l_1| + f(\mu_2)|r_2\rangle\langle l_2|. |r_i\rangle =$ right eigenvectors, $|l_i\rangle =$ left eigenvectors, normalized such that $\langle r_i|l_i\rangle = \delta_{ij}$. Here, $\sin(\frac{\pi}{3}x) \implies f(\mu_1) = \sin\pi = 0$, $f(\mu_2) = \sin 2\pi = 0 \therefore f(\hat{M}) = 0$.

15. $\frac{d\hat{\rho}}{dt} = \sum_i p_i\left[\left(\frac{d|\psi_i\rangle}{dt}\right)\langle\psi_i| + |\psi_i\rangle\frac{d}{dt}\langle\psi_i|\right] = \sum_i p_i\left[\left(\frac{\hat{\mathcal{H}}|\psi_i\rangle}{i\hbar}\right)\langle\psi_i| + |\psi_i\rangle\left(\frac{\langle\psi_i|\hat{\mathcal{H}}^\dagger}{-i\hbar}\right)\right]$

$= \frac{\hat{\mathcal{H}}}{i\hbar}\hat{\rho} - \hat{\rho}\frac{\hat{\mathcal{H}}}{i\hbar} = \frac{1}{i\hbar}[\hat{\mathcal{H}}, \hat{\rho}]$. ∎

Chapter 3

3.1 Exercises

1. (a) $|1111\rangle$, (b) $\frac{1}{\sqrt{3}}(|001\rangle + |010\rangle + |100\rangle)$, (c) $|01101110\rangle$.

3. $\hat{\rho} = \sum_i \lambda_i|\psi_i\rangle\langle\psi_i|$, where $\hat{\rho}|\psi_i\rangle = \lambda_i|\psi_i\rangle \therefore \hat{\rho}^2 = \sum_{ij}\lambda_i\lambda_j|\psi_i\rangle\underbrace{\langle\psi_i|\psi_j\rangle}_{\delta_{ij}}\langle\psi_j| = \sum_i\lambda_i^2|\psi_i\rangle\langle\psi_i|$

$\therefore Tr\{\hat{\rho}^2\} = \sum_i\lambda_i^2 \neq \sum_i\lambda_i$ unless $\lambda_i = 0, 1$.

An example is the maximally mixed state whose four eigenvalues are all the same and equal to 1/4.

5. (a) $\hat{\rho}_{AB} = \begin{bmatrix} |\alpha|^2 & \alpha\beta^* & \alpha\gamma^* & \alpha\delta^* \\ \beta\alpha^* & |\beta|^2 & \beta\gamma^* & \beta\delta^* \\ \gamma\alpha^* & \gamma\beta^* & |\gamma|^2 & \gamma\delta^* \\ \delta\alpha^* & \delta\beta^* & \delta\gamma^* & |\delta|^2 \end{bmatrix}$, $\hat{\rho}_A = Tr_B\{\hat{\rho}_{AB}\} = \begin{bmatrix} |\alpha|^2 + |\gamma|^2 & \alpha\beta^* + \gamma\delta^* \\ \beta\alpha^* + \delta\gamma^* & |\beta|^2 + |\delta|^2 \end{bmatrix}$, $\hat{\rho}_B =$

$Tr_A\{\hat{\rho}_{AB}\} = \begin{bmatrix} |\alpha|^2 + |\beta|^2 & \alpha\gamma^* + \beta\delta^* \\ \gamma\alpha^* + \delta\beta^* & |\gamma|^2 + |\delta|^2 \end{bmatrix}$ (same as $\hat{\rho}_A$ with γ and β swapped).

(b),(c) need eigenvalues of $\hat{\rho}_A$ and $\hat{\rho}_B$, $\begin{vmatrix} a-\lambda & b \\ c & d-\lambda \end{vmatrix}^2 = (a-\lambda)(d-\lambda) = (\underbrace{ad-bc}_{\Delta,\text{determinant}}) -$

$\underbrace{(a+d)}_{T,\text{trace}}\lambda + \lambda^2, Tr\{\hat{\rho}_A\} = 1.$ $\text{Det}\{\hat{\rho}_A\} = (|\alpha|^2 + |\gamma|^2)(|\beta|^2 + |\delta|^2) - |\alpha\beta^* + \gamma\delta^*|^2 = \cdots = |\alpha\delta -$

$\beta\gamma|^2 = (C/2)^2 \implies C = 2\sqrt{\text{Det}\{\hat{\rho}_A\}}$ $\therefore$ Eigenvalues of $\hat{\rho}_A$ are solutions of the equations: $\lambda^2 - \lambda + (C/2)^2 \implies \lambda = \frac{1 \pm \sqrt{1-C^2}}{2}$; $S(\rho) = -\lambda_1 \ln\lambda_1 - \lambda_2 \ln\lambda_2 = -\{\frac{1+\sqrt{1-C^2}}{2}\}\ln(\frac{1+\sqrt{1-C^2}}{2}) + \frac{1-\sqrt{1-C^2}}{2}\ln(\frac{1-\sqrt{1-C^2}}{2})$, function of C only.

7. (a) $C = 2|\frac{1}{\sqrt{6}}\frac{\sqrt{2}}{\sqrt{6}} - \frac{1}{\sqrt{6}}\frac{\sqrt{2}}{\sqrt{6}}| = 0$, separable: $\frac{(|0\rangle + \sqrt{2}|1\rangle)}{\sqrt{3}} \otimes \frac{(|0\rangle + |1\rangle)}{\sqrt{2}}$. (b) $C = 2|\frac{1}{\sqrt{2}}(\frac{-1}{\sqrt{2}}) - 0(0)| = 1$, max. entangled. (c) $C = 2|\frac{1}{\sqrt{5}}(\frac{2}{\sqrt{5}}) - 0(0)| = \frac{4}{5}$, partially entangled. (d) $C = 2|0(0) - 1(0)| = 0$, separable. (e) $C = 2|\frac{1}{2}\frac{1}{2} - \frac{1}{2}\frac{1}{2}| = 0$, separable: $\frac{(|0\rangle + |1\rangle)}{\sqrt{2}} \otimes \frac{(|0\rangle + |1\rangle)}{\sqrt{2}}$.

9. $C = 2\sqrt{\lambda_0 \lambda_1}$.

11. Let $\hat{U}_\chi = \begin{bmatrix} e^{i\chi_0} & 0 \\ 0 & e^{i\chi_1} \end{bmatrix}$, $\hat{U}^\dagger \hat{\rho}\hat{U} = \Lambda$ (diagonal), $\hat{U}_\chi^\dagger \hat{U}^\dagger \Lambda \hat{U}\hat{U}_\chi = \begin{bmatrix} e^{-i\chi_0} & 0 \\ 0 & e^{-i\chi_1} \end{bmatrix}\begin{bmatrix} \lambda_0 & 0 \\ 0 & \lambda_1 \end{bmatrix}\begin{bmatrix} e^{i\chi_0} & 0 \\ 0 & e^{i\chi_1} \end{bmatrix} =$

$\begin{bmatrix} \lambda_0 e^{-i\chi_0} & 0 \\ 0 & \lambda_1 e^{-i\chi_1} \end{bmatrix}\begin{bmatrix} e^{i\chi_0} & 0 \\ 0 & e^{i\chi_1} \end{bmatrix} = \begin{bmatrix} \lambda_0 & 0 \\ 0 & \lambda_1 \end{bmatrix} = \Lambda.$ ∎

3.2 Exercises

1. $\langle xy|\hat{I} \otimes \hat{U}|x'y'\rangle = \underbrace{\langle x|\hat{I}|x'\rangle}_{\delta_{xx'}} \langle u|\hat{U}|y'\rangle$. Using the $\{|00\rangle, |01\rangle, |10\rangle, |11\rangle\}$ basis, we get:

$$\hat{I} \otimes \hat{U} = \begin{bmatrix} a & b & 0 & 0 \\ -b^* & a^* & 0 & 0 \\ 0 & 0 & a & b \\ 0 & 0 & -b^* & a^* \end{bmatrix}, \quad \hat{U} \otimes \hat{I} = \begin{bmatrix} a & 0 & b & 0 \\ 0 & a & 0 & b \\ -b^* & 0 & a^* & 0 \\ 0 & -b^* & 0 & a^* \end{bmatrix}.$$

3. $|\psi_{AB}\rangle = \alpha|00\rangle + \beta|01\rangle + \gamma|10\rangle + \delta|11\rangle$. Let $\hat{U}_A = \begin{bmatrix} a & b \\ -b^* & a^* \end{bmatrix}$ and $\hat{U}_B = \begin{bmatrix} c & d \\ -d^* & c^* \end{bmatrix}$. $|0\rangle_A \rightarrow a|0\rangle_A - b^*|1\rangle_A$, $|1\rangle_A \rightarrow b|0\rangle_A + a^*|1\rangle_A$, $|0\rangle_B \rightarrow c|0\rangle_B - d^*|1\rangle_B$, $|1\rangle_B \rightarrow d|0\rangle_B + c^*|1\rangle_B$. Writing $|\psi_{AB}\rangle$ in terms on the new bases, gives $|\psi'_{AB}\rangle$, whose new probability amplitudes are labelled $\alpha', \beta', \gamma', \delta'$. With some algebraic manipulation, the relationship between the initial and final probability amplitudes in terms of the elements of the local unitaries can be expressed as follows using matrix notation:

$$\begin{bmatrix} \alpha' \\ \beta' \\ \gamma' \\ \delta' \end{bmatrix} = \begin{bmatrix} ac & ad & bc & bd \\ -ad^* & ac^* & -bd^* & bc^* \\ -b^*c & -b^*d & a^*c & a^*d \\ -b^*(-d^*) & -b^*c^* & a^*(-d^*) & a^*c^* \end{bmatrix}\begin{bmatrix} \alpha \\ \beta \\ \gamma \\ \delta \end{bmatrix}.$$

Working out the expression $\alpha'\delta' - \beta'\gamma'$ yields $\alpha'\delta' - \beta'\gamma' = \alpha\delta - \beta\gamma \implies C = C'$.

5. $|\psi\rangle = \frac{1}{2}(|00\rangle + |01\rangle + |10\rangle + |11\rangle) = \frac{1}{\sqrt{2}}(|0\rangle + |1\rangle) \otimes \frac{1}{\sqrt{2}}(|0\rangle + |1\rangle)$. $|0\rangle \rightarrow (|0\rangle + |1\rangle)/2$, $|1\rangle \rightarrow (|0\rangle - |1\rangle)/2$ $\therefore |\psi'\rangle = \hat{H} \otimes \hat{H}|\psi\rangle = \cdots = |00\rangle$, which is a lot easier to work with. $C(\psi) = 2|\frac{1}{2}\frac{1}{2} - \frac{1}{2}\frac{1}{2}| = 0$. $C(\psi') = 2|1(0) - 0(0)| = 0$.

7. $\text{CNOT}_{1,2}(a|00\rangle + b|01\rangle) = a|00\rangle + b|11\rangle$. $C = 2|ab|$.

9. $\hat{U}_1 = \frac{1}{\sqrt{2}}\begin{bmatrix} 1 & 1 \\ 1 & -1 \end{bmatrix} = \hat{H}$. $\text{CNOT}_{2,1}(\hat{H} \otimes \hat{I})|00\rangle = \text{CNOT}_{2,1}(|00\rangle + |10\rangle)/\sqrt{2} = (|00\rangle + |11\rangle)/\sqrt{2}$ $\therefore \lambda_0 = \lambda_1 = 1/\sqrt{2}, \hat{U}_2 = \hat{U}_3 = \hat{I}$.

11. $\hat{H}^2 = \hat{I}$,

13. $\text{SWAP} = \begin{bmatrix} 1 & 0 & 0 & 0 \\ 0 & 0 & 1 & 0 \\ 0 & 1 & 0 & 0 \\ 0 & 0 & 0 & 1 \end{bmatrix}$, $\sqrt{\text{SWAP}} = \begin{bmatrix} 1 & 0 & 0 & 0 \\ 0 & a & b & 0 \\ 0 & c & d & 0 \\ 0 & 0 & 0 & 1 \end{bmatrix}$. $\hat{X} = \begin{bmatrix} 0 & 1 \\ 1 & 0 \end{bmatrix} = $

$(+1)\underbrace{\frac{1}{2}\begin{bmatrix} 1 & 1 \\ 1 & 1 \end{bmatrix}}_{\hat{\Pi}_+} + (-1)\underbrace{\frac{1}{2}\begin{bmatrix} 1 & -1 \\ -1 & 1 \end{bmatrix}}_{\hat{\Pi}_-}$

$\therefore \sqrt{\hat{X}} = (\pm 1)\hat{\Pi}_+ + (\pm i)\hat{\Pi}_- = \pm\frac{1}{2}\begin{bmatrix} 1+i & 1-i \\ 1-i & 1+i \end{bmatrix}$ or $\pm\frac{1}{2}\begin{bmatrix} 1-i & 1+i \\ 1+i & 1-i \end{bmatrix}$; i.e.,

four possible roots, all equally valid. We choose $= \frac{1}{2}\begin{bmatrix} 1-i & 1+i \\ 1+i & 1-i \end{bmatrix}$ $\therefore \sqrt{\text{SWAP}} = $

$\begin{bmatrix} 1 & 0 & 0 & 0 \\ 0 & (1-i)/2 & (1+i)/2 & 0 \\ 0 & (1+i)/2 & (1-i)/2 & 0 \\ 0 & 0 & 0 & 1 \end{bmatrix}$.

15. $|\beta_0\rangle = (|00\rangle + |11\rangle)/\sqrt{2}$, $|\beta_k\rangle = i(\hat{I} \otimes \hat{\sigma}_k)|\beta_0\rangle$. $i\hat{I} \otimes \hat{X} = i\begin{bmatrix} 0 & 1 & 0 & 0 \\ 1 & 0 & 0 & 0 \\ 0 & 0 & 0 & 1 \\ 0 & 0 & 1 & 0 \end{bmatrix} \implies |\beta_1\rangle = $

$i(|01\rangle + |10\rangle)/\sqrt{2}$, $i\hat{I} \otimes \hat{Y} = \begin{bmatrix} 0 & 1 & 0 & 0 \\ -1 & 0 & 0 & 0 \\ 0 & 0 & 0 & 1 \\ 0 & 0 & -1 & 0 \end{bmatrix} \implies |\beta_2\rangle = (-|01\rangle + |10\rangle)/\sqrt{2}$, $i\hat{I} \otimes \hat{Z} = $

$i\begin{bmatrix} 1 & 0 & 0 & 0 \\ 0 & -1 & 0 & 0 \\ 0 & 0 & 1 & 0 \\ 0 & 0 & 0 & -1 \end{bmatrix} \implies |\beta_3\rangle = i(|00\rangle - |11\rangle)/\sqrt{2}$.

17. Let $|\psi_{AB}\rangle = \alpha|00\rangle + \beta|01\rangle + \gamma|10\rangle + \delta|11\rangle = A|\beta_0\rangle + B|\beta_1\rangle + C|\beta_2\rangle + D|\beta_3\rangle = \frac{A}{\sqrt{2}}(|00\rangle + |11\rangle) + \frac{B}{\sqrt{2}}i(|01\rangle + |10\rangle) + \frac{C}{\sqrt{2}}(-|01\rangle + |10\rangle) + i\frac{D}{\sqrt{2}}(|00\rangle - |11\rangle) = \frac{A+iD}{\sqrt{2}}|00\rangle + \frac{-C+iB}{\sqrt{2}}|01\rangle + \frac{C+iB}{\sqrt{2}}|10\rangle + \frac{A-iD}{\sqrt{2}}|11\rangle$. $2|\alpha\delta - \beta\gamma| = |(A+iD)(A-iD) + (C-iB)(C+iB)| = |A^2 + B^2 + C^2 + D^2|$. Hence, if A, B, C, and D are either real or share a common phase factor, the state has $C = 1$; i.e., is maximally entangled.

3.3 Exercises

1. $(\hat{X} \otimes \hat{X})|\beta_2\rangle = (\hat{X} \otimes \hat{I})(\hat{I} \otimes \hat{X})|\beta_2\rangle = (\hat{X} \otimes \hat{I})\mathrm{Det}\{\hat{X}\}(\hat{X}^{-1} \otimes \hat{I})|\beta_2\rangle$ (by fundamental theorem).
$\hat{X}^{-1} = \hat{X}, \mathrm{Det}\{\hat{X}\} = -1 \therefore (\hat{X} \otimes \hat{X})|\beta_2\rangle = (-1)(\hat{X} \otimes \hat{I})(\hat{X} \otimes \hat{I})|\beta_2\rangle = -(\hat{X}^2 \otimes \hat{I})|\beta_2\rangle =$
$-(\hat{I} \otimes \hat{I})|\beta_2\rangle = -|\beta_2\rangle$. ∎

3. $\hat{a} = \mathbf{a} \cdot \hat{\boldsymbol{\sigma}}, \hat{c} = \mathbf{c} \cdot \hat{\boldsymbol{\sigma}}, |\mathbf{a}| = |\mathbf{c}| = 1$, Eigenvalues $\pm 1 \implies \mathrm{Det} = -1, \mathrm{Det}\{\hat{c}\} = -1, \hat{a}\hat{c} =$
$\mathbf{a} \cdot \mathbf{c}\hat{I} + i(\mathbf{a} \times \mathbf{c}).\hat{\boldsymbol{\sigma}}, m = \langle\beta_2|\hat{a} \otimes \hat{c}|\beta_2\rangle = \langle\beta_2|(\hat{a} \otimes \hat{I})(\hat{I} \otimes \hat{c})|\beta_2\rangle = \langle\beta_2|(\hat{a} \otimes \hat{I})(\hat{c} \otimes \hat{I})|\beta_2\rangle =$
$\mathrm{Det}\{\hat{c}\}\langle\beta_2|\hat{a}\hat{c} \otimes \hat{I}|\beta_2\rangle = -\mathbf{a} \cdot \mathbf{c}\langle\beta_2|\hat{I} \otimes \hat{I}|\beta_2\rangle - i(\mathbf{a} \times \mathbf{c})\langle\beta_2|\hat{\boldsymbol{\sigma}} \otimes \hat{I}|\beta_2\rangle$. Prove $\langle\beta_2|\hat{\sigma}_i \otimes \hat{I}|\beta_2\rangle = 0$.
$|\beta_2\rangle = -(|01\rangle - |10\rangle)/\sqrt{2}, (\hat{X} \otimes \hat{I})|\beta_2\rangle = (-|11\rangle + |00\rangle)/\sqrt{2} \implies \langle\beta_2|\hat{X} \otimes \hat{I}|\beta_2\rangle = 0,$
$(\hat{Y} \otimes \hat{I})|\beta_2\rangle = (-i|11\rangle - i|00\rangle)/\sqrt{2} \implies \langle\beta_2|\hat{Y} \otimes \hat{I}|\beta_2\rangle = 0, (\hat{Z} \otimes \hat{I})|\beta_2\rangle = (-|01\rangle - |10\rangle)/\sqrt{2} \implies$
$\langle\beta_2|\hat{Z} \otimes \hat{I}|\beta_2\rangle = -1/2 + 1/2 = 0$. Hence, $\langle\beta_2|\hat{a} \otimes \hat{c}|\beta_2\rangle = -\mathbf{a} \cdot \mathbf{c}$. ∎

5. $a_2(b_1 - c_1) = \pm(1 + b_2 c_1)$:

- If $b_1 = c_1$, then LHS (left hand side) $= 0$; using $b_2 = -b_1$, RHS is $\pm(1 - c_1^2) = 0$ (since $c_1 = \pm 1$).

- If $b_1 \neq c_1$, then LHS $= \pm 2$; RHS $= 1 - b_1 c_1 = 2$ (since $b_1 c_1 = -1$ always).

Hence, since $1 + b_2 c_1 = 0$ or 2, we have $|a_2(b_1 - c_1)| = 1 + b_2 c_1$. Taking averages: $\overline{|a_2(b_1 - c_1)|} = 1 + \overline{b_2 c_1}$. $\overline{|x|} \geq |\bar{x}| \implies |\overline{a_2 b_1} - \overline{a_2 c_1}| \leq 1 + \overline{b_2 c_1} \implies |\overline{a_2 b_1} - \overline{a_2 c_1}| - \overline{b_2 c_1} \leq 1$. QM predictions: $\langle\hat{a} \otimes \hat{c}\rangle = -\mathbf{a} \cdot \mathbf{c}$ for singlet state. $|\mathbf{a} \cdot \mathbf{b} - \mathbf{a} \cdot \mathbf{c}| + \mathbf{b} \cdot \mathbf{c} \leq 1$. $\mathbf{a} \cdot \mathbf{b} = 1/2, \mathbf{a} \cdot \mathbf{c} = -1/2, \mathbf{b} \cdot \mathbf{c} = 1/2 \implies 1 + 1/2 = 3/2 > 1$ (! violation).

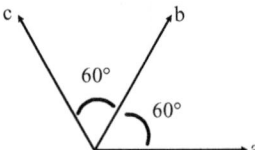

Figure S.3 Exercise 3.3.5.

3.4 Exercises

1. This solution is due to David Mermin. (QC S.5–QC S.11)
Parts (a) to (c):

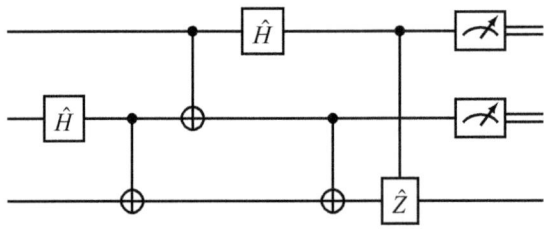

QC S.4 Exercise 3.4.1.

If we use the following identities for quantum gates:

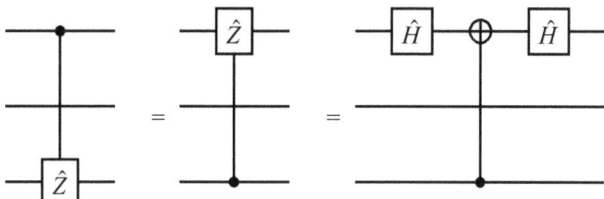

QC S.5 Exercise 3.4.1.

then the circuit becomes

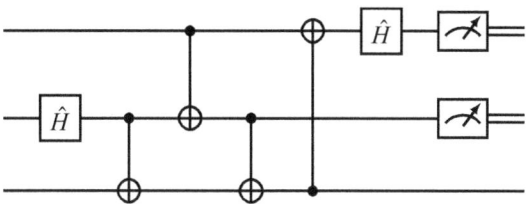

QC S.6 Exercise 3.4.1.

(d) to (g): The state $\hat{H}|0\rangle = \frac{1}{\sqrt{2}}(|0\rangle + |1\rangle)$ is the eigenstate of $\hat{H}$ with eigenvalue 1, i.e., $\hat{X}\hat{H}|0\rangle = \hat{H}|0\rangle$. Hence, we can add another CNOT changing nothing to the circuit without changing anything:

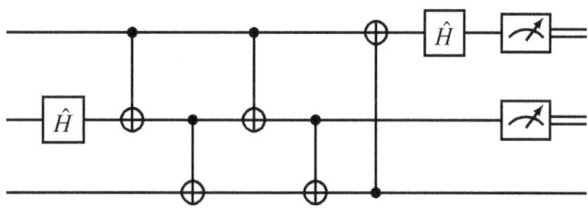

QC S.7 Exercise 3.4.1.

Now we employ another circuit identity:

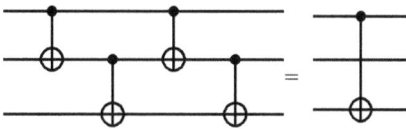

QC S.8 Exercise 3.4.1.

so the complete circuit is equivalent to the following:

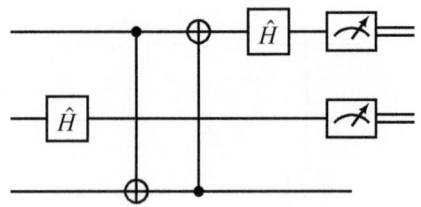

QC S.9 Exercise 3.4.1.

Since qubit #0 is in state $|0\rangle$, another CNOT with that qubit as the control will have no effect, so the circuit is equivalent to

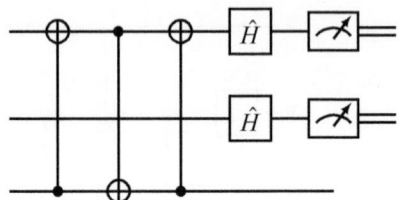

QC S.10 Exercise 3.4.1.

Let's analyze the combination of three CNOT gates on two qubits. The states evolve as follows:

$$|00\rangle \rightarrow |00\rangle \rightarrow |00\rangle \rightarrow |00\rangle,$$
$$|01\rangle \rightarrow |11\rangle \rightarrow |10\rangle \rightarrow |10\rangle,$$
$$|10\rangle \rightarrow |10\rangle \rightarrow |11\rangle \rightarrow |01\rangle,$$
$$|11\rangle \rightarrow |01\rangle \rightarrow |01\rangle \rightarrow |11\rangle;$$

i.e., SWAP gate. Thus, the circuit becomes

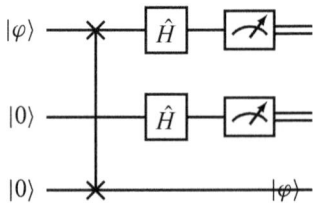

QC S.11 Exercise 3.4.1.

3. What's wrong is that $\hat{A}$ cannot exist.
$\hat{A}|00\rangle = |00\rangle \quad \hat{A}|10\rangle = |11\rangle$
$\hat{A}|+0\rangle = (|00\rangle + |11\rangle)/\sqrt{2} \quad |++\rangle = (|00\rangle + |01\rangle + |10\rangle + |11\rangle)/2\langle++|\hat{A}|+0\rangle = 1/\sqrt{2}$;
i.e., the probability the "Amplified" state is the state required for FLASH is 50%; you might as well toss a coin!

Chapter 3 Exercises

1. $\hat{U} = |\beta_0\rangle\langle 00| + |\beta_1\rangle\langle 01| + |\beta_2\rangle\langle 10| + |\beta_3\rangle\langle 11| = \frac{1}{\sqrt{2}}\begin{bmatrix}1\\0\\0\\1\end{bmatrix} \otimes \begin{bmatrix}1 & 0 & 0 & 0\end{bmatrix} +$

$\frac{i}{\sqrt{2}}\begin{bmatrix}0\\1\\1\\0\end{bmatrix} \otimes \begin{bmatrix}0 & 1 & 0 & 0\end{bmatrix} + \frac{1}{\sqrt{2}}\begin{bmatrix}0\\-1\\1\\0\end{bmatrix} \otimes \begin{bmatrix}0 & 0 & 1 & 0\end{bmatrix} + \frac{i}{\sqrt{2}}\begin{bmatrix}1\\0\\0\\-1\end{bmatrix} \otimes \begin{bmatrix}0 & 0 & 0 & 1\end{bmatrix} =$

$\frac{1}{\sqrt{2}}\begin{bmatrix}1 & 0 & 0 & i\\0 & i & -1 & 0\\0 & i & 1 & 0\\1 & 0 & 0 & -i\end{bmatrix}.$

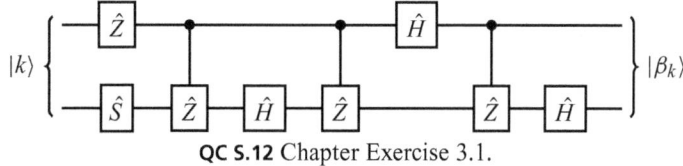

QC S.12 Chapter Exercise 3.1.

3. $\hat{U}$ and $\hat{V}$ are local unitaries that have no effect on entanglement. $CZ(\hat{H} \otimes \hat{H})|00\rangle = CZ\frac{1}{2}(|00\rangle + |01\rangle + |10\rangle + |11\rangle) = \frac{1}{2}(|00\rangle + |01\rangle + |10\rangle - |11\rangle)$. Concurrence $= 2|\frac{1}{2}\frac{-1}{2} - \frac{1}{2}\frac{1}{2}| = 1$. $C_{\text{in}} = 0$, $C_{\text{out}} = 1$.

5. (a) $\alpha|00\rangle + \beta|10\rangle \xrightarrow{\text{CNOT}} \alpha|00\rangle + \beta|11\rangle = |\varphi_1\rangle \xrightarrow{\hat{H}\otimes\hat{I}} \frac{1}{\sqrt{2}}\{\alpha|00\rangle + \alpha|10\rangle + \beta|01\rangle - \beta|11\rangle\} = \frac{1}{\sqrt{2}}\{|0\rangle(\alpha|0\rangle + \beta|1\rangle) + |1\rangle(\alpha|0\rangle - \beta|1\rangle)\} = |\varphi_2\rangle$.

(b) $\hat{U} = \hat{Z}$.

7. (a) 2×2 Hermitian matrix with eigenvalues ±1 $\therefore \hat{M} = \mathbf{a} \cdot \hat{\boldsymbol{\sigma}}, \mathbf{a} \cdot \mathbf{a} = 1$, Trace $= 0$, Determinant $= -a_z^2 - a_y^2 - a_x^2 = -1$. $\mathbf{a} \cdot \mathbf{b} = \cos\theta/2, |\mathbf{a} \times \mathbf{b}| = \sin\theta/2, \mathbf{u} = (\mathbf{a} \times \mathbf{b})/|\mathbf{a} \times \mathbf{b}|$. $(\mathbf{a} \cdot \hat{\boldsymbol{\sigma}})(\mathbf{b} \cdot \hat{\boldsymbol{\sigma}}) = (\mathbf{a} \cdot \mathbf{b})\hat{I} + i(\mathbf{a} \times \mathbf{b}) \cdot \hat{\boldsymbol{\sigma}} = \cos\theta/2\,\hat{I} + i\sin\theta/2\,\mathbf{u} \cdot \hat{\boldsymbol{\sigma}}$ (arbitrary unitary)
$\hat{U} = \cos\theta/2\,\hat{I} + i\sin\theta/2\,\mathbf{u} \cdot \hat{\boldsymbol{\sigma}}, \mathbf{a}$ and $\mathbf{b} \perp \mathbf{u}, \mathbf{a} \cdot \mathbf{b} = \cos\theta/2$.

(b) If qubit #2 is $|0\rangle$, $\hat{U}\hat{U}^\dagger\hat{V}\hat{V}^\dagger = I$. If qubit #2 is $|1\rangle$, $\hat{W} = \hat{V}^\dagger\hat{X}\hat{V}\hat{U}^\dagger\hat{X}\hat{U}, \hat{W}^\dagger = \hat{U}^\dagger\hat{X}\hat{U}\hat{V}^\dagger\hat{X}\hat{V},$
$\hat{W}^\dagger\hat{W} = \hat{I}$.

(c) $\hat{X} = \mathbf{e_x} \cdot \hat{\boldsymbol{\sigma}}, \hat{U}^\dagger\mathbf{A} \cdot \hat{\boldsymbol{\sigma}}\hat{U} = \mathbf{A'} \cdot \hat{\boldsymbol{\sigma}}$, where $\mathbf{A'}$ is the vector $\mathbf{A}$ rotated by an angle θ about an axis $\mathbf{n}$ (where θ and $\mathbf{n}$ are determined by the form of $\hat{U}$).
Let $\hat{V}^\dagger\hat{X}\hat{V} = \mathbf{a} \cdot \hat{\boldsymbol{\sigma}}, \hat{U}^\dagger\hat{X}\hat{U} = \mathbf{b} \cdot \hat{\boldsymbol{\sigma}}, \hat{W} = (\mathbf{a} \cdot \hat{\boldsymbol{\sigma}})(\mathbf{b} \cdot \hat{\boldsymbol{\sigma}}) = (\mathbf{a} \cdot \mathbf{b})\hat{I} + i(\mathbf{a}\ x\ \mathbf{b}) \cdot \hat{\boldsymbol{\sigma}}, \mathbf{a} \cdot \mathbf{b} = \cos\theta/2,$
$\mathbf{a} \times \mathbf{b} = \mathbf{n}\sin\theta/2$ (where $|\mathbf{n}| = 1$ and $\mathbf{a} \cdot \mathbf{n} = \mathbf{b} \cdot \mathbf{n} = 0$) $\therefore \hat{W} = \cos\theta/2\,\hat{I} + i\sin\theta/2\,\mathbf{n} \cdot \hat{\boldsymbol{\sigma}}$ (arbitrary unitary).

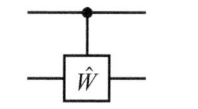

QC S.13 Chapter Exercise 3.7.

Chapter 4

4.1 Exercises

1. Four-bit integer: values 0 to $2^4 - 1 = 15$. $0 : 0000 \; 1 : 0001 \; 2 : 0010, \ldots, 15 : 1111 = 2^3 + 2^2 + 2^1 + 2^0 = 8 + 4 + 2 + 1 = 15$.

3.

Table S.1 Exercise 4.1.1.

x	x_2	#bits	$\log_2 x$	$\lfloor \log_2 x \rfloor$	$\lfloor \log_2 x \rfloor + 1$
0	00000	1	—	—	—
$1 = 2^0$	00001	1	0	0	1
$2 = 2^1$	00010	2	1	1	2
3	00011	2	1.58	1	2
$4 = 2^2$	00100	3	2	2	3
5	00101	3	2.32	2	3
6	00110	3	2.58	2	3
7	00111	3	2.81	2	3
$8 = 2^3$	01000	4	3	3	4
9	01001	4	3.17	3	4
10	01010	4	3.32	3	4
11	01011	4	3.46	3	4
12	01100	4	3.58	3	4
13	01101	4	3.7	3	4
14	01110	4	3.81	3	4
15	01111	4	3.91	3	4
$16 = 2^4$	10000	5	4	4	5

5. (a) $111_2 = 7_{10}$, (b) $1111_2 = 15_{10}$, (c) $0110_2 = 6_{10}$, (d) $1000000_2 = 64_{10}$, (e) $11001101_2 = 205_{10}$, (f) $111111111_2 = 511_{10}$.

7. $|x\rangle = |x_{L-1}\rangle \otimes |x_{L-2}\rangle \otimes \cdots \otimes |x_0\rangle$, $\langle y|x\rangle = \langle y_{L-1}|x_{L-1}\rangle \langle y_{L-2}|x_{L-2}\rangle \cdots \langle y_0|x_0\rangle$. $\langle 0|0\rangle = \langle 1|1\rangle = 1$, $\langle 0|1\rangle = \langle 1|0\rangle = 0$. If $y_p \neq x_p$ for any bit p, then $\langle y|x\rangle = 0$. If $y_p = x_p$, $\forall p = 0, 1, \ldots, L-1$, then $\langle y|x\rangle = 1 \therefore \langle y|x\rangle = \delta_{y_{L-1}x_{L-1}}, \ldots, \delta_{y_0 x_0} = \delta_{xy}$.

9. $\hat{Z}^{\otimes 7}|1100101\rangle = (-1)^4|1100101\rangle = |1100101\rangle$

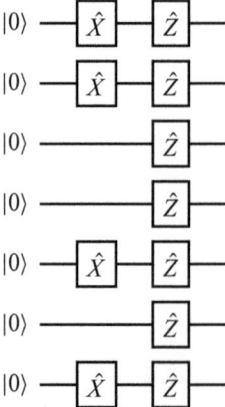

QC S.14 Exercise 4.1.9.

4.2 Exercises

1. (a) $(101)_2 \oplus (111)_2 = (010)_2 = 2_{10}$, (b) $(1111)_2 \oplus (0101)_2 = (1010)_2 = 10_{10}$, (c) $(01001)_2 \oplus (10010)_2 = (11011)_2 = 27_{10}$.

3. (a) $x, y, z = 1$ bit numbers:

Table S.2 Exercise 4.2.3.

xyz	$x \oplus y$	$(x \oplus y) \oplus z$	$y \oplus z$	$x \oplus (y \oplus z)$
000	0	0	0	0
001	0	1	1	1
010	1	1	1	1
011	1	0	0	0
100	1	1	0	1
101	1	0	1	0
110	0	0	1	0
111	0	1	0	1

$\implies (x \oplus y) \oplus z = x \oplus (y \oplus z)$.

(b) Since for L-bit numbers we have L independent operations, $(x \oplus y) \oplus z = \{(x_{L-1} \oplus y_{L-1}) \oplus z_{L-1}, \ldots\} = \{x_{L-1} \oplus (y_{L-1} \oplus z_{L-1}), \ldots\} = x \oplus (y \oplus z)$.

(c) $y \oplus y = \{\ldots, y_m \oplus y_n, \ldots\} = 0$. $0 \oplus 0 = 0$ $1 \oplus 1 = 0$, $y \oplus (y \oplus z) = (y \oplus y) \oplus z = 0 \oplus z = z$.

5.

Table S.3 Exercise 4.1.5.

x	y	f(x)	$y \oplus f$
00	0	0	0
01	0	1	1
10	0	0	0
11	0	1	1
00	1	0	1
01	1	1	0
10	1	0	1
11	1	1	0

Table S.4 Exercise 4.1.5.

$\lvert xy \rangle \rightarrow \lvert x\, y \oplus f(x) \rangle$
$\lvert 000 \rangle \rightarrow \lvert 000 \rangle$
$\lvert 010 \rangle \rightarrow \lvert 011 \rangle$
$\lvert 100 \rangle \rightarrow \lvert 100 \rangle$
$\lvert 110 \rangle \rightarrow \lvert 111 \rangle$
$\lvert 001 \rangle \rightarrow \lvert 001 \rangle$
$\lvert 011 \rangle \rightarrow \lvert 010 \rangle$
$\lvert 101 \rangle \rightarrow \lvert 101 \rangle$
$\lvert 111 \rangle \rightarrow \lvert 110 \rangle$

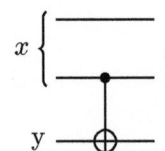

QC S.15 Exercise 4.2.5.

4.3 Exercises

1. (a) $|000\rangle$, one term.
   (b) $(|000\rangle + |010\rangle)/\sqrt{2}$, two terms.
   (c) $(|000\rangle + |010\rangle + |100\rangle + |110\rangle)/2$, four terms.
   (d) Eight terms.

3. $f(x) = \sin^2(\pi x/2)$, $f(0) = 0$, $f(1) = 1$.
   $\hat{U}_f(|00\rangle + |10\rangle)/\sqrt{2} = (|00\rangle + |11\rangle)/\sqrt{2}$.
   Measurement of both qubits: $p = 50\%$, $x = 0$, $f(0) = 0$; $p = 50\%$, $x = 1$, $f(1) = 1$ (not a very good calculator: you're rolling the dice to decide what the value of x is).

4.4 Exercises

1. (a) $f(0) = 0$, $f(1) = 0$, $y \oplus f(x) = y$, $x = 0, 1$.

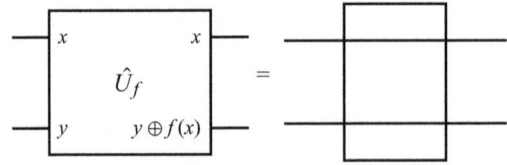

QC S.16 Exercise 4.4.1.

   (b) $f(0) = 0$, $f(1) = 1$, $y \oplus f(x) = y \oplus x$.

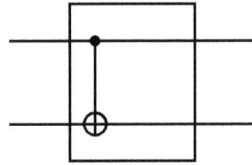

QC S.17 Exercise 4.4.1.

   (c) $f(0) = 1$, $f(1) = 0$, $y \oplus f(x) = y \oplus \bar{x}$.

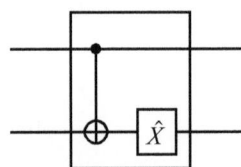

QC S.18 Exercise 4.4.1.

   (d) $f(0) = 1$, $f(1) = 1$, $y \oplus f(x) = \bar{y}$.

QC S.19 Exercise 4.4.1.

3. Measurement of first qubit yields 0 or 1 with 50% probability each.

5. $f(x) = \sin^2(\pi x/2)$, $f(0) = 0$, $f(1) = 1$, $f(2) = 0$, $f(3) = 1$.

(a) $|\varphi_2\rangle$ state, output of oracle:
$\frac{1}{2\sqrt{2}} (|00\rangle(|0\rangle - |1\rangle) + |01\rangle(|1\rangle - |0\rangle) + |10\rangle(|0\rangle - |1\rangle) + |11\rangle(|1\rangle - |0\rangle))$ $= \frac{1}{2\sqrt{2}}(|00\rangle - |01\rangle + |10\rangle - |11\rangle)(|0\rangle - |1\rangle) = \frac{1}{2\sqrt{2}}(|0\rangle + |1\rangle)(|0\rangle - |1\rangle)(|0\rangle - |1\rangle)$.

(b) $|\varphi_3\rangle$ state after final Hadamards:
$\hat{H} \otimes \hat{H} \otimes \hat{I}|\varphi_2\rangle = \frac{1}{\sqrt{2}}|01\rangle(|0\rangle - |1\rangle)$.

(c) 100% probability outcome of measurement of argument register gives $x = 1$. $x \neq 0 \implies$ function is balanced.

(d) $f(4) = f(6) = 0$, $f(5) = f(7) = 1$. $L = 3$: $|\varphi_2\rangle = \frac{1}{4}\{|000\rangle(|0\rangle - |1\rangle) + |001\rangle(|1\rangle - |0\rangle) + |010\rangle(|0\rangle - |1\rangle) + |011\rangle(|1\rangle - |0\rangle) + |100\rangle(|0\rangle - |1\rangle) + |101\rangle(|1\rangle - |0\rangle) + |110\rangle(|0\rangle - |1\rangle) + |111\rangle(|1\rangle - |0\rangle)\} = \frac{1}{4}\{|000\rangle - |001\rangle + |010\rangle - |011\rangle + |100\rangle - |101\rangle + |110\rangle - |111\rangle\}(|0\rangle - |1\rangle) = \frac{1}{4}\{(|0\rangle + |1\rangle)(|0\rangle + |1\rangle)(|0\rangle - |1\rangle)(|0\rangle - |1\rangle)\}$

$|\varphi_3\rangle = \hat{H}^{\otimes L} \otimes \hat{I}|\varphi_2\rangle = \frac{1}{\sqrt{2}}|001\rangle(|0\rangle - |1\rangle)$, measure $x = 1 \implies$ balanced function.

7. $a = \sum_{n=0}^{L-1} x_n y_n$, $b = x \cdot y = a \pmod 2$.

(a)

Table S.5 Exercise 4.4.7.

n	x_n	y_n	$x_n y_n$
0	0	1	0
1	1	1	1

$a = 1, b = 1 \implies (-1)^a = (-1)^b = -1$.

(b)

Table S.6 Exercise 4.4.7.

n	x_n	y_n	$x_n y_n$
0	0	1	0
1	0	1	0
2	0	0	0
3	1	0	0

$a = b = 0 \implies (-1)^a = (-1)^b = 1$.

(c)

Table S.7 Exercise 4.4.7.

n	x_n	y_n	$x_n y_n$
0	1	1	1
1	1	1	1
2	1	0	0
3	1	1	1

$a = 3, b = 1 \implies (-1)^a = (-1)^b = -1$.

4.5 Exercises

1. $|a_\perp\rangle = (|a\rangle - \sin\epsilon|all_L\rangle)/\cos\epsilon$, $\sin\epsilon = \langle all_L|a\rangle$. $\langle all_L|a_\perp\rangle = (\langle all_L|a\rangle - \sin\epsilon)/\cos\epsilon = (\sin\epsilon - \sin\epsilon)/\cos\epsilon = 0$. $\langle a_\perp|a_\perp\rangle = (\langle a| - \sin\epsilon\langle all_L|)(|a\rangle - \sin\epsilon|all_L\rangle)/\cos^2\epsilon = \{\langle a|a\rangle - \sin\epsilon(\langle a|all_L\rangle + \langle all_L|a\rangle) + \sin^2\epsilon\langle all_L|all_L\rangle\}/\cos^2\epsilon = \{1 - 2sin^2\epsilon + \sin^2\epsilon\}/\cos^2\epsilon = (1 - \sin^2\epsilon)/\cos^2\epsilon = 1$.

3. Proof by induction: $c_m = \cos(2m\epsilon)$, $s_m = \sin(2m\epsilon)$. $\begin{bmatrix} c_1 & -s_1 \\ s_1 & c_1 \end{bmatrix}^2 = \begin{bmatrix} c_1 & -s_1 \\ s_1 & c_1 \end{bmatrix}\begin{bmatrix} c_1 & -s_1 \\ s_1 & c_1 \end{bmatrix} =$

$\begin{bmatrix} c_1^2 - s_1^2 & -c_1 s_1 - s_1 c_1 \\ s_1 c_1 + c_1 s_1 & -s_1^2 + c_1^2 \end{bmatrix}$, $c_1^2 - s_1^2 = \cos^2 2\epsilon - \sin^2 2\epsilon = \cos(4\epsilon) = c_2$, $2c_1 s_1 =$

$2\sin(2\epsilon)\cos(2\epsilon) = \sin(4\epsilon) = s_2$ $\therefore \begin{bmatrix} c_1 & -s_1 \\ s_1 & c_1 \end{bmatrix}^2 = \begin{bmatrix} c_2 & -s_2 \\ s_2 & c_2 \end{bmatrix}$ ✓ for $m = 2$.

$\begin{bmatrix} c_m & -s_m \\ s_m & c_m \end{bmatrix}\begin{bmatrix} c_1 & -s_1 \\ s_1 & c_1 \end{bmatrix} = \begin{bmatrix} c_m c_1 - s_m s_1 & -c_m s_1 - s_m c_1 \\ s_m c_1 + c_m s_1 & -s_m s_1 + c_m c_1 \end{bmatrix}$, $\cos(2m\epsilon)\cos(2\epsilon) - \sin(2m\epsilon)$

$\sin(2\epsilon) = \cos(2(m+1)\epsilon)$, $\sin(2m\epsilon)\cos(2\epsilon) + \cos(2m\epsilon)\sin(2\epsilon) = \sin(2(m+1)\epsilon)$

$\therefore \begin{bmatrix} c_m & -s_m \\ s_m & c_m \end{bmatrix}^2\begin{bmatrix} c_1 & -s_1 \\ s_1 & c_1 \end{bmatrix} = \begin{bmatrix} c_{m+1} & -s_{m+1} \\ s_{m+1} & c_{m+1} \end{bmatrix}$, which completes the inductive step. ∎

5. $p = \sin^2\big((2m_{\text{opt}} + 1)\epsilon\big) \simeq \sin^2\big((2m_{\text{opt}} + 1)/2^{L/2}\big)$.

    (a) $L = 3$, $p = 0.945$.

    (b) $L = 5$, $p = 0.9991$.

    (c) $L = 10$, $p = 0.9994$.

    (d) $L = 24$, $p = 1 - \delta$ $\delta \sim 5.7 \times 10^{-8}$.

Chapter 4 Exercises

1. (a) $|11\rangle$ is the phase reference: we still have an arbitrary global phase.

    (b) $|\psi(x)\rangle = \frac{1}{2^{L/2}}\left(|2^L - 1\rangle + \sum_{y=0}^{2^L-2}(-1)^{x_y}|y\rangle\right)$

$\therefore \langle\psi(x')|\psi(x)\rangle = \frac{1}{2^L}\left(\langle 2^L - 1| + \sum_{y'=0}^{2^L-2}(-1)^{x'_{y'}}\langle y'|\right)\left(|2^L - 1\rangle + \sum_{y=0}^{2^L-2}(-1)^{x_y}|y\rangle\right) =$

$\frac{1}{2^L}\left(1 + \sum_{y,y'=0}^{2^L-2}(-1)^{x_y + x'_{y'}}\underbrace{\langle y'|y\rangle}_{\delta_{y'y}}\right) = \frac{1}{2^L}\left(1 + \sum_{y=0}^{2^L-2}(-1)^{x_y + x'_y}\right) \neq \delta_{xx'}$ in general.

    (c) (i) $|\psi(7)\rangle = \frac{1}{4}(|11\rangle - |10\rangle - |01\rangle - |00\rangle)$, $7_{10} = 111_2$.

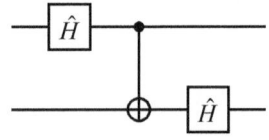

QC S.20 Chapter Exercise 4.1.

    (ii) $|\psi(3)\rangle = \frac{1}{4}(|11\rangle + |10\rangle - |01\rangle - |00\rangle) = \frac{1}{4}(|1\rangle - |0\rangle)(|1\rangle + |0\rangle)$, $3_{10} = 11_2$

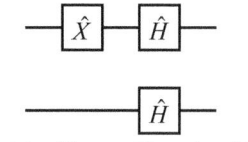

QC S.21 Chapter Exercise 4.1.

(iii) $|\psi(4)\rangle = \frac{1}{4}(|11\rangle - |10\rangle + |01\rangle + |00\rangle)$, $x = 4_{10} = 100_2$

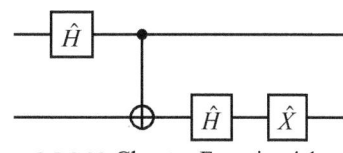

QC S.22 Chapter Exercise 4.1.

(d) Lack of orthogonality: Different numbers are not linearly independent; hence, any calculation and any result will be non-unique.

3. $0 \cdot 0 = 0$, $0 \cdot 1 = 0$, $1 \cdot 0 = 0$, $1 \cdot 1 = 1$. For $L = 1$, $2^L - 1 = 1$. $\hat{H}|0\rangle = \frac{1}{\sqrt{2}}(|0\rangle + |1\rangle)$, $\hat{H}|1\rangle = \frac{1}{\sqrt{2}}(|0\rangle - |1\rangle)$. $\underbrace{\text{sum}}_{x=0} = \frac{1}{\sqrt{2}}\sum_{y=0}^{1}(-1)^{0 \cdot y}|y\rangle = \frac{1}{\sqrt{2}}(|0\rangle + |1\rangle) = \hat{H}|0\rangle$. $\underbrace{\text{sum}}_{x=1} = \frac{1}{\sqrt{2}}\sum_{y=0}^{1}(-1)^{1 \cdot y}|y\rangle = \frac{1}{\sqrt{2}}(|0\rangle - |1\rangle) = \hat{H}|1\rangle$. For $L = 2$ (x is a two-bit number with bit values $x_1 x_0$). $\hat{H}^{\otimes 2}|x\rangle \equiv \hat{H}|x_1\rangle \otimes \hat{H}|x_0\rangle = \frac{1}{\sqrt{2^2}}\left(\sum_{y_1=0}^{1}(-1)^{y_1 x_1}|y_1\rangle\right) \otimes \left(\sum_{y_0=0}^{1}(-1)^{y_0 x_0}|y_0\rangle\right)$. $|y_1\rangle \otimes |y_0\rangle \equiv |y\rangle$ (y is a two-bit number with bit values $y_1 y_0$). $(-1)^{y_1 x_1 + y_0 x_0} = (-1)^{y_1 x_1 \oplus y_0 x_0} = (-1)^{y \cdot x}$ (second term is modulo 2 addition since $(-1)^2 = 1$) $\therefore \hat{H}^{\otimes 2}|x\rangle = \frac{1}{\sqrt{2^2}}\sum_{y_1=0}^{1}\sum_{y_0=0}^{1}(-1)^{y \cdot x}|y\rangle = \frac{1}{\sqrt{2^2}}\sum_{y=0}^{2^2-1}(-1)^{y \cdot x}|y\rangle$. $L = L + 1$: x, y L-bit numbers with bit values $x_{L-1}x_{L-2}\ldots x_1 x_0$ and $y_{L-1}y_{L-2}\ldots y_1 y_0$. x' is an $L+1$-bit number with bit values $x_L x_{L-1}\ldots x_0$ (where $x_{L-1}\ldots x_0$ are the same as the bit values for x) (similarly y'). $\hat{H}|x_L\rangle \otimes \hat{H}^{\otimes L}|x\rangle = \frac{1}{\sqrt{2}}\sum_{y_L=0}^{1}(-1)^{x_L y_L}|y_L\rangle \otimes \frac{1}{\sqrt{2^L}}\sum_{y=0}^{2^L-1}(-1)^{x \cdot y}|y\rangle = $

$\frac{1}{\sqrt{2^{L+1}}}\sum_{y_L=0}^{1}\sum_{y=0}^{2^L-1}\underbrace{(-1)^{x_L y_L + x \cdot y}}_{(-1)^{x_L y_L \oplus x \cdot y} = (-1)^{x' \cdot y'}}\underbrace{|y_L\rangle \otimes |y\rangle}_{\equiv |y'\rangle}$ $\therefore \hat{H}^{\otimes L+1}|x'\rangle = \frac{1}{\sqrt{2^{L+1}}}\sum_{y'=0}^{2^{L+1}-1}(-1)^{x' \cdot y'}|y'\rangle$, which

$\underbrace{\quad}_{\equiv \sum_{y'=0}^{2^{L+1}-1}}$

completes the induction step. ∎

Chapter 5

5.1 Exercises

1. $f(t) = \int_{-\infty}^{\infty}\tilde{f}(v)\exp(-2\pi i v t)dv = \sqrt{2\pi(\Delta t)^2}\int_{-\infty}^{\infty}\exp\left[-\frac{(2\pi\Delta t)^2}{2}(v - v_0)^2 - 2\pi i v t\right]dv$, change variable: $v - v_0 = v'$, $= \sqrt{2\pi(\Delta t)^2}\exp(-2\pi i v_0 t)\int_{-\infty}^{\infty}\exp\left[-\frac{(2\pi\Delta t)^2}{2}v_0^2 - 2\pi i v' t\right]dv'$. With some algebraic manipulation, the argument in the exponent can be shown to be $-\frac{(2\pi\Delta t)^2}{2}\left(v' - \frac{it}{2\pi\Delta t^2}\right)^2 -$ $\frac{t^2}{2\Delta t^2}$ $\therefore f(t) = e^{-2\pi iv_0 t}e^{-t^2/2\Delta t^2}\frac{\sqrt{2\pi\Delta t^2}}{2\pi\Delta t}\int_{-\infty}^{\infty}\exp(-x^2/2)dx$ $(x = 2\pi\Delta t v' - it/\Delta t, \, dx = (2\pi\Delta t)dv')$. $f(t) = \exp(-2\pi iv_0 t)\exp(-t^2/2\Delta t^2)$.

3. (a) $r = 4$; $N = 12$; plot has peaks at $l = 3, l = 9$, weak peak at $l = 6$. We would expect peaks at integer multiples of $N/r = 3, 6, 9$ (Fig. S.4).

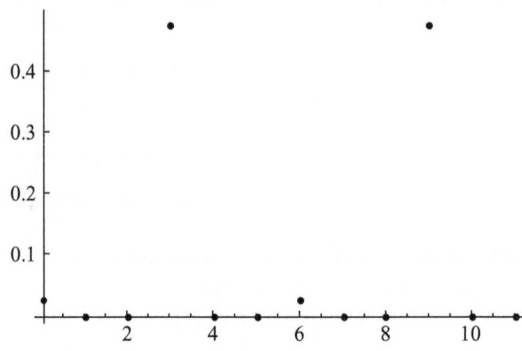

Figure S.4 Exercise 5.1.3.

(b) $r = 5$; $N = 18$; plot has peaks at $l = 4, 7, 11, 14$ (Fig. S.5). We expect peaks at integer multiples of $N/r = 3.6$. Round[3.6] = 4; Round[2 × 3.6] = 7; Round[3 × 3.6] = 11; Round[4 × 3.6] = 14.

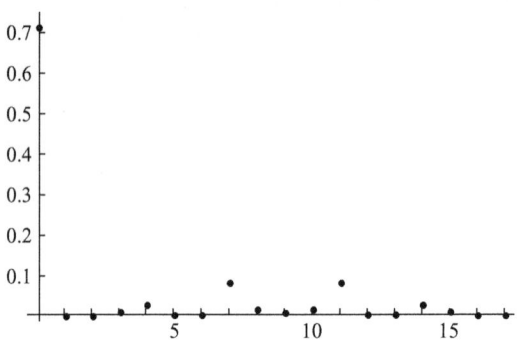

Figure S.5 Exercise 5.1.3.

5.2 Exercises

1. $|f\rangle = \sum_{l=0}^{2^L-1} f_l|l\rangle$. $\hat{U}_{\text{QFT}}|l\rangle = \frac{1}{\sqrt{2^L}}\sum_{k=0}^{2^L-1}\exp(2\pi i\frac{kl}{2^L})|l\rangle$ $\therefore \hat{U}_{\text{QFT}}|f\rangle = \sum_{l=0}^{2^L-1} f_l\hat{U}_{\text{QFT}}|l\rangle =$

$\sum_{k=0}^{2^L-1}\underbrace{\{\frac{1}{\sqrt{2^L}}\sum_{l=0}^{2^L-1} f_l\exp(\frac{2\pi ikl}{2^L})\}}_{=\tilde{f}_k}|k\rangle = \sum_{k=0}^{2^L-1}\tilde{f}_k|k\rangle = |\tilde{f}\rangle$.

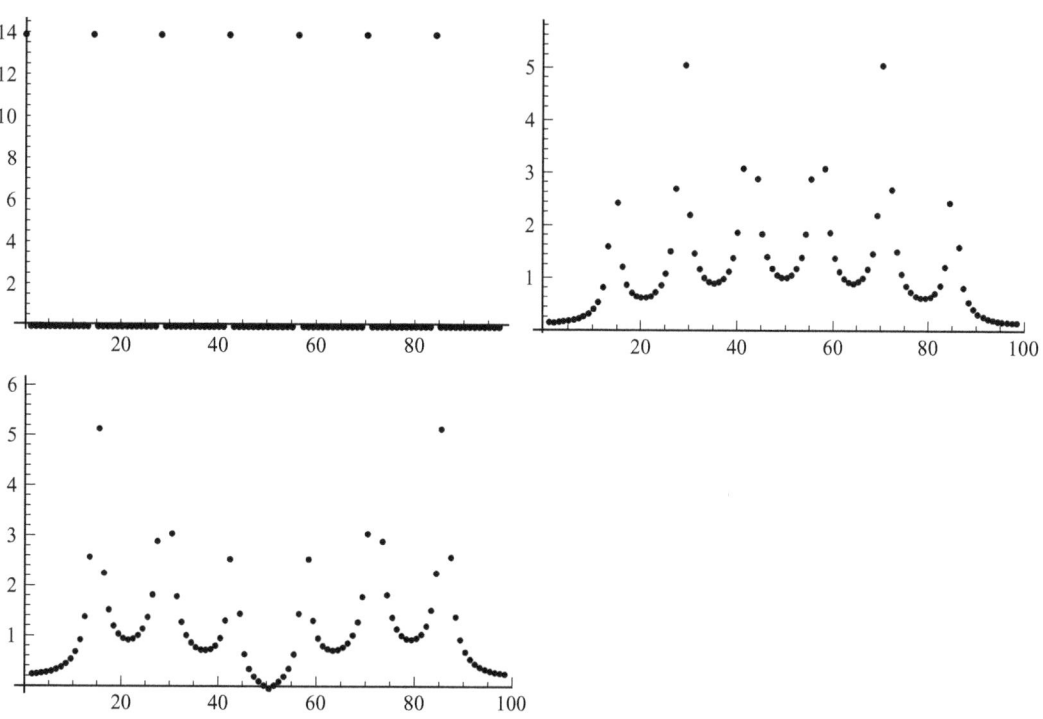

Figure S.6 Exercise 5.2.1.

3. $\hat{U}_{\text{QFT}}|l\rangle = \frac{1}{2}\{|00\rangle + \exp(\frac{2\pi il}{4})|01\rangle + \exp(\frac{2\pi i2l}{4})|10\rangle + \exp(\frac{2\pi i3l}{4})|11\rangle\}$. $C = 2|\alpha\delta - \beta\gamma| = 2|\frac{1}{2}\frac{e^{\frac{2\pi i3l}{4}}}{2} - \frac{e^{\frac{2\pi il}{4}}}{2}\frac{e^{\frac{2\pi i2l}{4}}}{2}| = 2|\frac{1}{4}e^{2\pi i3l/4} - \frac{e^{2\pi i3l/4}}{4}| = 0$.

5. (a) $1.85714 = 1 + \frac{1}{1.16667} = 1 + \frac{1}{1+\frac{1}{6}} = 1 + \frac{1}{\frac{7}{6}} = 1 + \frac{6}{7} = \frac{13}{7}$.

    (b) $0.73913 = \frac{1}{1.35294} = \frac{1}{1+\frac{1}{2.8333}} = \frac{1}{1+\frac{1}{2+\frac{1}{1.2}}} = \frac{1}{1+\frac{1}{2+\frac{1}{1+\frac{1}{5}}}} = \frac{1}{1+\frac{1}{2+\frac{5}{6}}} = \frac{1}{1+\frac{6}{17}} = \frac{17}{23}$.

    (c) $\sqrt{2} = 1.41421 = 1 + \frac{1}{2.41421} = 1 + \frac{1}{2+\frac{1}{2.41421}} = 1 + \frac{1}{2+\frac{1}{2+\frac{1}{2.41421}}}$ and so on. $1 + \frac{1}{2} = \frac{3}{2} = 1.5$,
    $1 + \frac{1}{2+\frac{1}{2}} = \frac{7}{5} = 1.4$, $1 + \frac{1}{2+\frac{1}{2+1/2}} = \frac{17}{12} = 1.41667$, $1 + \frac{1}{2+\frac{1}{2+\frac{1}{2+1/2}}} = \frac{41}{29} = 1.41379 \ldots$.

7. If one evaluates the function over (say) ~ 5 periods, then $2^L - 1 = 5T \implies L \simeq \log_2(5T + 1)$.

5.3 Exercises

1. (a) $13 - 3 = 10$ and 10 is divisible by 5.
    (b) $18 - 60 = -42$ and -42 is divisible by 2.
    (c) $46 - 2 = 44$ and 44 is divisible by 11.

3. $a \equiv b \pmod{m} \leftarrow n = 1$ case taken care of. Assume proposition is true for k such that $a^k \equiv b^k \pmod{m}$. Using result from Exercise 2, $aa^k \equiv bb^k \pmod{m} \implies a^{k+1} \equiv b^{k+1} \pmod{m} \leftarrow k+1$ case proved $\therefore a^n \equiv b^n \pmod{m}$. ∎

5. $p = 3, q = 5, N = 15, e = 7, M = 10, \phi(N) = 8, ed \equiv 1 \pmod{\phi(N)} \implies 7d \equiv 1 \pmod 8$, choose $d = 7$. $M^e \equiv R \pmod N$, $10^7 \equiv R \pmod{15}$, $10 \equiv -5 \pmod{15}$ (*). $(-5)^7 \equiv -5$ (mod 15), $(-5)^2 \equiv 10 \pmod{15}$, $(-5)^4 \equiv 10 \pmod{15}$, $(-5)^6 \equiv 10 \pmod{15}$, $(-5)^7 \equiv -50$ (mod 15), $(-5)^7 \equiv 10 \pmod{15}$ (**). Combining (*) and (**), $10^7 \equiv 10 \pmod{15} \implies R = 10$.
   $R^d \equiv M \pmod N$, $10^7 \equiv M \pmod{15}$, $10^7 \equiv 10 \pmod{15} \implies M = 10$.

7. (a) $1 = 29(-4) + 9(13)$, $2 = 38(-5) + 12(16)$, $1 = 60(-15) + 17(53)$.

   (b) $1 + 29(4) = 9(13)$, $2 + 38(5) = 12(16)$, $1 + 60(15) = 17(53)$. Note: The aim of this exercise will become apparent in Exercise 8 when the Euclidean algorithm is used to find $\text{GCD}(\phi(N), e)$. Then, the equation that has the last non-zero remainder is algebraically manipulated to be of the form in part (b) of this exercise, from which the value of d can be found.

5.4 Exercises

1. $C^x \equiv f_{N,C}(x) \pmod N$,
   $C^r \equiv f_{N,C}(r) = 1 \pmod N$,
   $C^{x+r} \equiv f_{N,C}(x + r) \pmod N$
   $\implies f_{N,C}(x)\underbrace{f_{N,C}(r)}_{=1} \equiv f_{N,C}(x + r) \pmod N \therefore f_{N,C}(x) \equiv f_{N,C}(x + r) \pmod N$.

3. $N = p_1 p_2 \cdots p_n$. If $N, N - 1$ are not coprime then one of the p_i must be a factor of $N - 1$; i.e., $\frac{N-1}{p_i} = $ integer. $\underbrace{\frac{p_1 p_2 \cdots p_n}{p_i}}_{\text{integer}} - \underbrace{\frac{1}{p_i}}_{\text{NOT an integer when } p_i = 1} \implies N, N - 1$ are coprime.

5. That is because when $x = x_f$, that is the first occurrence for $f_{N,C}(x) = 1$ (after the initial $f_{N,C}(0) = 1$); i.e., $x_f = r$, the period. That corresponds to the case when C^{x_f} is exactly 1 more than N; i.e., it equals $N + 1$. That is why the equality holds here.

Chapter 5 Exercises

1. p, q prime. $\phi(m) = $ # elements in $\{1, 2, 3, \ldots, m - 1\}$ that are relatively prime to $m, m \in \mathbb{Z}$. If $m = pq$ and $p \neq q$, in the set $\{1, 2, 3, \ldots, p, \ldots, \ldots, q, \ldots, \ldots, pq\}$, any number that is a multiple of either p and q is not relatively prime to pq. The number of multiples of p in the set is q, and the number of multiples of q in the set is p. Excluding pq, this means there are $p + q - 1$ numbers that are not relatively prime to $m = pq$, so subtracting this from the total number of numbers in the set, we obtain: $\phi(m) = \phi(pq) = pq - (p + q - 1) = pq - p - q + 1 = (p - 1)(q - 1)$. ∎

3. (a) $\lfloor \log_2 221 \rfloor + 1 = \lfloor 7.788 \rfloor + 1 = 8$ bits. $221_{10} = 11011101_2$. (b) $\text{GCD}(221, 11)$. Euclid's algorithm: $221 = 20 \times 11 + 1$, $11 = 11 \times 1 + 0 \implies 1$ is the GCD $\implies 221, 11$ are coprime. (c) Period of function is difficult to discern from plot. (d) $\tilde{f}_l$ has maxima at $l = 0, 21, 43, 64, 85, 107$. (e) $l_0 = 21 \simeq 2^{10}/r \implies 1/r = \frac{21}{1024} = 0.0205078$. Continued fractions: $\frac{21}{1024} = 0 + \frac{1}{48.7619} = 0 + \frac{1}{48 + \frac{1}{1.31251}} = 0 + \frac{1}{48 + \frac{1}{1 + \frac{1}{3.2}}} = 0 + \frac{1}{48 + \frac{1}{1 + \frac{1}{3 + \frac{1}{5}}}}$. Convergents: $\frac{1}{48}$, $\frac{1}{48 + \frac{1}{1}} = \frac{1}{49}$, $\frac{1}{48 + \frac{1}{1 + \frac{1}{3}}} = \frac{4}{195}$,

   $\frac{1}{48 + \frac{1}{1 + \frac{1}{3 + \frac{1}{5}}}} = \frac{21}{1024}$. We know that the ratio we seek is $1/r$ (since we looked at the first peak). Hence, we think: $r = 48$ or 49. Since C is not a perfect square (cf. Exercise 4), $r = 48$ seems the best to try. (f) $\text{GCD}(11^{24} + 1, 221) = 17$. $\text{GCD}(11^{24} - 1, 221) = 13$. $17 \times 13 = 221$.

5. (a)

Table S.5 Chapter Exercise 5.1.

k_{10}	k_2	$f_{15,4}(k)$
0	000	1=001
1	001	4=100
2	010	1=001
3	011	4=100
4	100	1=001
5	101	4=100
6	110	1=001
7	111	4=100

(b) Before modular exponentiation: $\underbrace{\frac{1}{2\sqrt{2}}(|000\rangle + |001\rangle + |010\rangle + |100\rangle + |101\rangle + |110\rangle + |000\rangle + |111\rangle)}_{|\text{all}\rangle_3}|000\rangle$. $\hat{X}$ on

qubit #0: $|\text{all}\rangle_3 \otimes |001\rangle$. Apply CNOTs: $|\varphi_1\rangle = \frac{1}{2}(|00\rangle + |01\rangle + |10\rangle + |11\rangle)\frac{1}{\sqrt{2}}(|0001\rangle +$

$|1100\rangle) = \frac{1}{2\sqrt{2}}\sum_{x,y=0}^{1}\left| \underbrace{xy0}_{\text{argument}}, \underbrace{001}_{\text{function}} \right\rangle + \left| \underbrace{xy1}_{\text{argument}}, \underbrace{100}_{\text{function}} \right\rangle$ (exactly what you'd expect from

part (a)).

(c) Qubits #4 and #5 are spectators; $\hat{H}$ is cancelled, and controlled R_k gates have no effect since control qubits are in $|0\rangle$. State before final $\hat{H}$ on qubit #3: $(|000001\rangle + |001100\rangle)1/\sqrt{2}$. $\hat{H}$ on #3: $(|000001\rangle + |001001\rangle + |000100\rangle - |001100\rangle +)1/2$. SWAP: $|\varphi_{\text{out}}\rangle = \frac{1}{2}(|000001\rangle + |100001\rangle + |000100\rangle - |100100\rangle)1/2$. $|\varphi_{\text{out}}\rangle = \frac{1}{2}(|000\rangle \otimes |001\rangle + |100\rangle \otimes |001\rangle + |000\rangle \otimes |100\rangle - |100\rangle \otimes |100\rangle)$. Denote $|000\rangle = |0\rangle$ and $|100\rangle = |4\rangle$ in argument register, and $|001\rangle = |1\rangle$ and $|100\rangle = |4\rangle$ in function register $\therefore |\varphi_{\text{out}}\rangle = \frac{1}{2}(|0\rangle \otimes |1\rangle + |4\rangle \otimes |1\rangle + |0\rangle \otimes |4\rangle - |4\rangle \otimes |4\rangle)$. Since both registers' states only involve two computational basis states, both are analogous to qubits. The concurrence is thus: $C = 2|\frac{1}{2}\frac{-1}{2} - \frac{1}{2}\frac{1}{2}| = 2|-\frac{2}{4}| = 1$. Yes, there is entanglement.

(d) Two possible values: $x = 0$ and $x = 4$ (50% probability each) (e) $x = 4 = \frac{2^L}{r} = \frac{8}{r} \implies r = \frac{8}{4} = 2$ (same as from table in part (a)). $C^{r/2} - 1 = 3$ and $C^{r/2} + 1 = 5 \implies$ factors of 15 are 3 and 5.

Chapter 6

6.1 Exercises

1. 2×10^7 qubits. $h\nu = \frac{hc}{\lambda} = \Delta E \implies \lambda = \frac{hc}{\Delta E}$. 2×10^7 qubits $\implies$ qubits on each row (with $\sqrt{20} \times 10^3$ rows). $\sqrt{20} \times 10^3 \times 10\lambda \simeq 4.5 \times 10^4 hc/\Delta E$. (a) 1.8 cm×1.8 cm. (b) 1.8 km×1.8 km. (c) 1859 km×1859 km.

3. $(1 - p_{\text{crit}})^N = 0.5$, $N\underbrace{\log(1 - p_{\text{crit}})}_{\simeq -p_{\text{crit}}} = -\log(2)$ $\therefore N = \log(2)/p_{\text{crit}} = 69$.

6.2 Exercises

1. $\hat{U}_0(t) = \cos(\omega t/2)\hat{I} + i\sin(\omega t/2)\hat{Z} = \begin{bmatrix} e^{i\omega t/2} & 0 \\ 0 & e^{-i\omega t/2} \end{bmatrix}$. $\hat{H} = \frac{1}{\sqrt{2}}(\hat{X} + \hat{Z}) \implies \hat{H}(t) = \frac{1}{\sqrt{2}}(\hat{X}(t) +$

$\hat{Z}(t))$. Use Eqs. (6.17) and (6.18) $\hat{H}(t) = \frac{1}{\sqrt{2}} \begin{bmatrix} 1 & e^{i\omega t} \\ e^{-i\omega t} & -1 \end{bmatrix}$.

3. (a) $(\hat{H} \otimes \hat{H})|00\rangle = \frac{1}{2}(|00\rangle + |01\rangle + |10\rangle + |11\rangle)$. $|\psi_{\text{out}}\rangle = \text{CNOT}(\hat{H} \otimes \hat{H})|00\rangle = \frac{1}{2}(|00\rangle + |01\rangle +$
$|10\rangle + |11\rangle)$. $C = 0$. (b) $|\psi_{\text{out}}\rangle = \text{CNOT}\hat{U}_0(t_1) \otimes \hat{U}_0(t_1)(\hat{H} \otimes \hat{H})|00\rangle = \text{CNOT}\frac{1}{2}(e^{i\omega t_1}|00\rangle + |01\rangle +$
$|10\rangle + e^{-i\omega t_1}|11\rangle) = \text{CNOT}\frac{1}{2}(e^{i\omega t_1}|00\rangle + |01\rangle + |11\rangle + e^{-i\omega t_1}|10\rangle)$. $C = 2|\frac{e^{i\omega t_1}}{4} - \frac{e^{-i\omega t_1}}{4}| = \sin(\omega t_1)$.

6.3 Exercises

1. (a) Let $A(t) = \alpha(t)\exp(i\omega_1 t/2)$, $B(t) = \beta(t)\exp(-i\omega_1 t/2)$ $\therefore \dot{A} = (\dot{\alpha} + i\frac{\omega_1}{2}\alpha)\exp(i\omega_1 t/2)$,
$\dot{B} = (\dot{\beta} - i\frac{\omega_1}{2}\beta)\exp(-i\omega_1 t/2)$. Substituting into equations for A and B: $i(\dot{\alpha} + \frac{i\omega_1}{2}\alpha)$
$\exp(i\omega_1 t/2) = -(\frac{\omega}{2})\alpha\exp(i\omega_1 t/2) - \Omega_0\cos(\omega_1 t)\beta\exp(-i\omega_1 t/2)$, $i(\dot{\beta} - \frac{i\omega_1}{2}\beta)\exp(-i\omega_1 t/2) =$
$-\Omega_0\cos(\omega_1 t)\alpha\exp(i\omega_1 t/2) + (\frac{\omega}{2})\beta\exp(-i\omega_1 t/2)$ $\therefore i\dot{\alpha} = -\frac{(\omega - \omega_1)}{2}\alpha - \Omega_0\cos(\omega_1 t)\beta\exp(-i\omega_1 t)$,
$i\dot{\beta} = -\Omega_0\cos(\omega_1 t)\alpha\exp(i\omega_1 t) + \frac{(\omega - \omega_1)}{2}\beta$. $\omega - \omega_1 = \Delta$, $\cos(\omega_1 t)\exp(\pm i\omega_1 t) = \frac{1 + \exp(\pm i 2\omega_1 t)}{2} \approx$
$\frac{1}{2}$ (RWA: rotating wave approximation). $\therefore i\dot{\alpha} = -\frac{\Delta}{2}\alpha - \frac{\Omega_0}{2}\beta$, $i\dot{\beta} = -\frac{\Omega_0}{2}\alpha - \frac{\Delta}{2}\beta$.

(b) Re-writing our equations: $\dot{\alpha} = i\frac{\Delta}{2}\alpha + i\frac{\Omega_0}{2}\beta$ (*), $\dot{\beta} = i\frac{\Omega_0}{2}\alpha - i\frac{\Delta}{2}\beta$. Differentiating: $\ddot{\alpha} =$
$i\frac{\Delta}{2}\dot{\alpha} + i\frac{\Omega_0}{2}\dot{\beta} = i\frac{\Delta}{2}\dot{\alpha} + i\frac{\Omega_0}{2}(\frac{i\Omega_0}{2}\alpha - i\frac{\Delta}{2}\beta) = i\frac{\Delta}{2}\dot{\alpha} - (\frac{\Omega_0}{2})^2\alpha + \frac{\Omega_0\Delta}{4}\beta$. (*) $\implies \beta = (\dot{\alpha} -$
$i\frac{\Delta}{2}\alpha)/(i\Omega_0/2)$ $\therefore \ddot{\alpha} = i\frac{\Delta}{2}\dot{\alpha} - \frac{\Omega_0^2}{4}\alpha + \frac{\Omega_0\Delta}{4}\frac{(\dot{\alpha} - i\frac{\Delta}{2}\alpha)}{i\Omega_0/2} = \cdots = \left(\frac{\Omega_0^2 + \Delta^2}{4}\alpha\right)$. Let $\Omega_0^2 + \Delta^2 = \Omega^2$.

$\therefore \alpha(t) = c_1\cos(\frac{\Omega t}{2}) + c_2\sin(\frac{\Omega t}{2})$, $\alpha(0) = c_1$, $\dot{\alpha}(0) = \frac{\Omega}{2}c_2 = i\frac{\Delta}{2}\alpha(0) + i\frac{\Omega_0}{2}\beta(0)$. $\therefore \alpha(t) =$
$\alpha(0)\cos(\frac{\Omega t}{2}) + i\frac{\{\Delta\alpha(0) + \Omega_0\beta(0)\}}{\Omega}\sin(\frac{\Omega t}{2})$. Similarly, $\beta(t) = \beta(0)\cos(\frac{\Omega t}{2}) + i\frac{\{\Omega_0\alpha(0) - \Delta\beta(0)\}}{\Omega}\sin(\frac{\Omega t}{2})$.

(c) $\begin{bmatrix} \alpha(t) \\ \beta(t) \end{bmatrix} = \begin{bmatrix} \cos(\Omega t/2) + i\frac{\Delta}{\Omega}\sin(\Omega t/2) & i\frac{\Omega_0}{\Omega}\sin(\Omega t/2) \\ i\frac{\Omega_0}{\Omega}\sin(\Omega t/2) & \cos(\Omega t/2) - i\frac{\Delta}{\Omega}\sin(\Omega t/2) \end{bmatrix} \begin{bmatrix} \alpha(0) \\ \beta(0) \end{bmatrix} = \{\cos(\Omega t/2)\hat{I} +$

$i\frac{\Delta}{\Omega}\sin(\Omega t/2)\hat{Z} + i\frac{\Omega_0}{\Omega}\sin(\Omega t/2)\hat{X}\} \begin{bmatrix} \alpha(0) \\ \beta(0) \end{bmatrix}$.

(d) $\alpha(0) = 1$, $\beta(0) = 0$ $\therefore \beta(t) = i\frac{\Omega_0}{\Omega}\sin(\Omega t/2) \implies |\beta(t)|^2 = \frac{\Omega_0^2}{\Omega^2}\sin^2(\Omega t/2)$. Maximum value:
$(\Omega_0/\Omega)^2 = \Omega_0^2/(\Omega_0^2 + \Delta^2) < 1$ $\therefore$ we can never do perfect X gates with $\Delta \neq 0$.

Chapter 6 Exercises

1. $T_{\text{crit}} = -\frac{\Delta E}{k_B \ln(p_{\text{crit}})}$. $T_{\text{crit}} = 77$ K $\implies \Delta E \gg -k_B T_{\text{crit}} \ln(p_{\text{crit}}) = 0.03$ eV. $\lambda = \frac{c}{\nu} = \frac{ch}{\Delta E} =$
4×10^{-5} m (terahertz range fields). $(N_{\text{qubits}})^{1/3}10\lambda \simeq 2700\lambda = 10$ cm $\implies$ 10 cm cube.

Chapter 7

7.1 Exercises

1. (a) $\lambda = 3$ cm, $hc/\lambda = 6.62 \times 10^{-24}$ J $= 4.1 \times 10^{-5}$ eV. (b) $\lambda = 1\mu$m, $hc/\lambda = 1.98 \times 10^{-19}$ J
$= 1.24$ eV. (c) $\lambda = 10^{-12}$ m $= 1$ pm, $hc/\lambda = 1.98 \times 10^{-13}$ J $= 1.24$ MeV.

3. Rotated HWP: $\hat{U}_{\text{HWP}}(\theta) = \begin{bmatrix} \cos 2\theta & \sin 2\theta \\ \sin 2\theta & -\cos 2\theta \end{bmatrix}$. Rotated QWP: $\hat{U}_{\text{QWP}}(\theta) = \frac{\exp(-i\pi/4)}{\sqrt{2}}$

$\begin{bmatrix} i + \cos 2\theta & \sin 2\theta \\ \sin 2\theta & i - \cos 2\theta \end{bmatrix}$. (a) Y-gate, $\hat{U}_{\text{HWP}}(\pi/4) = \hat{X}$, $\hat{U}_{\text{HWP}}(0) = \hat{Z}$. $\hat{Z}\hat{X} = i\hat{Y}$, where i is a

global phase. (b) $S = R_2$ gate $= \begin{bmatrix} 1 & 0 \\ 0 & \exp(2\pi i/4) \end{bmatrix} = \hat{U}_{\text{QWP}}(0)$. (c) H-gate $= \frac{1}{\sqrt{2}} \begin{bmatrix} 1 & 1 \\ 1 & -1 \end{bmatrix} =$

$\hat{U}_{\text{HWP}}(\pi/8)$.

7.2 Exercises

1. For electrons, $\Delta E = g\mu_B B_0$ and $g = 2$. For nuclei, $\Delta E = g\mu_B B_0 \left(\frac{m_e}{m_n}\right)$ and $g = 5.6$ (for ^{1}H), $g = 2.4$ (for ^{13}C), and $g = 39$ (for ^{31}P). (a) Electrons. $\Delta E = 1.8548 \times 10^{-23}$ J $= 1.16 \times 10^{-4}$ eV. (b) Proton. $g = 5.6$. $\Delta E = 2.83 \times 10^{-26}$ J $= 1.77 \times 10^{-7}$ eV. (c) ^{13}C, $g = 2.4$. $\Delta E = 1.2 \times 10^{-26}$ J $= 7.57 \times 10^{-8}$ eV.

3. $V_2 = V\cos\theta$, $\theta =$ angle of vector with z-axis. Some angles are more probable than others: there are more vectors pointing with $\theta = \pi/2$ than with $\theta \approx 0$ – there is only a single vector with $\theta = 0$. Number of vectors $\propto \sin\theta$. $\therefore$ $\overline{V_2} = \frac{\int_0^\pi V\cos\theta \sin\theta\,d\theta}{\int_0^\pi \sin\theta\,d\theta} = 0$. $\overline{V_2^2} = V^2 \int_0^\pi \cos^2\theta \sin\theta\,d\theta / \int_0^\pi \sin\theta\,d\theta = V^2 \left[-\frac{\cos^3\theta}{3}\right]_0^\pi / [-\cos\theta]_0^\pi = V^2 \frac{2}{3}/2 = \frac{V^3}{3} = \vec{V} \cdot \vec{V}/3$. ∎

7.3 Exercises

1. Use:

$$E(n) = \frac{-Z^2 E_I}{(n - \delta_n(l))^2} \begin{cases} +\frac{l\xi(n,l)}{2} & \text{if } j = l + 1/2, \\ -\frac{(l+1)\xi(n,l)}{2} & \text{if } j = l - 1/2, \end{cases}$$

$Z = 2$, $E_I = 13.59844$ eV. Use data from Table 7.1. The lowest levels are: $4s\,^2S_{1/2} \to -11.871745$ eV, $3d\,^2D_{3/2} \to -10.179318$ eV, $3d\,^2D_{5/2} \to -10.171796$ eV, $4p\,^2P_{1/2} \to -8.7483618$ eV, $4p\,^2P_{3/2} \to -8.7207273$ eV.

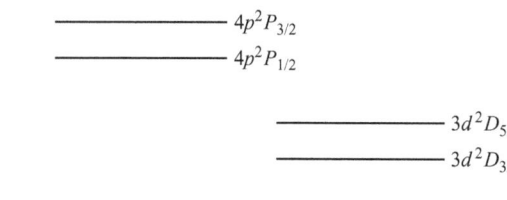

Figure S.7 Exercise 7.3.1.

3. Zeeman splitting: $\Delta\nu = \frac{g_J \mu_B B}{h}$. $3\,^2D_{5/2} \implies g_J = \frac{2(l+1)}{2l+1} = \frac{6}{5}$, $\Delta\nu = \frac{6}{5}\frac{\mu_B 0.01}{h} = 168$ MHz.

7.4 Exercises

1.

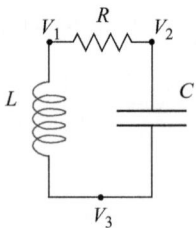

Figure S.8 Exercise 7.4.1.

Let Q be the charge on the capacitor. $V_2 - V_1 = -\dot{Q}R$, $V_3 - V_2 = -L\ddot{Q}$, $V_1 - V_3 = -Q/C$. $(V_2 - V_1) + (V_3 - V_2) + (V_1 - V_3) = 0$ ∴ $-\dot{Q}R - L\ddot{Q} - Q/C = 0 \implies \ddot{Q} + (R/L)\dot{Q} + (1/LC)Q = 0$. Denote: $\omega_0^2 = 1/LC$, $1/\tau = R/L$. To solve, find roots of $r^2 + r/\tau + \omega_0^2 = 0$ ∴ $r = \frac{-1/\tau \pm \sqrt{1/\tau^2 - 4\omega_0^2}}{2} = -\frac{1}{2\tau} \pm i\sqrt{\omega_0^2 - 1/(2\tau)^2}$ ∴ $Q(t) = Q_0 e^{-t/2\tau} \cos\left(\sqrt{\omega_0^2 - (1/2\tau)^2}\, t + \phi\right)$.

$R/L = 1/\tau$, $\tau =$ damping time; large $R \implies$ small τ.

3. (a) $\hat{a}^p \hat{n} = (\hat{n} + p)\hat{a}^p$. $p = 1$, $\hat{a}\hat{n} = \hat{a}\hat{a}^\dagger \hat{a} = (\hat{n} + 1)\hat{a}$, as required. For $p + 1$: $\hat{a}\hat{a}^p\hat{n} = \hat{a}(\hat{n} + p)\hat{a}^p = \hat{a}\hat{n}\hat{a}^p + p\hat{a}^{p+1} = (\hat{n} + 1)\hat{a}\hat{a}^p + p\hat{a}^{p+1} = (\hat{n} + p + 1)\hat{a}^{p+1}$, which completes the inductive step. ∎

(b) $p = 1$: $\hat{a}\hat{a}^\dagger = \hat{n} + 1$, as required. For $p+1$, $\hat{a}^{p+1}\hat{a}^{\dagger p+1} = \hat{a}(\hat{a}^p\hat{a}^{\dagger p})\hat{a}^\dagger = \hat{a}(\hat{n} + p)(\hat{n} + p - 1)\cdots(\hat{n} + 1)\hat{a}^\dagger$. From (a) we know $\hat{a}\hat{n} = \hat{a}\hat{a}^\dagger\hat{a} = (\hat{n} + 1)\hat{a}$, hence $\hat{a}(\hat{n} + p) = (\hat{n} + p + 1)\hat{a}$. Applying this recursively, $\hat{a}(\hat{n} + p)(\hat{n} + p - 1)\cdots(\hat{n} + 1)\hat{a}^\dagger = (\hat{n} + p + 1)(\hat{n} + p)(\hat{n} + p - 1)\cdots(\hat{n} + 2)\hat{a}\hat{a}^\dagger = (\hat{n} + p + 1)(\hat{n} + p)\cdots(\hat{n} + 1)$, which completes the inductive step.

(c) $p = 1$, $\hat{a}^\dagger \hat{a} = \hat{n}$. For $p + 1$: $\hat{a}^{\dagger p+1}\hat{a}^{p+1} = \hat{a}^\dagger(\hat{a}^{\dagger p}\hat{a}^p)\hat{a} = \hat{a}^\dagger \hat{n}(\hat{n} - 1)(\hat{n} - 2)\cdots(\hat{n} - p + 1)\hat{a}$, $\hat{n}\hat{a} = (\hat{a}\hat{a}^\dagger - 1)\hat{a} = (\hat{a}\hat{a}^\dagger\hat{a} - \hat{a}) = \hat{a}(\hat{n} - 1)$ ∴ $(\hat{n} - p)\hat{a} = \hat{a}(\hat{n} - p - 1)$. Apply recursively: $\hat{a}^\dagger \hat{n}(\hat{n} - 1)\cdots(\hat{n} - p + 1)\hat{a} = \hat{a}^\dagger \hat{n}(\hat{n} - 1)\cdots\hat{a}(\hat{n} - p - 1 + 1) = \hat{a}^\dagger\hat{a}(\hat{n} - 1)(\hat{n} - 2)\cdots(\hat{n} - p) = \hat{n}(\hat{n} - 1)(\hat{n} - 2)\cdots(\hat{n} - p)$, which completes the inductive step.

Chapter 7 Exercises

1. SHO, classically: $x(t) = x_0 \cos(\omega t) + (p_0/m\omega)\sin(\omega t) = \zeta \exp(-i\omega t) + \zeta^* \exp(i\omega t)$.

(a) $p(t) = m\dot{x}(t) - im\omega\zeta \exp(-i\omega t) + im\omega\zeta^* \exp(i\omega t)$. $\mathcal{E} = p^2/2m + m\omega^2 x^2/2 = \cdots = 2m\omega^2|\zeta|^2$.

(b) $\hat{x} = l(\hat{a} + \hat{a}^\dagger)$, $\hat{p} = -ilm\omega(\hat{a} - \hat{a}^\dagger)$, $l = \sqrt{\hbar/2m\omega}$. $\frac{1}{2m}\hat{p}^2 + \frac{1}{2}m\omega^2\hat{x}^2 = \frac{1}{2m}\left(-l^2 m^2 \omega^2\right)(\hat{a} - \hat{a}^\dagger)^2 + \frac{1}{2}m\omega^2 l^2(\hat{a} + \hat{a}^\dagger)^2 = \cdots = \hbar\omega(\hat{n} + 1/2)$. ∎

(c) $|\varphi(t)\rangle = \sum_n c_n \exp[-i(n + 1/2)\omega t]|\mathsf{n}\rangle$ ∴ $x(t) = l\langle\varphi|\hat{a}^\dagger + \hat{a}|\varphi\rangle = l\sum_{nm} c_n^* c_m \langle\mathsf{n}|\hat{a}^\dagger + \hat{a}|\mathsf{m}\rangle e^{i\omega(n-m)t} = l\sum_{nm} c_n^* c_m\{\sqrt{m + 1}\langle\mathsf{n}|\mathsf{m} + 1\rangle + \sqrt{m}\langle\mathsf{n}|\mathsf{m} - 1\rangle\}e^{i\omega(n-m)t} = \underbrace{l\sum_n c_n^* c_{n-1}\sqrt{n}\, e^{i\omega t}}_{\zeta^* e^{i\omega t}} + \underbrace{l\sum_m c_m c_{m-1}^* \sqrt{m}\, e^{-i\omega t}}_{\zeta e^{-i\omega t}}$. Let $m = n + 1$, $\zeta = l\sum_n c_{n+1}c_n^* \sqrt{n + 1}$.

(d) $\hat{\mathcal{H}} = \hbar\omega(\hat{n} + 1/2) - \hbar\Omega_0 \cos(\omega t)(\hat{a} + \hat{a}^\dagger)$, $i\hbar\frac{d}{dt}|\varphi(t)\rangle = \hat{\mathcal{H}}|\varphi(t)\rangle$. Let $|\varphi(t)\rangle = \sum_n c_n(t) \exp[-i(n + 1/2)\omega t]|\mathsf{n}\rangle$ (note that c_n is now time varying because of force) ∴ $\sum_n\{i\hbar\dot{c}_n(t) + \hbar\omega(n + 1/2)c_n(t)\}\exp[-i(n + 1/2)\omega t]|\mathsf{n}\rangle = \sum_n \hbar\omega(n + 1/2)c_n(t)\exp[-i(n + 1/2)\omega t]|\mathsf{n}\rangle - \hbar\Omega_0 \cos(\omega t)\sum_n c_n(t)\exp[-i(n + 1/2)\omega t](\hat{a} + \hat{a}^\dagger)|\mathsf{n}\rangle$. Multiply by $\langle\mathsf{m}|$ from the left and use

$\langle m|n\rangle = 0$ unless $n = m$, as well as the relations: $\langle m|\hat{a}|n\rangle = \sum_{p=0}^{\infty} \sqrt{p+1}\langle m|p\rangle\langle p+1|n\rangle = \delta_{n,m+1}\sqrt{m+1}$, $\langle m|\hat{a}^{\dagger}|n\rangle = \sum_{p=0}^{\infty} \sqrt{p+1}\langle m|p+1\rangle\langle p|n\rangle = \delta_{n,m-1}\sqrt{m}$ to get $\dot{c}_m(t) = i\Omega_0 \cos(\omega t)c_{m+1}(t)\sqrt{m+1}\exp(-i\omega t) + i\Omega_0 \cos(\omega t)c_{m-1}(t)\sqrt{m}\exp(i\omega t)$ $(m > 0)$. Rotating wave approximation: $\dot{c}_m(t) = \frac{i\Omega_0}{2}c_{m+1}(t)\sqrt{m+1} + \frac{i\Omega_0}{2}c_{m-1}(t)\sqrt{m}$ $(m > 0)$. Solution: If $c_0(0) = 1$, $c_m(0) = 0$, $m > 0$.

(e) Proof: Let $\Omega_0 t/2 = \tau$, $\frac{d}{dt} = \frac{d\tau}{dt}\frac{d}{d\tau} = \frac{\Omega_0}{2}\frac{d}{d\tau}$ $\therefore \dot{c}_m(t) = \frac{\Omega_0}{2}\frac{d}{d\tau}\left(\frac{(i\tau)^m}{\sqrt{m!}}\exp[-\tau^2/2]\right)_{\tau=\Omega_0 t/2} = \frac{\Omega_0}{2}\{\frac{im(i\tau)^{m-1}}{\sqrt{m!}}\exp(-\tau^2/2) + \frac{(i\tau)^m}{\sqrt{m!}}(-\tau)\exp(-\tau^2/2)\}_{\tau=\Omega_0 t/2} = \frac{\Omega_0}{2}\{i\sqrt{m}\frac{(i\tau)^{m-1}}{\sqrt{(m-1)!}}\exp(-\tau^2/2) + i\frac{(i\tau)^{m+1}}{\sqrt{(m+1)!}}\sqrt{m+1}\exp(-\tau^2/2)\}_{\tau=\Omega_0 t/2} = \frac{\Omega_0}{2}\{i\sqrt{m}c_{m-1}(t) + i\sqrt{m+1}c_{m+1}(t)\}$. ∎

$P_n(t) = $ Poisson distribution, with $\lambda = |\Omega_0 t/2|^2$. Trying to drive a two-level system on-resonance to perform quantum gates only results in all levels of the harmonic oscillator getting populated.

Chapter 8

8.1 Exercises

1. (a) $i\hbar\frac{\partial}{\partial t}\begin{bmatrix}\alpha\\\beta\\\gamma\\\delta\end{bmatrix} = \hbar J\begin{bmatrix}0\\-\beta+\gamma\\\beta-\gamma\\0\end{bmatrix}$, $\dot{\alpha} = \dot{\delta} = 0 \implies \alpha(t) = \alpha(0)$, $\delta(t) = \delta(0)$ $\therefore i\dot{\beta} = -J(\beta-\gamma)$,

$i\dot{\gamma} = J(\beta-\gamma) = -i\dot{\beta}$ $\therefore i(\dot{\beta}+\dot{\gamma}) = 0 \implies \beta(t) + \gamma(t) = \beta(0) + \gamma(0)$. $i(\dot{\beta}-\dot{\gamma}) = -2J(\beta-\gamma) \implies \beta(t) - \gamma(t) = \exp(i2Jt)\{\beta(0)-\gamma(0)\}$ $\therefore \beta(t) = \frac{1}{2}\{\beta(0)+\gamma(0)+e^{i2Jt}(\beta(0)-\gamma(0))\} = \beta(0)\left(\frac{1+e^{i2Jt}}{2}\right) + \gamma(0)\left(\frac{1-e^{i2Jt}}{2}\right) = e^{iJt}\{\beta(0)\cos(Jt) - i\gamma(0)\sin(Jt)\}$, $\gamma(t) = \frac{1}{2}\{\beta(0) + \gamma(0) - e^{i2Jt}\{\beta(0)-\gamma(0)\}\} = \beta(0)\left(\frac{1-e^{i2Jt}}{2}\right) + \gamma(0)\left(\frac{1+e^{i2Jt}}{2}\right) = e^{iJt}\{i\beta(0)\sin(Jt) + \gamma(0)\cos(Jt)\}$

$\therefore \begin{bmatrix}\alpha(t)\\\beta(t)\\\gamma(t)\\\delta(t)\end{bmatrix}\begin{bmatrix}1 & 0 & 0 & 0\\0 & e^{iJt}\cos(Jt) & ie^{iJt}\sin(Jt) & 0\\0 & ie^{iJt}\sin(Jt) & e^{iJt}\cos(Jt) & 0\\0 & 0 & 0 & 1\end{bmatrix}\begin{bmatrix}\alpha(0)\\\beta(0)\\\gamma(0)\\\delta(0)\end{bmatrix}$.

(b) If $t = \pi/2J$, then this is a SWAP gate. $\therefore t = \pi/4J$ gives a $\sqrt{\text{SWAP}}$.

(c) $|\varphi\rangle = \frac{1}{2}\{|00\rangle + e^{-2iJt}(|01\rangle + |10\rangle) + |11\rangle\}$. $C = \frac{2}{4}|1 - e^{-4iJt}| = \sin(2Jt)$.

3. $\hat{U}(t) = \exp\left[-i\{\frac{\omega_A t}{2}(\hat{Z}\otimes\hat{I}) + \frac{\omega_B t}{2}(\hat{I}\otimes\hat{Z}) + Jt(\hat{Z}\otimes\hat{Z})\}\right]$, $\hat{X}\hat{Z}\hat{X} = -\hat{Z}$ $\therefore (\hat{X}\otimes\hat{I})\hat{U}(t/2)(\hat{X}\otimes\hat{I})\hat{U}(t/2) = \exp\left[\frac{-i\omega_B t}{2}(\hat{I}\otimes\hat{Z})\right]$. $(\hat{I}\otimes\hat{X})\exp\left[\frac{-i\omega_B t}{2}(\hat{I}\otimes\hat{Z})\right](\hat{I}\otimes\hat{X})\exp\left[\frac{-i\omega_B t}{2}(\hat{I}\otimes\hat{Z})\right] = \hat{I}$. $\therefore (\hat{I}\otimes\hat{X})\{(\hat{X}\otimes\hat{I})\hat{U}(t/2)(\hat{X}\otimes\hat{I})\hat{U}(t/2)\}(\hat{I}\otimes\hat{X})\{(\hat{X}\otimes\hat{I})\hat{U}(t/2)(\hat{X}\otimes\hat{I})\hat{U}(t/2)\} = \hat{I}$, $(\hat{X}\otimes\hat{X})\hat{U}(t/2)(\hat{X}\otimes\hat{I})\hat{U}(t/2)(\hat{X}\otimes\hat{X})\hat{U}(t/2)(\hat{X}\otimes\hat{I})\hat{U}(t/2) = \hat{I}$.

8.2 Exercises

1. (a) Substituting the given Hamiltonian $\hat{\mathcal{H}}$ and the given state $|\varphi(t)\rangle$ into the Schrödinger equation, it can be shown that $\dot{\alpha}_n = g\sqrt{n}\beta_n e^{i(\omega-\omega_0)t}$, $\dot{\beta}_n = -g\sqrt{n}\alpha_n e^{-i(\omega-\omega_0)t}$, where $\omega - \omega_0 = \Delta$.

(b) Let $\alpha_n e^{-i\Delta t/2} = a_n \implies \dot{\alpha}_n = \dot{a}_n + i\frac{\Delta}{2}a_n = g\sqrt{n}b_n$ and $\beta_n e^{i\Delta t/2} = b_n \implies \dot{\beta}_n = \dot{b}_n - i\frac{\Delta}{2}b_n = -g\sqrt{n}a_n$ ∴ $\dot{a}_n = \frac{-i\Delta}{2}a_n + g\sqrt{n}b_n$, $\dot{b}_n = \frac{i\Delta}{2}b_n - g\sqrt{n}a_n \implies \ddot{a}_n = \frac{-i\Delta}{2}\dot{a}_n + g\sqrt{n}\dot{b}_n = \cdots = -\sqrt{(\frac{\Delta}{2})^2 + g^2 n}\, a_n = -(\frac{\Omega_n}{2})^2 a_n$, where $\Omega_n^2 = \Delta^2 + 4ng^2$. $a_n(t) = A\cos(\Omega_n t/2) + B\sin(\Omega_n t/2)$, $a_n(0) = A$, $\dot{a}_n(0) = B\Omega_n/2 = \frac{-i\Delta}{2}a_n(0) + g\sqrt{n}b_n(0)$ ∴ $a_n(t) = \alpha_n(0)\cos(\Omega_n t/2) + \{\frac{-i\Delta}{2}\alpha_n(0) + g\sqrt{n}\beta_n(0)\}\frac{\sin(\frac{\Omega_n t}{2})}{\Omega_n/2}$ ∴ $\alpha_n(t) = e^{i\Delta t/2}\alpha_n(0)\cos(\frac{\Omega_n t}{2}) + e^{i\Delta t/2}\frac{\{\frac{-i\Delta}{2}\alpha_n(0) + g\sqrt{n}\beta_n(0)\}}{\Omega_n/2}\sin(\frac{\Omega_n t}{2})$, $\beta_n(t) = e^{-i\Delta t/2}\beta_n(0)\cos(\frac{\Omega_n t}{2}) + e^{-i\Delta t/2}\frac{\{\frac{i\Delta}{2}\beta_n(0) - g\sqrt{n}\alpha_n(0)\}}{\Omega_n/2}\sin(\frac{\Omega_n t}{2})$.

(c) To perform pulses needed for a CZ gate, $\Delta = 0$, $n = 1$, $\alpha_1(0) = 0$, $\beta_1(0) = 1$, $\beta_1(t) = 0$, $\alpha_1(t) = 1$. $\alpha_1(t) = \beta_1(0)\sin(\frac{\Omega_n t}{2})$, $\frac{\Omega_n t}{2} = \frac{\pi}{2} \implies t = \pi/2g$.

(d) n random, $\Delta = 0$, $n = 1, 2, \ldots$ with probabilities p_n. $\bar{n} = \sum_n p_n n$. $\alpha_n(0) = 0$, $\beta_n(0) = 1$ (with probability p_n). Desired final state: $|0, 1\rangle$, $\langle 0, \text{n}|\varphi(t)\rangle = \sin(\Omega_n t/2)$ (with probability p_n). Fidelity $F = \langle 0, 1|\varphi(t)\rangle = \sin^2(t/2g)p_0$ – multiple excitations in quantum data bus (QDB) degrades fidelity of gate.

Chapter 8 Exercises

1. $\frac{1}{2}(\hat{Z} - \hat{I}) = -|1\rangle\langle 1| \implies \frac{1}{4}(\hat{I} - \hat{Z}) \otimes (\hat{I} - \hat{Z}) = |11\rangle\langle 11|$ ∴ $\exp\left[\frac{i\pi}{4}(\hat{Z} - \hat{I}) \otimes (\hat{Z} - \hat{I})\right] = e^{i\pi}|11\rangle\langle 11| + e^{i0}(|00\rangle\langle 00| + |01\rangle\langle 01| + |10\rangle\langle 10|) = (|00\rangle\langle 00| + |01\rangle\langle 01| + |10\rangle\langle 10| - |11\rangle\langle 11|) =$ CZ gate. ∎

3. (a) Denote $\hat{\Pi}_s = (\hat{I}\otimes\hat{I} - \hat{X}\otimes\hat{X} - \hat{Y}\otimes\hat{Y} - \hat{Z}\otimes\hat{Z})/4$. This is a rank-1 projector $\hat{\Pi}_s = |\beta_2\rangle\langle\beta_2|$, $|\beta_2\rangle = \frac{-1}{\sqrt{2}}(|01\rangle - |10\rangle)$ (singlet). $\hat{W}(\theta + \phi) = \hat{W}(\theta)\hat{W}(\phi)$. $\hat{W}(\theta) = \exp(i\theta\hat{\Pi}_s) = \exp(i\theta)|\beta_2\rangle\langle\beta_2| +$

$$|\beta_0\rangle\langle\beta_0| + |\beta_1\rangle\langle\beta_1| + |\beta_3\rangle\langle\beta_3| = \begin{bmatrix} 1 & 0 & 0 & 0 \\ 0 & (1+e^{i\theta})/2 & (1-e^{i\theta})/2 & 0 \\ 0 & (1-e^{i\theta})/2 & (1+e^{i\theta})/2 & 0 \\ 1 & 0 & 0 & 1 \end{bmatrix}, 1 - e^{i\theta} = 2 \implies e^{i\theta} =$$

$-1 \implies \theta = \pi$ ∴ $\hat{W}(\pi) = \text{SWAP} \implies \hat{W}(\pi/2) = \sqrt{\text{SWAP}}$.

(b) $\hat{Z}\hat{X}\hat{Z} = -\hat{X}, \hat{Z}\hat{Y}\hat{Z} = -\hat{Y}, \hat{Z}\hat{Z}\hat{Z} = \hat{Z}, \hat{Z}\hat{I}\hat{Z} = \hat{I}$ ∴ $(\hat{I}\otimes\hat{Z})\hat{\Pi}_s(\hat{I}\otimes\hat{Z}) = (\hat{I}\otimes\hat{I} + \hat{X}\otimes\hat{X} + \hat{Y}\otimes\hat{Y} - \hat{Z}\otimes\hat{Z})/4$ ∴ $(\hat{I}\otimes\hat{Z})\hat{\Pi}_s(\hat{I}\otimes\hat{Z}) + \hat{\Pi}_s = (\hat{I}\otimes\hat{I} - \hat{Z}\otimes\hat{Z})/2$ ∴ $(\hat{I}\otimes\hat{Z})\hat{W}(\pi/2)(\hat{I}\otimes\hat{Z})\hat{W}(\pi/2)$. Two useful facts: (i) if $\hat{a}^2 = \hat{I}$, then $\hat{a}f(\hat{b})\hat{a} = f(\hat{a}\hat{b}\hat{a})$. Proof: Use power-series form for $f(\hat{b})$: $f(\hat{b}) = f(0)\hat{I} + f'(0)\hat{b} + \frac{f''(0)}{2}\hat{b}^2 + \cdots + \frac{f^n(0)}{n!}\hat{b}^n + \cdots$, $(\hat{a}\hat{b}\hat{a})^n = \hat{a}\hat{b}\underbrace{\hat{a}\hat{a}}_{\hat{I}}\hat{b}\hat{a}\ldots\hat{a}\hat{b}\hat{a} = \hat{a}\hat{b}^n\hat{a}$ ∴ $\hat{a}f(\hat{b})\hat{a} = f(\hat{a}\hat{b}\hat{a})$.

(ii) For any operator, $\exp(\hat{A} + \hat{B}) = \exp(\hat{A})\exp(\hat{B})$ if $[\hat{A}, \hat{B}] = 0$. Thus, $(\hat{I} \otimes \hat{Z})\hat{W}(\theta/2)(\hat{I} \otimes \hat{Z})\hat{W}(\theta/2) = \exp(i\theta/2)\{\hat{I}\otimes\hat{I} + \hat{X}\otimes\hat{X} + \hat{Y}\otimes\hat{Y} - \hat{Z}\otimes\hat{Z}\}\exp(i\theta/8)\{\hat{I}\otimes\hat{I} - \hat{X}\otimes\hat{X} - \hat{Y}\otimes\hat{Y} - \hat{Z}\otimes\hat{Z}\} = \exp\left(i\frac{\theta}{2}\left[\frac{(\hat{I}\otimes\hat{I} - \hat{Z}\otimes\hat{Z})}{2} - \hat{\Pi}_s\right]\right)\exp\left(\frac{i\theta}{2}\hat{\Pi}_s\right)$, $\left[\left(\frac{(\hat{I}\otimes\hat{I} - \hat{Z}\otimes\hat{Z})}{2} - \hat{\Pi}_s\right), \hat{\Pi}_s\right] = \frac{-1}{2}[\hat{Z}\otimes\hat{Z}, \hat{\Pi}_s] = \frac{1}{2}[\hat{Z}\otimes\hat{Z}, \hat{X}\otimes\hat{X} + \hat{Y}\otimes\hat{Y}] = \frac{1}{8}(\hat{Z}\hat{X}\otimes\hat{Z}\hat{X} + \hat{Z}\hat{Y}\otimes\hat{Z}\hat{Y} - \hat{X}\hat{Z}\otimes\hat{X}\hat{Z} - \hat{Y}\hat{Z}\otimes\hat{Y}\hat{Z}) = \frac{1}{8}(-\hat{Y}\otimes\hat{Y} - \hat{X}\otimes\hat{X} + \hat{Y}\otimes\hat{Y} + \hat{X}\otimes\hat{X}) = 0$. ∴ $(\hat{I}\otimes\hat{Z})\hat{W}(\theta/2)(\hat{I}\otimes\hat{Z})\hat{W}(\theta/2) = \exp\left(\frac{i\theta}{8}2\{\hat{I}\otimes\hat{I} - \hat{Z}\otimes\hat{Z}\}\right) = e^{i\theta/4}\exp(\frac{-i\theta}{4}\hat{Z}\otimes\hat{Z})$. From Exercise 1 we know $\exp\left[\frac{i\pi}{4}(\hat{Z} - \hat{I}) \otimes (\hat{Z} - \hat{I})\right] =$ CZ. Since CZ is real and Hermitian, this is the same as CZ $= \exp\left[-\frac{i\pi}{4}(\hat{Z} - \hat{I}) \otimes (\hat{Z} - \hat{I})\right] = \exp(\frac{-i\pi}{4})\{\hat{Z} \otimes \hat{Z} - \hat{Z} \otimes \hat{I} - \hat{I} \otimes \hat{Z} + \hat{I} \otimes \hat{I}\} = e^{-i\pi/4}\left(\exp(\frac{i\pi}{4}\hat{Z}) \otimes \hat{I}\right)\left(\hat{I} \otimes \exp(\frac{i\pi}{4}\hat{Z})\right)\exp\left(\frac{-i\pi}{4}(\hat{Z} \otimes \hat{Z})\right) =$

$$e^{-i\pi/4}e^{i\pi/4}\hat{R}_2^\dagger \otimes \hat{I}e^{i\pi/4}(\hat{I} \otimes \hat{R}_2^\dagger)\exp\left(\frac{-i\pi}{4}(\hat{Z} \otimes \hat{Z})\right), \text{where } \hat{R}_k = \begin{bmatrix} 1 & 0 \\ 0 & \exp(i2\pi/2^k) \end{bmatrix} \therefore \text{ CZ} =$$

$$\exp(i\pi/4)(\hat{R}_2^\dagger \otimes \hat{R}_2^\dagger)\exp\left(\frac{-i\pi}{4}(\hat{Z} \otimes \hat{Z})\right) = \cdots = (\hat{R}_2 \otimes \hat{R}_2)(\hat{Z} \otimes \hat{I})\hat{W}(\pi/2)(\hat{I} \otimes \hat{Z})\hat{W}(\pi/2) \text{ (using }$$

$$\hat{R}_2^\dagger = \hat{R}_2\hat{Z}).$$

Chapter 9

9.1 Exercises

1.

Table S.8 Exercise 9.1.1.

	1	2	3	4	5	6	Score, n	p
1	2	3	4	5	6	7	2	1/36
2	3	4	5	6	7	8	3	2/36
3	4	5	6	7	8	9	4	3/36
4	5	6	7	8	9	10	5	4/36
5	6	7	8	9	10	11	6	5/36
6	7	8	9	10	11	12	7	6/36
							8	5/36
							9	4/36
							10	3/36
							11	2/36
							12	1/36

(a) 7 is the most probable score from two dice. $p_7 = 6/30 = 1/6$. (b) $\bar{n} \sum_{n=2}^{12} np_n = 7$. $V[n] = \overline{(n - \bar{n})^2} = \overline{n^2} - \bar{n}^2 = \frac{329}{6} - 7^2 = \frac{35}{6} \implies \sigma[n] = \sqrt{35/6} = 2.41$. (c) $1 - p_7 = 30/36 = 5/6$.
(d) New probability table:

Table S.9 Exercise 9.1.1(d).

n	2	3	4	5	6	7	8	9	10	11	12
p_n	$\frac{2}{42}$	$\frac{2}{42}$	$\frac{4}{42}$	$\frac{4}{42}$	$\frac{6}{42}$	$\frac{6}{42}$	$\frac{6}{42}$	$\frac{4}{42}$	$\frac{4}{42}$	$\frac{2}{42}$	$\frac{2}{42}$

6, 7, and 8 equally likely.
(e) For one dice:

Table S.10 Exercise 9.1.1(e).

n	1	2	3	4	5	6
p	$\frac{2}{8}$	$\frac{1}{8}$	$\frac{1}{8}$	$\frac{1}{8}$	$\frac{1}{8}$	$\frac{2}{8}$

$$p(n + m) = p_n \times p_m, p(k) = \sum_{n=1}^{6} p_n p_{k-n}$$

Table S.11 Exercises 9.1.1(e).

n	2	3	4	5	6	7	8	9	10	11	12
p	$\frac{1}{16}$	$\frac{1}{16}$	$\frac{5}{64}$	$\frac{3}{32}$	$\frac{7}{64}$	$\frac{3}{16}$	$\frac{7}{64}$	$\frac{3}{32}$	$\frac{5}{64}$	$\frac{1}{16}$	$\frac{1}{16}$

most likely result: 7 $p_7 = 3/16 = 0.1875$.

9.2 Exercises

1. (a) $S = (0, 0, 1)$. (b) $S = (0, 0, -1)$. (c) $S = (\cos\phi, \sin\phi, 0)$. (d) $S = (0, 0, 0)$.

3. $S_i = (2n_i/N) - 1$, $S_x = \frac{22}{25}$, $S_y = \frac{-7}{50}$, $S_z = \frac{-12}{25}$, $\hat{\rho} = \frac{1}{100}\begin{bmatrix} 26 & 44 + 7i \\ 44 - 7i & 74 \end{bmatrix}$. Eigenvalues:

$\lambda = 50 \pm \sqrt{2561}$. One eigenvalue is less than 0, and one is greater than 1. State is nearly pure. Small error due to Shor noise caused recovered state to be unphysical.

9.3 Exercises

1. $\mathcal{R} = \sum_i (y_i - mx_i - b)^2$, $\frac{\partial \mathcal{R}}{\partial m} = \sum_i -2x_i(y_i - mx_i - b) = -2(\overline{xy} - m\overline{x^2} - b\overline{x})\mathcal{N}$, $\frac{\partial \mathcal{R}}{\partial b} = \sum_i -2(y_i - mx_i - b) = -2(\overline{y} - m\overline{x} - b)\mathcal{N}$ $\therefore$ $\frac{\partial \mathcal{R}}{\partial m} = 0$ and $\frac{\partial \mathcal{R}}{\partial b} = 0$ imply: $m\overline{x^2} + b\overline{x} = \overline{xy}$, $m\overline{x} + b = \overline{y}$ $\implies$ $m\overline{x^2} + (\overline{y} - m\overline{x})\,\overline{x} = \overline{xy}$ $\implies$ $m(\overline{x^2} - \overline{x}^2) = \overline{xy} - \overline{x}\,\overline{y}$ $\implies$ $m = (\overline{xy} - \overline{x}\,\overline{y})/(\overline{x^2} - \overline{x}^2)$ and $b = \overline{y} - m\overline{x}$.

3. $S_i = \text{Tr}\hat{\rho}\hat{\sigma}_i$, $\rho = \frac{1}{2}\begin{bmatrix} 1 + S_x & S_x - iS_y \\ S_x + iS_y & 1 - S_z \end{bmatrix}$, Cholesky decomposition: $\hat{T} = \begin{bmatrix} u_1 & 0 \\ u_3 + iu_4 & u_2 \end{bmatrix}$. $\hat{T}^\dagger \hat{T} =$

$\begin{bmatrix} u_1 & u_3 - iu_4 \\ 0 & u_2 \end{bmatrix}\begin{bmatrix} u_1 & 0 \\ u_3 + iu_4 & u_2 \end{bmatrix} = \begin{bmatrix} u_1^2 + u_3^2 + u_4^2 & u_2(u_3 - iu_4) \\ u_2(u_3 + iu_4) & u_2^2 \end{bmatrix}$ $\therefore$ $S_z = u_1^2 - u_2^2 + u_3^2 + u_4^2 = 1 - 2u_2^2$, $S_x = 2u_2u_3$, $S_y = 2u_2u_4$, $u_2 = \sqrt{(1 - S_z)/2}$, $u_3 = S_x/\sqrt{2(1 - S_z)}$, $u_4 = S_y/\sqrt{2(1 - S_z)}$, $u_2^2 + u_3^2 + u_4^2 = \cdots = \frac{(1 - S_z)^2 + S_x^2 + S_y^2}{2(1 - S_z)}$ $\therefore$ $u_1^2 = 1 - (u_2^2 + u_3^2 + u_4^2) = \cdots = \frac{1 - S_x^2 - S_y^2 - S_z^2}{2(1 - S_z)}$

$\therefore$ $\hat{T} = \frac{1}{\sqrt{2(1 - S_z)}}\begin{bmatrix} \sqrt{1 - S_x^2 - S_y^2 - S_z^2} & 0 \\ S_x + iS_y & (1 - S_z) \end{bmatrix}$.

Chapter 10

10.1 Exercises

1. $\hat{\rho}_L = \frac{\exp(-\hat{\mathcal{H}}/k_BT)}{\text{Tr}\exp(-\hat{\mathcal{H}}/k_BT)}$, $\exp(\frac{-\hat{\mathcal{H}}}{k_BT}) \simeq \hat{I} - \frac{\hat{\mathcal{H}}}{k_BT} + \frac{1}{2}\frac{\hat{\mathcal{H}}^2}{(k_BT)^2} + \cdots$ $\therefore$ $\text{Tr}\exp(\frac{-\hat{\mathcal{H}}}{k_BT}) = \text{Tr}\hat{I} - \frac{1}{k_BT}\text{Tr}\hat{\mathcal{H}} +$

$O(\frac{1}{T^2}) = 2^L - \frac{1}{k_BT}\text{Tr}\hat{\mathcal{H}} + O(\frac{1}{T^2})$. $\therefore$ $\hat{\rho}_L = \frac{1}{2^L}\left(\hat{I} - \frac{\hat{\mathcal{H}}}{k_BT}\right)\left(1 - \frac{1}{2^Lk_BT}\text{Tr}\hat{\mathcal{H}}\right)^{-1} + O(\frac{1}{T^2}) =$

$\frac{1}{2^L}\left(\hat{I} - \frac{\hat{\mathcal{H}}}{k_BT}\right)\left(1 + \frac{1}{2^Lk_BT}\text{Tr}\hat{\mathcal{H}}\right) + O(\frac{1}{T^2}) = \frac{1}{2^L}\{\hat{I} - (k_BT)^{-1}(\hat{\mathcal{H}} - \text{Tr}\mathcal{H}/2^L) + O((k_BT)^2)\}$.

10.2 Exercises

1. Hadamard gate: $\hat{H}|0\rangle = (|0\rangle+|1\rangle)/\sqrt{2}, \hat{H}|1\rangle = (|0\rangle-|1\rangle)/\sqrt{2}$. Notation: $|1,0\rangle = 1$ photon in left–right propagating mode; 0 photon in the up–down mode. $|1,0\rangle = r|0,1\rangle + t|1,0\rangle, |0,1\rangle = r'|1,0\rangle + t'|0,1\rangle$. 50-50 beamsplitter $\implies r = t = t' = 1/\sqrt{2}, r' = -1/\sqrt{2}. |1,0\rangle \rightarrow 1/\sqrt{2}|0,1\rangle + |1,0\rangle$, $|0,1\rangle \rightarrow 1/\sqrt{2}|0,1\rangle - |1,0\rangle$; i.e., two modes represent one qubit. $|1,0\rangle = |0\rangle, |0,1\rangle = |1\rangle$.

3. Best analyzed in the Heisenberg picture: $\hat{a}_0, \hat{a}_1$ are the two modes before the beamsplitter, $\hat{a}_2, \hat{a}_3$ are the two modes after the beamsplitter. $r = -r' = t = t' = 1/\sqrt{2}. \hat{a}_2 = \frac{1}{\sqrt{2}}(\hat{a}_0 + \hat{a}_1)$ and $\hat{a}_3 = \frac{1}{\sqrt{2}}(-\hat{a}_0 + \hat{a}_1) \implies \hat{a}_0 = \frac{1}{\sqrt{2}}(\hat{a}_2 - \hat{a}_3)$ and $\hat{a}_1 = \frac{1}{\sqrt{2}}(\hat{a}_2 + \hat{a}_3). |1,1\rangle$ state before the beamsplitter is $\hat{a}_0^\dagger \hat{a}_1^\dagger |0,0\rangle$, after the beamsplitter: $\frac{1}{2}(\hat{a}_2 - \hat{a}_3)(\hat{a}_2 + \hat{a}_3)|0;0\rangle = \frac{1}{2}(\hat{a}_2^{\dagger 2} - \hat{a}_3^{\dagger 2})|0,0\rangle = \frac{1}{\sqrt{2}}(|2,0\rangle - |0,2\rangle). |2,2\rangle$ state before the beamsplitter is $|2,2\rangle = \frac{(\hat{a}_0^\dagger \hat{a}_1^\dagger)^2|0,0\rangle}{2} = \frac{(\hat{a}_2^{\dagger 2} - 2\hat{a}_2^\dagger \hat{a}_3^\dagger + \hat{a}_3^{\dagger 2})(\hat{a}_2^{\dagger 2} + 2\hat{a}_2^\dagger \hat{a}_3^\dagger + \hat{a}_3^{\dagger 2})|0,0\rangle}{8} = \cdots = (\sqrt{3}|4,0\rangle - \sqrt{2}|2,2\rangle + \sqrt{3}|0,4\rangle)/\sqrt{8}$.

10.3 Exercises

1. $(\hat{H} \otimes \hat{H} \otimes \hat{H})|000\rangle = \frac{1}{2\sqrt{2}}\{|000\rangle + |001\rangle + |010\rangle + |011\rangle + |100\rangle + |101\rangle + |110\rangle + |111\rangle\}$. CZ gates on (2,3) and (2,1). $|\varphi_{\text{cluster}}\rangle = \frac{1}{2\sqrt{2}}\{|000\rangle + |001\rangle + |010\rangle - |011\rangle + |100\rangle + |101\rangle - |110\rangle + |111\rangle\}$.

3. 1. Here is the circuit:

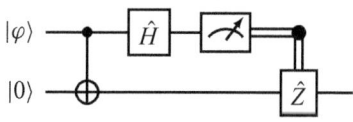

QC S.23 Exercise 10.3.3.

2. Apply the Griffiths–Niu theorem:

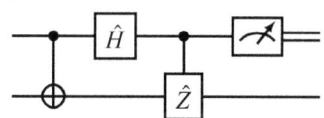

QC S.24 Exercise 10.3.3.

3. Reverse the controlled-Z gate and insert a pair of Hadamards:

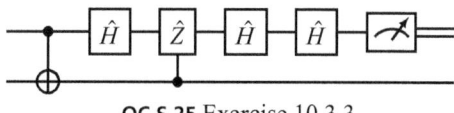

QC S.25 Exercise 10.3.3.

4. Use identity for controlled-NOT gates:

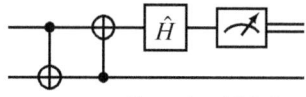

QC S.26 Exercise 10.3.3.

5. Qubit #0 is in state $|0\rangle$, hence another CNOT will have no effect:

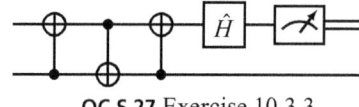

QC S.27 Exercise 10.3.3.

6. Three CNOTs make a SWAP:

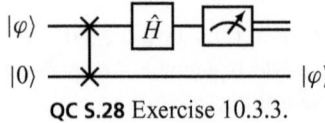

QC S.28 Exercise 10.3.3.

Chapter 10 Exercises

1.

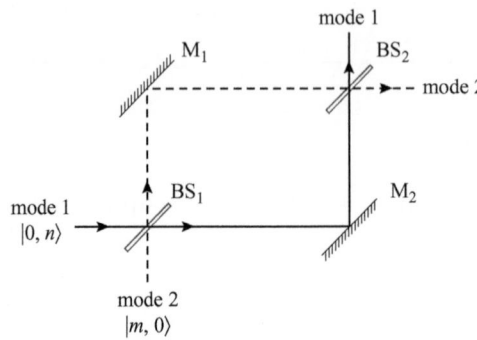

Figure S.9 Chapter Exercise 10.1.

Solid line: mode #1, dashed line: mode #2. $|m, n\rangle = |m\rangle_2 \otimes |n\rangle_1$. BS: sign change on reflection from the left.

(a) $|0, 1\rangle \xrightarrow{\text{BS}_1} \frac{1}{\sqrt{2}}(-|1, 0\rangle + |0, 1\rangle) \xrightarrow{\text{BS}_2} \frac{1}{\sqrt{2}}\left(\frac{-1}{\sqrt{2}}\{-|0, 1\rangle + |1, 0\rangle\} + \frac{1}{\sqrt{2}}\{|0, 1\rangle + |1, 0\rangle\}\right) = |0, 1\rangle$, 1 photon in mode 1 (solid line).

(b) M_2: reflection coefficient r, absorption/transmission coefficient a, $|r|^2 + |a|^2 = 1$. $|0, 1\rangle \xrightarrow{\text{BS}_1}$
$\frac{1}{\sqrt{2}}(-|1, 0\rangle + |0, 1\rangle) \xrightarrow{M_1} \frac{1}{\sqrt{2}}(-r|1, 0\rangle + |0, 1\rangle + a|0, 0\rangle) \xrightarrow{\text{BS}_2} \frac{1}{\sqrt{2}}(-r\frac{1}{\sqrt{2}}\{-|0, 1\rangle + |1, 0\rangle\}\frac{1}{\sqrt{2}}\{|0, 1\rangle + |1, 0\rangle\} + a|0, 0\rangle)$. $|\psi_{\text{out}}\rangle = \frac{1}{2}((1 + r)|0, 1\rangle + (1 - r)|1, 0\rangle) + \frac{a}{\sqrt{2}}|0, 0\rangle$.

(c) $r \to 0$, $|\psi_{\text{out}}\rangle = \frac{1}{2}(|0, 1\rangle + |1, 0\rangle) + \frac{1}{\sqrt{2}}|0, 0\rangle$. Detection of a photon in mode #2 implies there is a probability amplitude associated with photon absorption in either one of the paths: the photon has "sensed" the presence of an absorber without being scattered by it.

Chapter 11

11.1 Exercises

1. $\hat{\rho} = \begin{bmatrix} |\alpha|^2 & \alpha\beta^* e^{-t/T} \\ \alpha^*\beta e^{-t/T} & |\beta|^2 \end{bmatrix}$. $\text{Tr}\left\{\begin{bmatrix} a & b \\ c & d \end{bmatrix}\begin{bmatrix} a & b \\ c & d \end{bmatrix}\right\} = a^2 + d^2 + 2bc = a^2 + 2ad + d^2 - 2ad + bc = (a + d)^2 - 2(ad - bc)$. $\therefore \text{Tr}\hat{\rho}^2 = (|\alpha|^2 + |\beta|^2)^2 - 2(|\alpha|^2|\beta|^2 - |\alpha|^2|\beta|^2 e^{-t/T}) = 1 - 2|\alpha|^2|\beta|^2(1 - e^{-t/T})$.

11.2 Exercises

1. Qubit #1: $|\varphi\rangle \to \hat{U}_z(\phi)|\varphi\rangle$, Qubit #2: $|\varphi\rangle \to \hat{U}_z(-\phi)|\varphi\rangle \therefore \hat{U}_z(\phi) \otimes \hat{U}_z(-\phi)|00\rangle = |00\rangle, \hat{U}_z(\phi) \otimes \hat{U}_z(-\phi)|01\rangle = e^{i\phi}|01\rangle, \hat{U}_z(\phi) \otimes \hat{U}_z(-\phi)|10\rangle = e^{-i\phi}|10\rangle, \hat{U}_z(\phi) \otimes \hat{U}_z(-\phi)|11\rangle = |11\rangle$, hence the DFS should be $\{|00\rangle, |11\rangle\}$.

11.3 Exercises

1. Suppose we use n copies. The probability that m of them flip during transmission is $p(m) = \frac{m!}{n!(n-m)!}p^m(1-p)^{n-m}$. Majority voting fails after $m > n/2$, hence probability of a failed message is $\sum_{m=\lceil n/2 \rceil}^{n} p(m)$. $\Pr(m > \lceil n/2 \rceil) = \sum_{m=\lceil n/2 \rceil}^{n} \frac{m!}{n!(n-m)!}p^m(1-p)^{n-m}$ – sum evaluated numerically – smallest n for which $\Pr(m > \lceil n/2 \rceil) < 1/10$ is $n = 15$.

3. (a) Without error correction, state is $\hat{\rho} = (1-p)|\varphi_0\rangle\langle\varphi_0| + p\hat{X}|\varphi_0\rangle\langle\varphi_0|\hat{X}$. Assuming $\langle\varphi_0|\hat{X}|\varphi_0\rangle = 0$, $F_1 = \langle\varphi_0|\hat{\rho}|\varphi_0\rangle = 1 - p$. (b) Error-correcting code does not care if one or two errors occurred; all it does is transform three-qubit states as follows: $|000\rangle, |001\rangle, |010\rangle, |100\rangle \to |000\rangle$, $|111\rangle, |110\rangle, |101\rangle, |011\rangle \to |111\rangle$. Thus, if two qubits flipped, the final error-corrected state will be $\hat{X}|\varphi_0\rangle$. (c) Probability of two or three errors is $p' = 3p^2(1-p) + p^3 = p^2(3-2p)$. Thus, with three-qubit error corrections, the state is $\hat{\rho} = (1-p')|\varphi_0\rangle\langle\varphi_0| + p'\hat{X}|\varphi_0\rangle\langle\varphi_0|\hat{X}$. $F_3 = 1 - p' = 1 - 3p^2 + 2p^3 = (1-p)^2(1+2p) \therefore \frac{F_3}{F_1} = (1-p)(1+2p)$. This function has a maximum at $p = 1/4$. $(1 - \frac{1}{4})(1 + \frac{2}{4}) = \frac{9}{8} = 1.125$.

Chapter 11 Exercises

1. (a) $\hat{\rho}' = \sum_m \hat{K}_m^\dagger \hat{\rho} \hat{K}_m$. $\text{Tr}\hat{\rho}' = \sum_m \text{Tr}\hat{K}_m^\dagger \hat{\rho}\hat{K}_m = \text{Tr}\left(\sum_m \hat{K}_m \hat{K}_m^\dagger\right)\hat{\rho} = 1$. Let $\hat{E} = \sum_m \hat{K}_m \hat{K}_m^\dagger$. This must be true for any state. Let $\{|n\rangle\}$ be an orthonormal basis for the space. $\hat{\rho} = |n\rangle\langle n|$ is a valid state, hence, $\langle n|\hat{E}|n\rangle = 1$. $\hat{\rho} = \frac{1}{2}(|n\rangle + |m\rangle)(\langle n| + \langle m|)$ is also a valid state ($|n\rangle \neq |m\rangle$), hence $\frac{1}{2}(\langle n|\hat{E}|n\rangle + \langle m|\hat{E}|m\rangle + \langle n|\hat{E}|m\rangle + \langle m|\hat{E}|n\rangle) = 1 \therefore \langle n|\hat{E}|m\rangle + \langle m|\hat{E}|n\rangle = 0$ ($n \neq m$). Finally, $\hat{\rho} = \frac{1}{2}(|n\rangle + i|m\rangle)(\langle n| - i\langle m|)$ is also a state ($n \neq m$), hence $\frac{1}{2}(\langle n|\hat{E}|n\rangle + \langle m|\hat{E}|m\rangle + i\langle n|\hat{E}|m\rangle - i\langle m|\hat{E}|n\rangle) = 1$, hence $\langle n|\hat{E}|m\rangle - \langle m|\hat{E}|n\rangle = 0 \implies \langle n|\hat{E}|m\rangle = 0$ ($n \neq m$) $\therefore \hat{E} = \sum_m \hat{K}_m \hat{K}_m^\dagger = |i\rangle$.

(b) $\hat{Z}\begin{bmatrix} a & b \\ c & d \end{bmatrix}\hat{Z} = \begin{bmatrix} a & -b \\ -c & d \end{bmatrix} \therefore \alpha\hat{I}\begin{bmatrix} a & b \\ c & d \end{bmatrix}\hat{I} + \beta\hat{Z}\begin{bmatrix} a & b \\ c & d \end{bmatrix}\hat{Z} = \begin{bmatrix} (\alpha + \beta)a & (\alpha - \beta)b \\ (\alpha - \beta)c & (\alpha + \beta)d \end{bmatrix}$. Let $\alpha + \beta = 1, \alpha - \beta = e^{-t/T}, \alpha = \frac{1 + e^{-t/T}}{2}, \beta = \frac{1 - e^{-t/T}}{2} \therefore \hat{\rho}(t) = \sum_{m=1}^{2} \hat{K}_m^\dagger \hat{\rho}\hat{K}_m, \hat{K}_1 = \hat{I}\sqrt{\frac{1 + e^{-t/T}}{2}}, \hat{K}_2 = \hat{Z}\sqrt{\frac{1 - e^{-t/T}}{2}}$.

3. No known method to do error correction with two qubits, but errors can be *detected* using the following circuit:

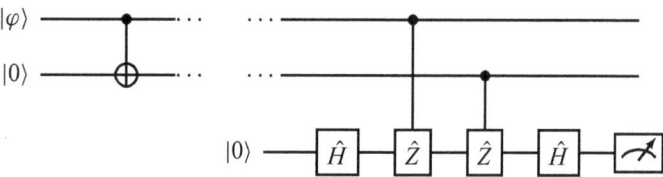

QC S.29 Chapter Exercise 11.3.

Let $|\varphi\rangle = \alpha|0\rangle + \beta|1\rangle$. With no errors $(\alpha|00\rangle + \beta|11\rangle)\otimes(|0\rangle + |1\rangle)/\sqrt{2} = 1/\sqrt{2}(\alpha|000\rangle + \alpha|001\rangle + \beta|110\rangle + \beta|111\rangle) \xrightarrow{CZ_{3,1}} 1/\sqrt{2}(\alpha|000\rangle + \alpha|001\rangle + \beta|110\rangle - \beta|111\rangle) \xrightarrow{CZ_{2,1}} 1/\sqrt{2}(\alpha|000\rangle + \alpha|001\rangle + \beta|110\rangle + \beta|111\rangle) \xrightarrow{\hat{I}\otimes\hat{I}\otimes\hat{H}} \frac{1}{\sqrt{2}}(\alpha|00\rangle + \beta|11\rangle)|0\rangle$.

If there is a bit flip: $(\alpha|01\rangle + \beta|10\rangle)(|0\rangle + |1\rangle)/\sqrt{2} = (\alpha|010\rangle + \alpha|011\rangle + \beta|100\rangle + \beta|101\rangle)/\sqrt{2} \xrightarrow{CZ_{3,1}} (\alpha|010\rangle + \alpha|011\rangle + \beta|100\rangle - \beta|101\rangle)/\sqrt{2} \xrightarrow{CZ_{2,1}} (\alpha|010\rangle - \alpha|011\rangle + \beta|100\rangle - \beta|101\rangle)/\sqrt{2} = (\alpha|01\rangle + \beta|10\rangle)(|0\rangle - |1\rangle)/\sqrt{2} \xrightarrow{\hat{I}\otimes\hat{I}\otimes\hat{H}} (\alpha|01\rangle + \beta|10\rangle)|1\rangle$. $|1\rangle \implies$ a bit-flip error occurred (but there is no way to know which qubit it was).

Chapter 12

12.1 Exercises

1. $k = $ one-time pad, (a) $m_1 = 0$, $m_2 = 1$, $d_1 = m_1 \oplus k = 0$, $d_2 = m_2 \oplus k = 1$. Assuming the eavesdropper is aware that the one-time pad is re-used, she concludes $m_1 \neq m_2$. In other words, if one is 0, the other is 1, and vice versa. (b) $m_1 = 00$, $m_2 = 10$, $k = 11$, $d_1 = 11$, $d_2 = 01$. $d_1 \oplus d_2 = m_1 \oplus k \oplus m_2 \oplus k = m_1 \oplus m_2$, which allows one bit to be deduced from the other. (c) $m_1 = 010$, $m_2 = 101$, $k = 101$, $d_1 = 111$, $d_2 = 000$, $m_1 \oplus m_2 = 111$, $d_1 \oplus d_2 = 111$.

3. (a) $|0\rangle$ measurement outcome 0 with 100% probability. (b) $|1\rangle$ measurement outcome 1 with 100% probability. (c) $|+\rangle$ measurement outcome 0 with 50% probability, measurement outcome 1 with 50% probability. (d) $|-\rangle$: identical to $|+\rangle$.

5. (a) Measurement in $\{|0\rangle, |1\rangle\}$ basis yields 0 if in state $|0\rangle$ and 1 if in state $|1\rangle$.

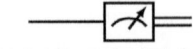

QC S.30 Exercise 12.1.5(a).

(b) Apply $\hat{H}$ followed by measurement yields 0 if in state $|+\rangle$ and 1 if in state $|1\rangle$.

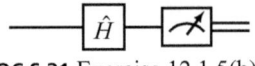

QC S.31 Exercise 12.1.5(b).

(c) Apply $\hat{U} = (\hat{Z} + \zeta\hat{X})/\sqrt{1 + \zeta^2}$, where $\zeta = 1 - \sqrt{2}$, followed by measurement: Outcome "0" implies $|0\rangle$ and "1" implies $|+\rangle$ with 85.4% probability; 14.6% chance of wrong answer.

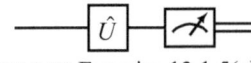

QC S.32 Exercise 12.1.5(c).

(d) Similar to (c) only with $\hat{U} = (\hat{Z} - \zeta\hat{X})/\sqrt{1 - \zeta^2}$: Outcome "0" implies $|1\rangle$ and "1" implies $|-\rangle$ with 85.4% probability; 14.6% chance of wrong answer.

7. If Inez measures in the $|+\rangle$, $|-\rangle$ basis, she gets $|+\rangle$; Bob gets $|+\rangle$ with 100% probability $\implies$ Inez and Bob both get the bit value of the key.

12.2 Exercises

1. $\hat{H}\hat{X}\hat{H} = \hat{Z}$, hence a bit flip can be converted into a phase flip (just like we did in error correction).

12.3 Exercises

1. We measured a different qubit that is not part of this entangled pair. The two photons were not measured in the same basis.

3. Need to distinguish photons transmitted then scattered from object from ambient photons.

12.4 Exercises

1. (a) $(\hat{A}_i^0 \equiv \hat{I})$, $\exp[\lambda(\hat{A}_1 + \hat{A}_2)] = \sum_{m=0}^{\infty} \frac{\lambda^m}{m!}(\hat{A}_1 + \hat{A}_2)^m = \sum_{m=0}^{\infty} \frac{\lambda^m}{m!} \sum_{n=0}^{m} \frac{m!}{n!(m-n)!} \hat{A}_1^n \hat{A}_2^{m-n} =$
$\sum_{m=0}^{\infty} \sum_{n=0}^{m} \left(\frac{\lambda^n \hat{A}_1^n}{n!}\right)\left(\frac{\lambda^{m-n}\hat{A}_2^{m-n}}{(m-n)!}\right) = \sum_{p,q=0}^{\infty} \left(\frac{\lambda^p \hat{A}_1^p}{p!}\right)\left(\frac{\lambda^q \hat{A}_2^q}{q!}\right) = \exp(\lambda\hat{A}_1)\exp(\lambda\hat{A}_2)$. ∎

(b) Prove: $\exp\left(\lambda \sum_{i=1}^{N} \hat{A}_i\right) = \prod_{i=1}^{N} \exp(\lambda\hat{A}_i)$ – true for $N = 2$ (see part (a)). $\exp\left(\lambda \sum_{i=1}^{N+1} \hat{A}_i\right) =$
$\exp\left(\lambda\left(\sum_{i=1}^{N}\hat{A}_i + \hat{A}_{N+1}\right)\right) = \exp\left(\lambda \sum_{i=1}^{N}\hat{A}_i\right)\exp(\lambda\hat{A}_{N+1}) = \left(\prod_{i=1}^{N}\exp(\lambda\hat{A}_i)\right)\exp(\lambda\hat{A}_{N+1}) =$
$\prod_{i=1}^{N-1}\exp(\lambda\hat{A}_{N+1})$, which completes the inductive step. ∎

3. $\hat{\mathcal{H}} = \frac{\hbar\Delta}{2}(\hat{Z} \otimes \hat{I} + \hat{I} \otimes \hat{Z}) + \hbar J(\hat{X} \otimes \hat{X})$. Let $J = \Delta \tan\chi$.

$$\hat{\mathcal{H}} = \frac{\hbar\Delta}{2}\begin{bmatrix} 1 & 0 & 0 & \tan\chi \\ 0 & 0 & \tan\chi & 0 \\ 0 & \tan\chi & 0 & 0 \\ \tan\chi & 0 & 0 & -1 \end{bmatrix}.$$

Eigenvalues and eigenvectors:
$\lambda_1 = -\sec\chi$, $|\varphi_1\rangle = \{-\sin(\chi/2), 0, 0, \cos(\chi/2)\}$
$\lambda_2 = \sec\chi$, $|\varphi_1\rangle = \{\cos(\chi/2), 0, 0, \sin(\chi/2)\}$
$\lambda_3 = -\tan\chi$, $|\varphi_1\rangle = \{0, 1, -1, 0\}/\sqrt{2}$
$\lambda_4 = \tan\chi$, $|\varphi_1\rangle = \{0, 1, 1, 0\}\sqrt{2}$
$\exp(-i\hat{\mathcal{H}}t/\hbar) = \sum_{n=1}^{4} e^{-i\Delta\, t\lambda_n}|\varphi_n\rangle\langle\varphi_n|$.

$\therefore P_{00} = |\sum_{n=1}^{4}\exp(-i\Delta\, t\lambda_n)(\langle 00|\varphi_n\rangle)^2|^2 = |\exp(i\Delta\, t\sec\chi)\sin^2(\chi/2) + \exp(-i\Delta\, t\sec\chi)$
$\cos^2(\chi/2)|^2 = \sin^4(\chi/2) + \cos^4(\chi/2) + \sin^2(\chi/2)\cos^2(\chi/2)2\cos(\Delta\, t2\sec\chi) = 1 - 2\sin^2(\chi/2)$
$\cos^2(\chi/2)(1 - 2\cos(2\Delta\, t\sec\chi)) = 1 - \sin^2\chi\sin^2(\Delta\, t\sec\chi) = 1 - \frac{J^2}{J^2+\Delta^2}\sin^2(\sqrt{J^2+\Delta^2}t)$
$(\tan\chi = \frac{J}{\Delta} \implies \sin\chi = \frac{J}{\sqrt{J^2+\Delta^2}}, \sec\chi = \frac{\sqrt{J^2+\Delta^2}}{\Delta})$ $(\omega_0 \to \Delta)$.

Chapter 12 Exercises

1. k = one-time pad. $m_1 = 0100$, $m_2 = 1111$, $k = 111 \implies d_1 = 1011$ and $d_2 = 0000$. $m_1 \oplus m_2 = 1011 \implies$ three different bits, one bit the same.

3. n = number of key bit-values adversary obtains: $p = 0.01$, $p(n = 0) = (1-p)^{20} = 81.7\%$, $p(n = 1) = 16.5\%$, $p(n = 2) = 0.016\%$, etc.

A Appendix A
Derivatives and Differential Equations

A.1 Derivatives

In science, we are interested in quantities that change as well as in the rates at which they change. For example, a policeman watching cars whizzing by is interested in the rate of change of the position of the cars with respect to time. This rate of change is known as velocity. Velocity is defined to be the derivative of position with respect to time. If the velocity is constant, the plot of position versus time is represented by a straight line. In this case, the derivative of the position with respect to time can be interpreted as *a rate of change*. On the other hand, if the position–time plot is not a straight line, but is represented by a curve, then the derivative is interpreted as *the slope of the tangent at a point on the curve*. In both cases, the derivative, using Leibniz's notation, is represented as shown in the left-hand side of these equations from Chapter 1:

$$\frac{d\mathbf{r}_i}{dt} = \mathbf{p}_i/m_i,$$

$$\frac{d\mathbf{p}_i}{dt} = \sum_{\substack{j=1 \\ j \neq i}}^{N} \mathbf{F}_{ij}. \tag{A.1}$$

Notes:

- When a function f depends on more than one variable, say x and y, the notation used for the derivative is $\frac{\partial f(x,y)}{\partial x}$ (read: partial $f(x,y)$ by partial x) or $\frac{\partial f(x,y)}{\partial y}$ instead of the straight ds as we saw in the preceding equations.
- Sometimes the derivative is represented simply by an apostrophe, instead of all these ds. For example, the first derivative of a function $f(x)$ is written as $f'(x)$, its second derivative as $f''(x)$, and so on. When making it unspecific, so for the n-th derivative, the representation is often written as $f^{(n)}(x)$. This notation is, for example, often used when writing the Taylor series expansion of a function around a point a, assuming it has a power series representation, as $f(x) = \frac{f'(a)}{1!}(x-a) + \frac{f''(a)}{2!}(x-a)^2 + \frac{f'''(a)}{3!}(x-a)^3 + \cdots$, or more compactly as $f(x) = \sum_{n=0}^{\infty} \frac{f^{(n)}(a)}{n!}(x-a)^n$.

A.2 Differential Equations

Equations that involve derivatives, such as the ones in (A.1) are called differential equations. There is a whole branch of mathematics that deals with solving them, given certain conditions. In this book, we will be dealing with very basic and simple ones, such as Schrödinger's *ordinary*, *homogeneous*, *linear*, and *first-order* equation as well as the harmonic oscillator's *ordinary*, *homogeneous*, *linear*, and *second-order* equation.

The sort of first-order differential equation we encounter in this book looks like this:

$$a\frac{d}{dt}y(t) + by(t) = 0,$$
(A.2)

where a and b are constants. A little algebra gives

$$\frac{d}{dt}y(t) = -\frac{b}{a}y(t).$$
(A.3)

What Eq. (A.3) tells us about $y(t)$ is that it is related to its first derivative; i.e., $\frac{d}{dt}y(t)$, by a constant. In other words, they are multiples of each other. This gives us a hint to what this function could possibly be: the exponential function. In fact, the general solution to such a differential equation is of the form:

$$y(t) = Ce^{-\frac{b}{a}t},$$
(A.4)

where C is a constant that can be found using the initial conditions; i.e., by using a given value for $y(0)$.

Now we move on to the kind of second-order differential equations that we encounter in this book, and they are of the form:

$$a\frac{d^2}{dt^2}y(t) + b\frac{d}{dt}y(t) + cy(t) = 0,$$
(A.5)

where a, b, and c are constants. This is called second order because the highest derivative that appears in the equation is a *second* derivative: $\frac{d^2}{dt^2}y(t)$. By examining this equation, again, we guess that the solution is of the form: e^{rt}, where r is a constant. If we substitute e^{rt} into (A.5), we will obtain the following quadratic equation:

$$ar^2 + br + c = 0,$$
(A.6)

which, in general, gives two solutions for r: r_1 and r_2. Both of them will give an e^{rt} that is a solution for (A.5), but to obtain the *general* solution, one has to take the linear combination of these two solutions so that

$$y(t) = C_1 e^{r_1 t} + C_2 e^{r_2 t},$$
(A.7)

where C_1 and C_2 are constants and can be found using the values for the initial conditions of the function and its first derivative; i.e., $y(0)$ and $\dot{y}(0) = \frac{d}{dt}y(0)$: $C_1 = (r_1 y(0) - \dot{y}(0))/(r_1 - r_2)$ and $C_2 = (r_2 y(0) - \dot{y}(0))/(r_2 - r_1)$.

B Appendix B
Singular Value Decomposition

We have seen how, for Hermitian matrices, the eigenvectors can be used to form a very useful *ortho-normal basis* for the vector space in question. But sometimes nature presents us with a non-Hermitian matrix: what can we do with it?

Let us call the matrix $\mathcal{M}$, with components M_{ij}, where the indices i and j take values $1, 2, \ldots, n$, N being the dimension of the space (here we will use the subscript/summation notation for the matrices to emphasize the analysis applies for *all* matrices, not just those representing quantum operations and quantum states, for which the ket-bra notation is reserved); thus, the product of $\mathcal{M}$ with another matrix $\mathcal{N}$ is

$$(\mathcal{M} \cdot \mathcal{N})_{ij} = \sum_{k=1}^{n} M_{ik} N_{kj}, \tag{B.1}$$

and so on. We assume that $\mathcal{M}$ is non-singular, so that it has an inverse and its determinant is not zero.

While $\mathcal{M}$ is not Hermitian, the matrix

$$\mathcal{A} = \mathcal{M}^{\dagger} \cdot \mathcal{M} \tag{B.2}$$

is Hermitian. It is also positive definite, since for any vector v_i,

$$\sum_{i,j=1}^{n} v_i^* A_{ij} v_j = \sum_{k=1}^{n} \left| \sum_{i=1}^{n} M_{ki} v_i \right|^2 \geq 0 \tag{B.3}$$

and $\mathrm{Det}\{\mathcal{A}\} = |\mathrm{Det}\{\mathcal{M}\}|^2 \neq 0$. In other words, the eigenvalues of $\mathcal{A}$ are positive real numbers; we will call them $s_1^2, s_2^2, \ldots, s_n^2$, where the s_n are called the *singular values* of $\mathcal{M}$. To keep track of the singular values, we define the diagonal matrix $\mathcal{S}$:

$$\mathcal{S} = \begin{bmatrix} s_1 & & \\ & \ddots & \\ & & s_n \end{bmatrix}. \tag{B.4}$$

Since $\mathcal{A}$ is Hermitian it can be diagonalized using a unitary matrix $\mathcal{V}$:

$$\mathcal{V}^{\dagger} \cdot \mathcal{A} \cdot \mathcal{V} = \mathcal{S}^2, \tag{B.5}$$

where $\mathcal{V}^{\dagger} \cdot \mathcal{V} = \mathcal{V} \cdot \mathcal{V}^{\dagger} = \mathcal{I}$. Now let's define another matrix $\mathcal{U}$ as follows:

$$\mathcal{U} = \mathcal{M} \cdot \mathcal{V} \cdot \mathcal{S}^{-1}. \tag{B.6}$$

The matrix $\mathcal{U}$ is also unitary:

$$
\begin{aligned}
\mathcal{U}^\dagger \cdot \mathcal{U} &= (\mathcal{S}^{-1} \cdot \mathcal{V}^\dagger \cdot \mathcal{M}^\dagger) \cdot (\mathcal{M} \cdot \mathcal{V} \cdot \mathcal{S}^{-1}) \\
&= \mathcal{S}^{-1} \cdot \mathcal{V}^\dagger \cdot \mathcal{A} \cdot \mathcal{V} \cdot \mathcal{S}^{-1} \\
&= \mathcal{S}^{-1} \cdot \mathcal{S}^2 \cdot \mathcal{S}^{-1} = \mathcal{I},
\end{aligned}
\tag{B.7}
$$

and

$$
\begin{aligned}
\mathcal{U} \cdot \mathcal{U}^\dagger &= (\mathcal{M} \cdot \mathcal{V} \cdot \mathcal{S}^{-1}) \cdot (\mathcal{S}^{-1} \cdot \mathcal{V}^\dagger \cdot \mathcal{M}^\dagger) \\
&= \mathcal{M} \cdot \mathcal{A}^{-1} \cdot \mathcal{M}^\dagger \\
&= \mathcal{M} \cdot \mathcal{A}^{-1} \cdot \mathcal{M}^\dagger \cdot \mathcal{M} \cdot \mathcal{M}^{-1} \\
&= \mathcal{M} \cdot \mathcal{A}^{-1} \cdot \mathcal{A} \cdot \mathcal{M}^{-1} = \mathcal{I}.
\end{aligned}
\tag{B.8}
$$

Thus, from Eq. (B.6), we find that any non-singular matrix $\mathcal{M}$ can be written as a simple product of a unitary, a positive diagonal matrix, and another unitary:

$$
\boxed{\mathcal{M} = \mathcal{U} \cdot \mathcal{S} \cdot \mathcal{V}^\dagger.}
\tag{B.9}
$$

This is the singular value decomposition. Our proof is much simplified by the assumption that $\mathcal{M}$ and hence $\mathcal{A}$ are non-singular, and so the inverses $\mathcal{M}^{-1}$ and $\mathcal{A}^{-1}$ exist: using a different approach, the theorem can in fact be proven for *any* matrix $\mathcal{M}$, even for ones that are not square.

Schmidt Decomposition

A pure quantum state of a two-qubit system can be written as

$$
\begin{aligned}
|\varphi\rangle &= \alpha|00\rangle + \beta|01\rangle + \gamma|10\rangle + \delta|11\rangle \\
&= \sum_{i,j=0}^{1} M_{ij}|i\rangle \otimes |j\rangle,
\end{aligned}
\tag{B.10}
$$

where

$$
\mathcal{M} = \begin{bmatrix} \alpha & \beta \\ \gamma & \delta \end{bmatrix}.
\tag{B.11}
$$

Applying the singular value decomposition to this matrix, we find the state $|\varphi\rangle$ can be written in the following form:

$$
|\varphi\rangle = \sum_{k=0}^{1} s_k \left\{ \sum_{i=0}^{1} U_{ik}|i\rangle \right\} \otimes \left\{ \sum_{j=0}^{1} V_{jk}^*|j\rangle \right\},
\tag{B.12}
$$

where the matrices $\mathcal{U}$ and $\mathcal{V}$ are unitary. Let's define two local unitary operators $\hat{U}_A$ and $\hat{U}_B$ as follows:

$$
\begin{aligned}
\hat{U}_A^\dagger &= \sum_{\ell,m=0}^{1} U_{m\ell}^*|\ell\rangle\langle m|, \\
\hat{U}_B^\dagger &= \sum_{\ell',m'=0}^{1} V_{m'\ell'}|\ell'\rangle\langle m'|.
\end{aligned}
\tag{B.13}
$$

We find

$$
\begin{aligned}
(\hat{U}_A^\dagger \otimes \hat{U}_B^\dagger)|\varphi\rangle &= \sum_{k=0}^{1} s_k \left\{ \sum_{i,\ell,m=0}^{1} U_{m\ell}^* U_{ik} |\ell\rangle \langle m|i\rangle \right\} \otimes \left\{ \sum_{j,\ell',m'=0}^{1} V_{m'\ell'} V_{jk}^* |\ell'\rangle \langle m'|j\rangle \right\} \\
&= \sum_{k=0}^{1} s_k \left\{ \sum_{i,\ell=0}^{1} U_{i\ell}^* U_{ik} |\ell\rangle \right\} \otimes \left\{ \sum_{j,\ell'=0}^{1} V_{j\ell'} V_{jk}^* |\ell'\rangle \right\} \\
&= \sum_{k=0}^{1} s_k \left\{ \sum_{\ell=0}^{1} \delta_{\ell k} |\ell\rangle \right\} \otimes \left\{ \sum_{j,\ell'=0}^{1} \delta_{\ell' k} |\ell'\rangle \right\} \\
&= \sum_{k=0}^{1} s_k |k\rangle \otimes |k\rangle.
\end{aligned}
\tag{B.14}
$$

And thus we have the Schmidt decomposition theorem for two qubits: any pure state can be written in the following form

$$
|\varphi\rangle = (\hat{U}_A \otimes \hat{U}_B)(s_0|00\rangle + s_1|11\rangle),
\tag{B.15}
$$

where s_0 and s_1 are positive real quantities and $s_0^2 + s_1^2 = 1$.

This theorem can be extended to multi-level bipartite systems by using a larger matrix $\mathcal{M}$: the singular value decomposition applies to matrices of any size. Unfortunately there is no easy way to extend the decomposition to more than two subsystems.

Exercises

1. Show that the unitary matrix $\mathcal{U}$ defined by Eq. (B.6) diagonalizes the matrix $\mathcal{B} = \mathcal{M} \cdot \mathcal{M}^\dagger$.
2. Work out the components of $\mathcal{A} = \mathcal{M}^\dagger \cdot \mathcal{M}$ for the matrix defined in Eq. (B.11). Show by direct calculation that it has a determinant equal to $(C/2)^2$, where C is the concurrence.

Appendix C
C
Probabilities and Probability Distributions

The probability of an event occurring is a quantification of the likelihood that it occurs. Unless otherwise specified, we assume that all outcomes of an experiment are equally likely. For example, if the experiment in question is tossing a coin, then we assume that obtaining Heads is as equally likely as obtaining Tails. In that case, if we want to calculate the probability of obtaining Heads, we obtain 1/2. In another experiment that involves throwing a die, the probability of obtaining an even number would be $3/6 = 1/2$. In other words, to obtain the probability, one has to divide the number of outcomes of the nature that we are interested in, n(event), by the number of all possible outcomes, n(all outcomes). In the die experiment we describe above, n(event) = 3 and n(all outcomes) = 6, since the event in this case is obtaining 2, 4, or 6, while all possible outcomes correspond to obtaining 1, 2, 3, 4, 5, or 6.

In science, we are often interested in finding the probability of obtaining a specific event if we repeat the experiment several times. Before we elaborate, let us first assume that these repeated experiments are *identical* and *independent*, such as in throwing a die multiple times. We call such experiments **Bernoulli trials**. In each experiment, if we obtain the event we are interested in, such as obtaining the number 5 when throwing the die, we call that a *success*. Otherwise, we say the experiment was a *failure*! We can label the probability of success p and the probability of failure $q = 1 - p$. In this case, the probability of obtaining a certain number of successes in a certain number of Bernoulli trials is given by the *binomial distribution*. If the number of trials is n, then the possible number of successes, x, is an element in the set: $\{0, 1, \ldots, n\}$. The binomial distribution is given by

$$p(x) = \binom{n}{x} p^x q^{n-x}, \tag{C.1}$$

where $\binom{n}{x} = n!/x!(n-x)!$. To comment on the rationale behind Eq. (C.1), let us look at an example. Say we are going to throw a die 10 times, where success is obtaining 5. In this case, $p = 1/6$ and $q = 5/6$. When p is small compared to q, as we have in this example, we expect that for high values of x, say $x = n = 10$, the probability is very low. That is why raising p to the power x takes care of this. At the same time we have to also take into account the probability of obtaining a failure, which is what the term q^{n-x} takes care of, using a similar reasoning. Moreover, we also need to take into account that there are many ways that x successes may occur in n Bernoulli trials, which is taken care of by the term $\binom{n}{x}$. What Eq. (C.1) gives us is the probability of obtaining x successes in n Bernoulli trials.

When dealing with the binomial distribution, an issue arises when the number of trials is large, as the expression becomes impractical to evaluate. Fortunately, one can avoid this situation by approximating the binomial distribution by a *normal distribution*. Of course, this is not always possible, but we will introduce a rule of thumb that we can use to help decide when it is okay to do so and how one can go about doing it.

A normal distribution is a distribution that can be defined for continuous as well as discrete values. It represents data that are symmetrically distributed about the mean (average), μ, where the probability of getting a result in a certain interval decreases as the distance of the interval from the mean increases. This distribution (probabilities versus data) is given by a *bell*-shaped curve. Hence, besides being called a *normal curve*, it is also know as the *bell curve* (other names for the normal curve that you might encounter are Gaussian curve and error curve). The probability that the score lies between the values x and $x + dx$ is $P(x)dx$ where

$$P(x) = \frac{1}{\sqrt{2\pi}\sigma} \exp\left(-\frac{(x - \mu)^2}{2\sigma^2}\right). \tag{C.2}$$

Since the probabilities have to add up to 1, the area under a normal curve has to equal 1. Standard deviation, σ, is a statistical property that is defined as the distance between μ and the point of inflection of the curve. A normal curve is completely determined by μ and σ and is often represented by the symbol $N(\mu, \sigma^2)$, where σ^2, the square of the standard deviation, is also known as the variance.

Since the area under a normal curve equals 1, it can be used to determine probabilities so that the probability that an observation is in a given interval is simply the area under the curve bound by that interval. To avoid calculating this area each time it is needed, one can refer to tables in the literature that already have the various areas under a normal curve calculated, but keep in mind that they are usually constructed for the normal curve $N(0, 1)$, the standard normal distribution. Therefore, to use these tables for other normal distributions, once has to first standardize them. This ends up being much simpler to calculate than the binomial distribution in cases with a large number of trials. That is not always possible, but a good rule of thumb is this: If both np and nq are ≥ 5, then it is acceptable to approximate the binomial distribution by a normal distribution with the properties that $\mu = np$ and $\sigma^2 = npq$.

Appendix D
Haar-Distributed Random Unitary Matrices

The **Haar measure** is the equivalent for random matrices of the uniform distribution for random variables, and numerical matrices distributed according to the Haar measure are used in applications such as random circuit sampling.

Generating a random matrix is quite simple to do numerically: just use a random number generator to fill in each element of the matrix! Generating a random *unitary* matrix is a bit more complicated: we have to use the specific model of a unitary (such as Eq. (2.60)) and choose random values of the parameters, appropriately normalized. But uniformly distributed random parameters – in which all allowed values of the parameters are equally likely – does not necessarily lead to a uniform distribution of the matrix: certain types of matrix might have higher probability than others. To resolve this issue when simulating random circuits, for example, mathematicians use a concept called the *Haar measure*. A probability measure is an alternative way of specifying probability distributions; the Haar measure is such a probability measure with specific symmetry properties, making it akin to a uniform distribution of angles on the Bloch sphere. (A more thorough introduction can be found in "Introduction to Haar measure tools in quantum information: A beginner's tutorial" by Antonio Anna Mele, arXiv:2307.08956.)

The following simple numerical recipe to generate Haar-distributed random unitary matrices is adapted from Francesco Mezzadri, "How to generate random matrices from the classical compact groups," *Notices of the American Mathematical Society*, **54** (2007), 592–604 (see also arXiv:math-ph/0609050v2).

1. Create an $N \times N$ complex matrix $\mathcal{Z}$ whose elements are all complex random numbers with both real and imaginary parts being standard normal random variables (i.e., they are normally distributed random numbers of mean 0 and variance 1).
2. Use a numerical **QR decomposition** routine to factor $\mathcal{Z} = \mathcal{QR}$, where $\mathcal{Q}$ is an $N \times N$ unitary matrix and $\mathcal{R}$ is an $N \times N$ upper triangular matrix. The QR decomposition is widely used in linear algebra, and routines to find the matrices $\mathcal{Q}$ and $\mathcal{R}$ are standard in many numerical libraries.
3. Using the matrix $\mathcal{R}$ create the diagonal matrix Λ whose elements are $R_{ii}/|R_{ii}|$ $(i = 1, 2, \ldots, N)$, where R_{ij} are the elements of $\mathcal{R}$.
4. The matrix $\mathcal{Q}' = \mathcal{Q}\Lambda$ is distributed with the Haar measure.

Glossary

Addition modulo 2: For *single*-bit numbers x and y, $x \oplus y$ is the remainder when $x + y$ is divided by 2. For two L-bit numbers, $z = x \oplus y$ is defined to be the L-bit number whose bits are the sum modulo 2 of the corresponding bits of x and y, i.e., $z_m = x_m \oplus y_m$, for $m = 0, 1, \ldots, L - 1$.

Alkali: A type of atom with a single valence electron; all the other electrons are in closed shells; this simplifies the atomic structure considerably.

Angular momentum: Formally, $\hat{\mathbf{L}} = \hat{\mathbf{r}} \times \hat{\mathbf{p}}$, where $\hat{\mathbf{r}}$ is the particle's position and $\hat{\mathbf{p}}$ is its momentum.

Annihilation operator: An operator that decreases the number of photons (or other excitation) by one when applied on a number state.

Anti-bunched light: Light in which the time interval between photons is on average longer than one would expect from Poisson statistics.

Anti-Hermitian: A matrix or operator that equals the negative of its Hermitian adjoint: $\hat{A}^\dagger = -\hat{A}$.

Argument register: A set of qubits whose overall state stores the bit values of a number representing the argument x of some function $f(x)$.

ASCII: *American Standard Code for Information Interchange*; a set of standardized 8-bit binary numbers denoting the letters and punctuation used in European and North American script.

Auto-correlation function: A function that quantifies how much a random variable $v(t)$ will change in time; formally $\Gamma_v(t_1, t_2) = \overline{v(t_1)v(t_2)}$, where the overbar indicates the average.

Avalanche photo-diode: An electronic photon detector that releases a large number of electrons in response to a single photon (just like a single snowflake can induce an avalanche of snow in the mountains).

Basis: A linearly independent set of vectors that can be used to describe any given vector in the space in question. For example, the horizontal and vertical directions in a two-dimensional space are independent; therefore, any vector in that space can be written as a linear combination of the vectors that describe the horizontal and vertical directions.

BB84: The Bennett and Brassard 1984 protocol for quantum key distribution which allows the sender and the receiver to construct the one-time pad securely by following a series of steps that involves sending and measuring photon states in conjunction with classical communication.

Beamsplitter: An optical device that partially reflects and partially transmits an incoming electromagnetic wave. In the case of a single photon, there is a certain probability associated with reflection and another with transmission. A polarizing beamsplitter reflects vertically polarized light and transmits horizontally polarized light.

Bell states: A set of four linearly independent and maximally entangled two-qubit states. They also make a basis called the *Bell basis*. There are a variety of notations in use: Bell himself used the following notation:

$$|\Phi^+\rangle = (|00\rangle + |11\rangle)/\sqrt{2}$$

$$|\Phi^-\rangle = (|00\rangle - |11\rangle)/\sqrt{2}$$
$$|\Psi^+\rangle = (|01\rangle + |10\rangle)/\sqrt{2}$$
$$|\Psi^-\rangle = (|01\rangle - |10\rangle)/\sqrt{2}.$$

The convention based on the circuit used to create Bell states is also popular (QC G.1):

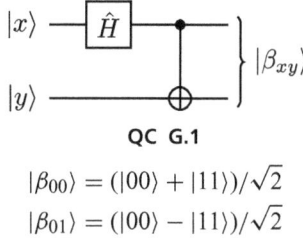

QC G.1

$$|\beta_{00}\rangle = (|00\rangle + |11\rangle)/\sqrt{2}$$
$$|\beta_{01}\rangle = (|00\rangle - |11\rangle)/\sqrt{2}$$
$$|\beta_{10}\rangle = (|01\rangle + |10\rangle)/\sqrt{2}$$
$$|\beta_{11}\rangle = (|01\rangle - |10\rangle)/\sqrt{2}.$$

The *magic basis* was introduced specifically for problems in mixed-state entanglement:

$$|e_1\rangle = (|00\rangle + |11\rangle)/\sqrt{2}$$
$$|e_2\rangle = i(|00\rangle - |11\rangle)/\sqrt{2}$$
$$|e_3\rangle = i(|01\rangle + |10\rangle)/\sqrt{2}$$
$$|e_4\rangle = (|01\rangle - |10\rangle)/\sqrt{2}.$$

In this book we adopted a constructionist approach as well: the state $|\beta_0\rangle = (|00\rangle + |11\rangle)/\sqrt{2}$ came from a natural maximally entangled Schmidt decomposition; and the other basis states from applying a Pauli gate to one qubit: $|\beta_k\rangle = i\hat{I} \otimes \hat{\sigma}_k |\beta_0\rangle$

$$|\beta_0\rangle = (|00\rangle + |11\rangle)/\sqrt{2}$$
$$|\beta_1\rangle = i(|01\rangle + |10\rangle)/\sqrt{2}$$
$$|\beta_2\rangle = -(|01\rangle - |10\rangle)/\sqrt{2}$$
$$|\beta_3\rangle = i(|00\rangle - |11\rangle)/\sqrt{2}.$$

Bell's inequality: In order to explain the randomness of quantum mechanics, the notion of *hidden variables* was hypothesized: the randomness of quantum mechanics is no more than a consequence of our ignorance about the true state of the system, quantified by these variables which are hidden from us due to the inadequacy of experimental technique. In 1964 John S. Bell discovered a test of hidden variable theories, based on the notion that their existence must constrain the possible correlations between measurements of two quantum systems. Experiments soon demonstrated the violation of these constraints, implying hidden variables were not consistent with nature.

Bernoulli trials: Repeated experiments that are identical, but independent.

Bit: A numerical digit that can only take two values, usually labelled 0 and 1. It is short for binary digit.

Bit-flip gate: A unitary operator that flips the two possible states of a qubit into each other. Another name for the $\hat{X}$-gate.

Bloch vector: A three-component real vector that specifies the quantum state of a qubit; analogous to the older Stokes parameters used for photon polarization.

Bohr magneton: $\mu_B = e\hbar/2m_e = 9.2740100657 \times 10^{-24}$ joules per tesla. This is the fundamental unit of magnetic dipole moment.

Born's rule: What happens when you measure a quantity (let's call it Ω) pertaining to a quantum system is that the value you obtain is random; it can have a variety of possible values ω_n (which are the eigenvalues of the operator $\hat{\Omega}$ that corresponds to the quantity you are measuring). The probability of obtaining the value ω_n is $P(\omega_n) = |\langle \omega_n | \varphi \rangle|^2$ where $|\omega_n\rangle$ is the normalized eigenvector of $\hat{\Omega}$ and $|\varphi\rangle$ is the state of the system. After the measurement, the system (if it has survived the measurement process) should be projected into the state $|\omega_n\rangle$.

Bra: The Hermitian adjoint of its corresponding ket.

Byte: A term used to describe a group of eight bits.

C* algebra: An algebra is a linear vector space that has a vector product; i.e., one can multiply two elements of the vector space to get another element. In quantum mechanics, the kets (or state vectors) are part of a linear vector space, but not part of an algebra: one cannot multiply two kets together to get another ket. However operators – including the density operator – can be multiplied together to produce another operator, and hence they form an algebra. Since this algebra is closed (i.e., you cannot multiply two operators together and get something that isn't an operator) and can have complex elements with complex conjugation, mathematically it's called a C* algebra.

CHSH inequality: A Bell inequality for an experimental scenario set up by Clauser, Horne, Shimony, and Holt. Eventually, Clauser and Freedman demonstrated its violation in an actual experiment.

Classical entanglement: Intermingling of the degrees of freedom of the same classical system with each other.

Classical randomness: Randomness that is due to our ignorance or lack of complete knowledge of the system or experimental limitations. In other words, it is a randomness that decreases or goes away the more we know about our system and the better experiments we can build.

Closed shells: Energy levels of an atom in which all the angular momentum states available to electrons are occupied: for every electron with a spin m_s and orbital angular momentum m_l there is a corresponding electron with spin $-m_s$ and orbital angular momentum $-m_l$.

Closed system: A system that is isolated from interactions with the outside world, and, therefore, exchanges nothing, such as energy, with it.

Cluster state: A massively entangled multi-qubit state which can be used for quantum computation simply by measurements and local one-qubit operations. This is a promising alternative paradigm for universal quantum computing in that it avoids a number of challenges posed by the circuit-based quantum computer.

Coherence time: Roughly speaking, the time interval over which a quantum state will remain pure.

Commutator bracket: In general, matrices and operators do not commute, i.e., $\hat{A}\hat{B} \neq \hat{B}\hat{A}$. The commutator bracket is defined by $[\hat{A}, \hat{B}] = \hat{A}\hat{B} - \hat{B}\hat{A}$.

Completely positive trace-preserving map: A linear operation which transforms a density operator to another operator that is Hermitian, positive, and unit trace. These requirements place certain constraints on the mathematical form of the map.

Computational basis: The basis that is made of the eigenvectors of the Pauli matrix $\hat{Z}$. It emulates classical bits in computer science in that its elements are made of single kets, as opposed to superposition of kets.

Concatenation: Adding a new layer of complexity to quantum error correction. For example, if one corrects for bit-flip errors using a 3-qubit code, one could decrease the effect further by replacing each of those three qubits by three more, for a total of 9; a further level of concatenation would replace each of those 9 qubits with 3 more, for at total of 27. Provided one can perform all the required quantum gates sufficiently accurately, repeated levels of concatenation enable fault-tolerant quantum computation of arbitrary accuracy.

Concurrence: A measure for quantum entanglement of two qubits. It is based on the idea that if two qubits are entangled, then the fact that the individual qubits cannot be described independently will lead to them becoming mixed if that is attempted. How mixed they become is used to quantify how strong the entanglement between them is.

Configuration: The actual and definite state that a system is in, with no probabilities associated with this information.

Control qubit: The qubit whose state determines whether an operation is to be performed on another qubit or not, namely the *target* qubit.

Convex hull decomposition: A way of writing the density matrix as a set of pure-state projectors with associated probabilities: $\hat{\rho} = \sum_k p_k |\varphi_k\rangle\langle\varphi_k|$. Since $|\varphi_k\rangle$ is not necessarily an eigenvector, the different pure states are not necessarily orthogonal, nor is the decomposition unique.

Cooper pair box: Electronic superconducting circuit consisting of an inductor and a Josephson junction in parallel which can act as a qubit.

CPTP maps: see Completely positive trace-preserving map.

Creation operator: An operator that increases the number of photons by one when applied on a number state.

Cryptography: The science of writing encoded messages that only the recipient can decode.

Decoherence: The decay of the quantum properties of a system due to its interactions with the environment.

Decoherence-free subspace: A subspace of a space of more than one qubit that is immune to the effects of decoherence, which can be used to create a single logical qubit.

Density matrix: A matrix representation that describes all the correlations between the various basis elements that describe the quantum state. For pure states, it is simply the outer product of the ket representing the state.

Dephasing: Decoherence that results in the off-diagonal elements of the density matrix of a system decaying due to the randomization of the phases of these elements. In this book, we label the dephasing time by T. It is more commonly labelled T_2 in the literature.

Depopulation: Decoherence that not only leads to the decay of the off-diagonal elements of the density matrix, but also to the flipping of the qubit state, affecting the diagonal elements as well. The depopulation time is longer than the dephasing time and often labelled by T_1 in the literature.

Detuning: The mismatch between the laser frequency and the frequency of the energy difference between two atomic levels of interest.

Deutsch's algorithm: The first quantum computing algorithm to demonstrate a computational advantage from quantum superposition.

Deutsch-Jozsa algorithm: A generalization of Deutsch's algorithm that was the first quantum computing algorithm to demonstrate a computational advantage that scales exponentially with the size of the problem.

Dichotomic observable: An observable that has two possible measurement outcomes, usually ± 1 (from the same root as the word *dichotomy*).

Discrete Fourier transform: A mathematical transformation of a list of values which reveals the strengths of various harmonic signals present in the data.

Doppler effect: A change in the vibrational frequency of light (or other wave field) due to relative motion of the source and observer.

Earnshaw's theorem: A mathematical theorem for electrostatic fields that shows there is no electrostatic field with a point of stable equilibrium.

Eigenvector decomposition: Decomposition of a matrix in terms of the outer products of its eigenvectors. A matrix is in effect equal to the sum of the outer product of its eigenvectors, where each term in the sum is multiplied by the corresponding eigenvalue.

Electron spin resonance (ESR): Resonant interaction between the magnetic moment of an electron and an oscillating magnetic field.

Entanglement: The inability to write a state of a system as a product, each pertaining to the system's degrees of freedom.

EPR dilemma: Named for Einstein, Podolsky, and Rosen, who pointed out the seeming incompatibility of instantaneous wave-function collapse upon measurement with special relativity.

Expectation value: The average value of a quantity.

Fault-tolerant quantum computing: see Concatenation.

Fidelity: A measure that tells how closely similar one quantum state is to another quantum state. For pure states $F = |\langle \varphi | \psi \rangle|^2$.

Fine structure: An atomic electron's spin interacts with the magnetic field induced by the electron's orbit to produce a splitting in the atomic energy levels.

Fourier transform: When expressing a function as a linear combination of complex exponential functions, the weighting factor in each term is called the Fourier transform.

Function register: A set of qubits whose overall state represents the binary representation of a function's value evaluated for a given number.

Fundamental theorem of entanglement: In a maximally entangled two-qubit state, applying a local operator on one qubit of the pair is equivalent to applying a related operator on the other qubit.

Gate: A device or an experimental setup that is designed to perform an operation on the building blocks of a computer.

Global phase: A phase that multiplies every element of the ket representation of a quantum state. It has no effect on the study of quantum systems, making it acceptable to ignore it.

Griffiths–Niu theorem: A two-qubit gate followed by a single-qubit measurement is equivalent to a single-qubit measurement followed by a conditional single-qubit gate (see QC 3.12).

Grover's search algorithm: A quantum algorithm that results in a polynomial speed-up of database search.

Haar measure: A symmetric probability distribution for unitary operators, roughly equivalent to a uniform distribution for variables.

Hamiltonian: A Hermitian matrix that describes the energy of a system and, paired with the Schrödinger equation, tells us what the quantum state evolves to at a later time.

Harmonic oscillator: A dynamic system in which the force on a body is always acting towards the body's rest position, and grows in magnitude in proportion to the distance from the rest position. The resultant motion is a sinusoidal oscillation.

Hermitian: A matrix is Hermitian if it is equal to its transposed complex conjugate. Hermitian matrices have real eigenvalues and their eigenvectors are orthogonal, making them ideal to represent observables.

Hermitian adjoint (of a matrix): The transpose and complex conjugate (of the matrix).

Hidden variables: Alleged unknown variables that if included in describing the quantum system of interest would remove the apparent inherent randomness imposed by quantum theory; in other words, they are variables that are claimed to make measurement results in quantum mechanics certain, as opposed to probabilistic.

Hilbert space: A vector space that can be infinite, if necessary, and whose elements are, in general, defined on the space of complex numbers. These attributes are what is needed for the purposes of quantum mechanics. Sometimes we have to deal with observables that have infinite possible outcomes, such as the position operator, and the coefficients in the ket representation are, in general, complex numbers. However, in this book, we only deal with finite-dimensional Hilbert spaces, such as the two-dimensional complex Hilbert space that describes one qubit, or the four-dimensional one describing two qubits, etc.

Hong–Ou–Mandel effect: A quantum interference effect whereby two photons arriving simultaneously, one on each input port of a beamsplitter, will always exit together.

Ion: An atom that has an unequal number of electrons and protons, consequently giving it an overall charge.

Interaction picture: The variant of the Schrödinger equation in which the Hamiltonian is simplified by incorporating complex-exponential time variation into the probability amplitudes.

Jaynes–Cummings model: A single qubit interacting with a quantum harmonic oscillator.

Josephson junction: A narrow interface in a superconducting circuit which acts as a barrier for the charge carriers (the "Cooper pairs"); there is a probability amplitude, dependent on their energy, for the charge carriers to pass through the barrier, giving rise to the Josephson effect.

Ket: A vector that describes a quantum state. In general, it is a linear combination of more than one vector, and its matrix representation is a column vector.

Kraus operator: A mathematical representation of a completely positive trace-preserving map (see above): $\hat{\rho}' = \sum_m \hat{K}_m^\dagger \hat{\rho} \hat{K}_m$, where $\sum_m \hat{K}_m \hat{K}_m^\dagger = \hat{I}$ (note the $\hat{K}_m^\dagger$ comes first in the first equation and second in the second equation). This is the generalization of the unitary time evolution obtained from the Schrödinger equation, so it also covers non-unitary evolution.

Laser: An acronym from Light Amplification by Simulated Emission of Radiation, lasers are powerful and coherent light sources used as an enabling technology for many quantum technologies.

Linear regression: A method for modelling the relationship between two variables by assuming that the relationship is given by a linear equation.

Local: Pertaining to a subsystem, which is part of the bigger system, only. For example, if we say that a local operator is applied on a two-qubit system, we mean it is applied on either of the qubits individually.

Logical qubit: A collection of physical two-level quantum systems; i.e., of multiple qubits, that together have two effective levels only and behave as a single qubit.

Master equation: A generalization of the Schrödinger equation that describes the quantum system to include mixed states, using density matrices, as well as generalizing the dynamics to include decay terms. The latter includes the case when the system under investigation is not closed, interacting with outside systems (such as the environment).

Maximally entangled state: As the name suggests, a state that has the maximum possible entanglement. Examples include the Bell states. The concurrence for such states equals 1.

Maximally mixed state: A mixed state in which all the possible states of the system are equally probable and whose density matrix representation is a diagonal. In other words, it is proportional to the identity matrix. It corresponds to states that are completely random.

Measurement: The projection of the quantum state into one of the possible states dictated by the measurement operator.

Measurement (of an observable): The projection of the quantum state into one of the eigenvectors (of the matrix representing the observable).

Mixed state: A state that has both inherent quantum randomness and classical randomness due to our ignorance or experimental limitations. A mixed state does not have a ket representation, only a density matrix representation.

Modular exponential function: A function that is defined using modular arithmetic (in number theory) so that it represents the remainder of an exponential function.

Nitrogen-vacancy colour centre: A localized defect in a diamond where a single nitrogen atom is bound close to a vacancy in the carbon lattice. The result is an artificial atom, fixed in place inside the diamond, which can be used as either a qubit or a light source.

No-cloning theorem: States that there is no linear device that can make an exact duplicate of an arbitrary quantum state.

Noisy intermediate-scale quantum (NISQ) devices: Quantum computers with dozens or even hundreds of qubits, with good decoherence properties, which allow the performance of various algorithms, mostly aimed at validating device physics, small-scale quantum simulation or testing some fundamental precepts of computer science.

Non-unitary operation: An operation that is irreversible. Measurement is an example of a non-unitary operation.

Normalized: A state is normalized when its inner product with itself equals 1: $\langle \varphi | \varphi \rangle = 1$.

Nuclear magnetic resonance (NMR): The resonant interaction of a nuclear magnetic moment with an oscillating magnetic field.

Number field sieve: An algorithm factoring integers on a conventional computer.

Number operator: The operator $\hat{n}$ corresponding to the number of excitations of a harmonic oscillator, photon mode or phonon mode.

Observables: Quantities that define properties of a system and can be measured. Examples include position, momentum, spin, charge, photon number, etc. In quantum mechanics, an observable is represented by a Hermitian operator.

Operationalist interpretation: The perspective that one does not need to interpret the axioms of quantum mechanics beyond the fact that they agree with experimental observations (also known as the "shut up and calculate" school of quantum interpretation).

Optical cavity: The region between two high-quality mirrors, which precludes propagation of any but a few frequencies of light.

Optical lattice: A trapping potential for atoms formed by the interference pattern of two (or more) lasers.

Oracle: A subroutine of a quantum algorithm which evaluates functions.

Orbitals: The eigenstates of the Hamiltonian of an atomic electron; the name stems from Bohr's original ideas.

Order (of a function): Non-dimensional period.

Orthogonal: Two vectors are said to be orthogonal if their inner product is 0.

Orthogonal matrix: A matrix whose transpose is equal to its inverse.

Parallelism: Performing various computational tasks involving separate sub-systems simultaneously instead of one after another.

Partial trace: A mathematical operation in which one qubit of a two-qubit state is discarded, leaving a single-qubit state (which will be mixed if the two-qubit state was entangled).

Pauli matrices: A set of three 2×2 matrices that along with the 2×2 identity matrix forms a complete basis to describe any 2×2 matrix.

Phase: The angle in the exponential representation of a complex number, also known as the argument of the number.

Phase-flip gate: A gate that changes the relative phase of the two levels of a qubit by π; equivalent to $\hat{Z}$.

Phonon: A quantum excitation of a vibrational mode of a crystal.

Photon: A term that is used to describe light when it displays particle-like properties, as opposed to a continuous wave.

Physical qubit: A qubit that is an actual two-level system in real life, such as a two-level atom or a polarized photon.

Poisson statistics: The probability of n events occurring in a given interval of time when the events are all independent of one another, and the average number of events in the interval, λ, is constant: $p(n) = \lambda^n e^{-\lambda}/n!$.

Polarization: Describes the direction in which a light wave, or more generally an electromagnetic wave, oscillates as it propagates in space.

Polarizer: An optical device that only allows certain polarization components to pass through, blocking the other components.

Polarizing beamsplitter: An optical element that reflects vertically polarized light while transmitting horizontally polarized light.

Positive definite matrix: An Hermitian matrix whose eigenvalues are all real and positive.

Principal quantum number: An integer that specifies the energy level of an atomic electron;

Probability: A quantification of the likelihood that an event will occur.

Probability amplitude: The complex number that specifies the amplitudes of the different basis states that make up a quantum state.

Probability density function: The function that describes the probability of a continuous variable: $p(x)dx$ is the probabilty that x takes a value in the range x to $x + dx$.

Projection (of a quantum state): see Born rule.

Projection-valued measure: Mathematical expression of the Born rule (q.v.).

Projector: A non-unitary operation that renders a quantum state into a specific state as determined by the projector.

Pseudo-pure state: A mixed quantum state for which almost all states are equally likely, except for one particular pure state that has a higher probability.

Pure states: States whose randomness is only due to the inherent quantum randomness. Pure states can be described by kets and, consequently, density matrices. Mixed states, on the other hand, do not have a ket representation.

Purity (of a density matrix): A measure for how close a state is to being pure; i.e., having a ket representation. If a state is pure, the purity is at its maximal value of 1. On the other hand, if the state is maximally mixed, then the purity is at its lowest value of 0.

PVM: see Projection-valued measure.

QR decomposition: A mathematical method of factoring a matrix into a product of a unitary matrix Q and a triangular matrix R (triangular matrices have all elements equal to 0 below and to the left of the leading diagonal).

Quantum algorithm: An ordered set of unitary operations, measurements, and classical communication whose aim is to obtain some desired numerical result.

Quantum circuit (Quantum circuit diagram): A visual representation of the operations performed on qubit systems to implement a quantum algorithm.

Quantum coin: see Qubit.

Quantum computer: An information processing device employing multiple qubits and which can reliably implement a quantum algorithm.

Quantum data bus: A quantum system that allows the implementation of quantum gates involving two or more distant qubits.

Quantum defect: In atomic physics, the quantum defect is a correction to Bohr's energy level formula that takes into account the presence of closed shells of electrons.

Quantum detection: The detection of an object using the advantages of entangled quantum systems.

Quantum dot: A microscopic semiconductor device that acts like an artificial atom, and thus could be used as a qubit.

Quantum entanglement: When the state of multiple quantum systems cannot be described as a product of states, each representing the state of the individual systems, we say that they are entangled or that there is quantum entanglement in the total system.

Quantum error correction: The technique of identifying and correcting any random error that might occur during quantum computations. Generally, it involves using multiple physical qubits to represent a single logical qubit.

Quantum Fourier transform: A unitary transform that reveals periodicities in quantum states.

Quantum gate: A unitary operator that is applied on one or more qubits to change their quantum state.

Quantum information: Broadly speaking, the application of quantum mechanics to problems of information technology. A more narrow definition is the study of the information content of quantum states and the optimal means of its storage and transmission.

Quantum key distribution (QKD): The use of quantum mechanics (in particular, the no-cloning theorem and the indivisibility of photons) to allow two people to share a common stream of random bits with eavesdropping.

Quantum metrology: The use of quantum mechanics to enhance experimental techniques for precision measurement.

Quantum operation/Qubit operation: see Quantum gate.

Quantum parallelism: The computational advantage afforded by quantum systems that can be in multiple configurations simultaneously.

Quantum randomness: A randomness that is due to the inherent probabilistic nature of quantum measurement results, and is not due to limitations imposed by our scientific ignorance or experimental limitations.

Quantum simulations: The use of quantum devices to simulate the behaviour of other quantum systems.

Quantum state: The fundamental description of a system as a superposition of distinct configurations weighted by probability amplitudes.

Quantum state tomography: Experimental method for deducing the quantum state of a system.

Quantum teleportation: The transfer of the state of one quantum system to another.

Quantum volume: If a quantum computer has a quantum volume of $\mathcal{V}_Q = 2^d$, it implies that it has reliably demonstrated quantum circuits involving d qubits and d sequential algorithmic layers (each involving several multi-qubit gates).

Qubit: A simple quantum system with two possible states; the building block of quantum computing hardware.

Qudit: A d-level quantum system, generalizing the notion of a qubit.

Qutrit: a three-level quantum system, extending the notion of a qubit.

Rabi frequency: The frequency at which probability amplitudes oscillate between $|0\rangle$ and $|1\rangle$ when a force is applied to the system.

Reduced density matrix: The reduced-dimension density matrix of a subsystem (in a larger system) that is left when the degrees of freedom of another subsystem are traced out (or discarded), shrinking the system's density matrix.

Regression analysis: A data analysis technique to compare a specific mathematical model (characterized by a set of parameters) with measured data, and obtain the optimal values of the model's parameters to best fit the data.

Resonant frequency: The natural frequency of oscillation of a physical system; if the system is excited by a force oscillating at this frequency, the system's response will be strongest.

Rotating wave approximation: An important approximation in the theory of qubit control, whereby rapidly oscillating terms in the equation of motion are neglected.

RSA encryption: A classical method of encryption whose strength depends on the difficulty in prime factoring a very large number. Quantum technologies, which promise making prime factorization of large numbers much easier, threaten the security of this encryption method.

Schmidt decomposition theorem: Any pure quantum state of two qubits can be written as a simple sum of tensor products of orthogonal states: $|\varphi\rangle = \sqrt{\lambda_1}|u_1\rangle \otimes |v_1\rangle + \sqrt{\lambda_2}|u_2\rangle \otimes |v_2\rangle$, where $\langle u_1|u_2\rangle = \langle v_1|v_2\rangle = 0$ and λ_1 and λ_2 are real and positive. The results can be generalized to two n-level systems. This is the starting point for understanding and exploiting quantum entanglement.

Schrödinger equation: A differential equation that, along with the initial state of the quantum system, gives the state of the system at a later time. This equation only predicts the evolution of the system if it is closed; i.e., if it is not interacting with an outside system.

Separable state: A state that can be written as a product of the states of the subsystems. For example, if system C is composed of two subsystems A and B, and if we are able to write down the state representing C as a product of a state representing A and a state representing B, then we say that the state of C is separable.

Shor's algorithm: The most important quantum computing algorithm discovered so far, which allows for finding the factors of integers with exponentially more efficiency than known conventional computation.

Singular value decomposition: A mathematical method of simplifying operators and matrices by writing them as a product of (i) a unitary matrix; (ii) a real, positive, diagonal matrix (i.e., a matrix whose elements are all zero except for those elements on the top-left to bottom-right diagonal); and (iii) another unitary matrix, in general different from the first.

Spectral decomposition: see Eigenvector decomposition

Spontaneous decay: Quantum systems in an excited state will spontaneously flip to a lower energy state.

Stabilizer formalism: A mathematical method of identifying and classifying different types of error-correcting codes.

Standard quantum limit: The limit on precision in a measurement due to the discrete and indivisible nature of photons.

Stationary (random function): A random function whose statistical properties (mean, standard deviation, probability distribution, etc.) do not vary in time; as a consequence, the auto-correlation function $\Gamma_v(t_1, t_2)$ is a function of $t_2 - t_1$ only: $\Gamma_v(t_1, t_2) = \Gamma(t_2 - t_1)$.

Stokes parameters: A set of four measured quantities which fully determine the polarization of light.

Symmetric matrix: A matrix that is its own transpose: $S_{mn} = S_{nm}$.

Target qubit: A qubit whose fate in whether an operation is to be performed on it or not is dependent on the state of another qubit, namely the *control* qubit.

Transpiler: The conventional computer software used to simplify quantum algorithms for implementation.

Unitary operator: An operator whose inverse equals its Hermitian adjoint: $\hat{U}^{-1} = \hat{U}^{\dagger}$.

Universal gate: A two-qubit gate that can be used to implement a CNOT gate, and thus can be used to perform any quantum operation.

von Neumann entropy: A quantity that indicates how mixed a quantum state is: $S = -\text{Tr}\{\rho \ln \hat{\rho}\}$; $S = 0$ if the state is pure.

Wave plate: An optical element that changes the relative phases of different polarization components of light; used to implement local unitary operations on photon qubits.

Zeeman effect: The change of energy of atomic electrons due to their interaction with an external magnetic field.

Suggestions for Further Reading

Here are suggestions for books, articles, pre-prints, web-pages, etc., which might help the reader delve more deeply into topics covered in this book. There are over 50,000 papers on quantum computing published since 2000, so we have to be very selective. Besides a selection of foundational papers, we have concentrated on works adopting a pedagogical approach accessible to the new student of the field. We have also included reference numbers on the arXiv, which is a free distribution service for scientific papers: https://arxiv.org/.

General Works on Quantum Computing and Quantum Information

- J. Preskill, *Lecture Notes for Physics 229: Quantum Information and Computation* (California Institution of Technology, Pasadena, 1998).
- M. A. Nielsen and I. L. Chuang, *Quantum Computation and Quantum Information* (Cambridge University Press, Cambridge, 2000).
- N. David Mermin, *Quantum Computer Science* (Cambridge University Press, Cambridge, 2007).
- B. Schumacher and M. Westmoreland, *Quantum Processes, Systems, & Information* (Cambridge University Press, Cambridge, 2010).
- M. M. Wilde, *Quantum Information Theory* (Cambridge University Press, Cambridge, 2013; 2nd ed., 2019).
- J. Watrous, *The Theory of Quantum Information* (Cambridge University Press, Cambridge, 2018).
- I. B. Djordjevic, *Quantum Information Processing, Quantum Computing, and Quantum Error Correction: An Engineering Approach* (Academic Press, 2021).
- Qiskit (on-line resource) www.ibm.com/quantum/qiskit.

Chapter 2: Quantum Mechanics for Quantum Computers

Mathematical Techniques: Matrices, Linear Algebra, Differential Equations

- D. B. Damiano and J. B. Little, *A Course in Linear Algebra* (Dover Publications, New York, 2011).
- G. Strang, *Linear Algebra and Its Applications* (Thomson, Brooks/Cole, Belmont, CA, 2006).
- W. E. Boyce and R. C. DiPrima, *Elementary Differential Equations and Boundary Value Problems* (Wiley, New York, 2001).

Quantum Mechanics: General Works

- D. J. Griffiths and D. F. Schroeter, *Introduction to Quantum Mechanics* (Cambridge University Press, Cambridge, 2018).
- R. Shankar, *Principles of Quantum Mechanics* (Plenum Press, New York, 1994).
- J. J. Sakurai, *Modern Quantum Mechanics* (Addison-Wesley, Reading, MA, 1994).
- N. Zettilli, *Quantum Mechanics: Concepts and Applications* (John Wiley & Sons, Chichester, 2006).

Chapter 3: Two Qubits and Entanglement

Seminal Papers

- A. Einstein, B. Podolsky, and N. Rosen, "Can quantum-mechanical description of physical reality be considered complete?" *Physical Review* **47 (10)**, 777–780 (1935).
- E. Schrödinger, "Discussion of probability relations between separated systems," *Mathematical Proceedings of the Cambridge Philosophical Society* **31**, 555–556 (1935).
- D. Bohm and Y. Aharonov, "Discussion of experimental proof for the paradox of Einstein, Rosen, and Podolsky," *Physical Review* **108**, 1070–1076 (1957).
- J. S. Bell, "On the Einstein Podolsky Rosen paradox," *Physics Physique Fizika* **1**, 195–200 (1964); see also *Speakable and Unspeakable in Quantum Mechanics* (Cambridge University Press, Cambridge, 2004).
- J. F. Clauser, M. A. Horne, A. Shimony, and R. A. Holt, "Proposed experiment to test local hidden-variable theories," *Physical Review Letters* **23**, 880–884 (1969).
- S. J. Freedman and J. F. Clauser, "Experimental test of local hidden-variable theories," *Physical Review Letters* **28**, 938–941 (1972).
- A. Aspect, P. Grangier, and G. Roger, "Experimental realization of Einstein-Podolsky-Rosen-Bohm Gedankenexperiment: A new violation of Bell's inequalities," *Physical Review Letters* **49**, 91–94 (1982).
- C. H. Bennett, G. Brassard, C. Crépeau, R. Jozsa, A. Peres, and W. K. Wootters, "Teleporting an unknown quantum state via dual classical and Einstein-Podolsky-Rosen channels," *Physical Review Letters* **70**, 1895 (1993).

Pedagogical Work

- J. H. Eberly, "Schmidt analysis of pure-state entanglement," *Laser Physics* **16**, 921–926, (2006); arXiv:quant-ph/0508019.

Chapter 4: Quantum Algorithms

Seminal Papers

- P. A. Benioff, "The computer as a physical system: A microscopic quantum mechanical Hamiltonian model of computers as represented by Turing machines," *Journal of Statistical Physics* **22**, 563–591 (1980).

- R. P. Feynman, "Simulating physics with computers," *International Journal of Theoretical Physics* **21**, 467–488 (1982).
- D. Deutsch, "Quantum theory, the Church–Turing principle and the universal quantum computer," *Proceedings of the Royal Society of London A* **400**, 97–117 (1985).
- D. Deutsch and R. Jozsa, "Rapid solutions of problems by quantum computation," *Proceedings of the Royal Society of London A* **439**, 553–558 (1992).
- E. Bernstein and U. Vazirani, "Quantum complexity theory," *SIAM Journal on Computing* **26**, 1411–1473 (1997).
- L. K. Grover, "A fast quantum mechanical algorithm for database search," *Proceedings of the 28th Annual ACM Symposium on the Theory of Computing*, pp. 212–219 (1996).

Recent Review

- A. M. Dalzell, S. McArdle, *et al.*, "Quantum algorithms: A survey of applications and end-to-end complexities," https://arxiv.org/abs/2310.03011 (2023).

Chapter 5: Period Finding and Shor's Algorithm

Seminal Papers

- P. W. Shor, "Algorithms for quantum computation: Discrete logarithms and factoring," *Proceedings of the 35th Annual Symposium on Foundations of Computer Science*, pp. 124–134 (1994).
- D. Coppersmith, "An approximate Fourier transform useful in quantum factoring," IBM Technical Report RC19642 (1994); https://arxiv.org/abs/quant-ph/0201067.
- C. Gidney and M. Ekerå, "How to factor 2048 bit RSA integers in 8 hours using 20 million noisy qubits,"*Quantum* **5**, 433 (2021).

Pedagogical Work

- R. L. Singleton, "Shor's factoring algorithm and modular exponentiation operators," *Quanta* **12**, 41–130 (2023); https://arxiv.org/abs/2306.09122.

Background Reading

- J. F. James, *A Student's Guide to Fourier Transforms* (Cambridge University Press, Cambridge, 2011).
- D. Rosenthal, D. Rosenthal, and P. Rosenthal, *A Readable Introduction to Real Mathematics* (Springer, 2018).
- G. H. Hardy and E. M. Wright, *An Introduction to the Theory of Numbers* (Oxford University Press, Oxford; 4th ed., 1968); (result cited in Section 5.2.3 on convergences of continued fractions is Theorem 184, p. 153).
- S. Singh, *The Code Book* (Anchor, New York, 2000).

Chapters 6, 7 and 8: Physical Realizations of Qubits

Seminal Papers

- D. P. DiVincenzo, "The physical implementation of quantum computation," *Progress of Physics* **48**, 771–783 (2000); arXiv:quant-ph/0002077v3.
- G. J. Milburn, "Quantum optical Fredkin gate," *Physical Review Letters* **62**, 2124–2127 (1989).
- J. I. Cirac and P. Zoller, "Quantum computations with cold trapped ions," *Physical Review Letters* **74**, 4091–4094 (1995).
- C. Monroe, D. M. Meekhof, B. E. King, W. M. Itano, and D. J. Wineland, "Demonstration of a fundamental quantum logic gate," *Physical Review Letters* **75**, 4714–4717 (1995).
- M. H. Devoret, "Quantum fluctuations in electrical circuits," in *Fluctuations quantiques: Les Houches, Session LXIII, 27 juin–28 juillet 1995* (S. Reynaud, S. Giacobino and J. Zinn-Justin, eds.).

Background Reading

- E. Hecht, *Optics* (Pearson, Boston; 5th ed., 2015).
- C. C. Gerry and P. L. Knight, *Introductory Quantum Optics* (Cambridge University Press, Cambridge; 2nd ed., 2023).
- L. Allen and J. H. Eberly, *Optical Resonance and Two-Level Atoms* (Dover, New York, 1987).
- H. M. Rosenberg, *The Solid State* (Oxford University Press, Oxford; 3rd ed., 1988).
- G. K. Woodgate, *Elementary Atomic Structure* (Oxford University Press, Oxford; 2nd ed., 1983).
- D. F. V. James, "Quantum dynamics of cold trapped ions, with application to quantum computation," *Applied Physics B* **66**, 181–190 (1998).

Chapter 9: Device Characterization

Seminal Papers

- D. F. V. James, P. G. Kwiat, W. J. Munro, and A. G. White, "Measurement of qubits," *Physical Review A* **64**, 052312 (2001).
- A. W. Cross *et al.*, "Validating quantum computers using randomized model circuits," *Physical Review A* **100**, 032328 (2019) (IBM Superconducting Qubit Group).

Chapter 10: Variations on the DiVincenzo Criteria

Seminal Papers

- D. G. Cory, A. F. Fahmy, and T. F. Havel, "Ensemble quantum computing by NMR spectroscopy," *Proceedings of the National Academy of Sciences USA* **94**, 1634 (1997) and, independently,
- N. A. Gershenfeld and I. L. Chuang, "Bulk spin-resonance quantum computation," *Science* **275**, 350 (1997).

- E. Knill, R. Laflamme, and G. J. Milburn, "A scheme for efficient quantum computation with linear optics," *Nature* **409** 46–52 (2001).
- B. P. Lanyon, *et al.*, "Experimental demonstration of Shor's algorithm with quantum entanglement," *Physical Review Letters* **99**, 250505 (2007) (Optical quantum computing group of A. G. White at the University of Queensland).
- R. Raussendorf and H. J. Briegel, "A one-way quantum computer," *Physical Review Letters* **86**, 5188 (2001).
- M. A. Nielsen, "Cluster-state quantum computation," preprint (2005); https://arxiv.org/abs/quant-ph/0504097.

Chapter 11: Decoherence and How to Defeat It

Seminal Papers

- D. A. Lidar, I. L. Chuang, and K. B. Whaley, "Decoherence-free subspaces for quantum computation," *Physical Review Letters* **81**, 2594–2597 (1998).
- P. W. Shor, "Scheme for reducing decoherence in quantum computer memory," *Physical Review A* **52**, R2493–R2496 (1995).
- J. Chiaverini, *et al.*, "Realization of quantum error correction," *Nature* **432**, 602–605 (2004) (Ion trapping group of D. J. Wineland at NIST, Boulder).

Pedagogical Work

- J. Roffe, "Quantum error correction: An introductory guide," *Contemporary Physics* **60**, 226–245 (2019).

Chapter 12: Quantum Technologies

- C. H. Bennett and G. Brassard, "Quantum cryptography: Public key distribution and coin tossing," *Proceedings of the IEEE International Conference on Computers, Systems and Signal Processing*, Bangalore, 175–179 (1984).
- I. Buluta and F. Nori, "Quantum simulators," *Science* **326**, 108 (2009).
- R. Blatt and C. F. Roos "Quantum simulations with trapped ions," *Nature Physics* **8**, 277 (2012).

Index